T0192732

Materials Forming, Machining and Tribology

Series editor

J. Paulo Davim, Aveiro, Portugal

More information about this series at http://www.springer.com/series/11181

J. Paulo Davim
Editor

Introduction to Mechanical Engineering

 Springer

Editor
J. Paulo Davim
Department of Mechanical Engineering
University of Aveiro
Aveiro
Portugal

ISSN 2195-0911   ISSN 2195-092X   (electronic)
Materials Forming, Machining and Tribology
ISBN 978-3-030-08711-1   ISBN 978-3-319-78488-5   (eBook)
https://doi.org/10.1007/978-3-319-78488-5

© Springer International Publishing AG, part of Springer Nature 2018
Softcover re-print of the Hardcover 1st edition 2018
This work is subject to copyright. All rights are reserved by the Publisher, whether the whole or part of the material is concerned, specifically the rights of translation, reprinting, reuse of illustrations, recitation, broadcasting, reproduction on microfilms or in any other physical way, and transmission or information storage and retrieval, electronic adaptation, computer software, or by similar or dissimilar methodology now known or hereafter developed.
The use of general descriptive names, registered names, trademarks, service marks, etc. in this publication does not imply, even in the absence of a specific statement, that such names are exempt from the relevant protective laws and regulations and therefore free for general use.
The publisher, the authors and the editors are safe to assume that the advice and information in this book are believed to be true and accurate at the date of publication. Neither the publisher nor the authors or the editors give a warranty, express or implied, with respect to the material contained herein or for any errors or omissions that may have been made. The publisher remains neutral with regard to jurisdictional claims in published maps and institutional affiliations.

Printed on acid-free paper

This Springer imprint is published by the registered company Springer International Publishing AG part of Springer Nature
The registered company address is: Gewerbestrasse 11, 6330 Cham, Switzerland

Preface

Currently, it is possible to define mechanical engineering, as the branch of engineering that "*which involves the application of principles of physics and engineering for the design, manufacturing, automation and maintenance of mechanical systems.*" Mechanical Engineering is much related to a number of other engineering disciplines. This book fosters information exchange and discussion on all aspects of introductory matters of mechanical engineering from a number of perspectives including materials, design and manufacturing (Part I-Chaps. 1–4), thermal engineering (Part II-Chaps. 5–6), robotics and automation (Part III-Chaps. 7–8) and advanced machining (Part IV-Chaps. 9–12).

The purpose of this book is to present a collection of chapters exemplifying theoretical and practical introductory aspects of mechanical engineering. Chapter 1 of the book provides mechanical properties of engineering materials (relevance in design and manufacturing). Chapter 2 is dedicated to analysis and material selection of a continuously variable transmission for a bicycle drivetrain. Chapter 3 describes coin minting. Chapter 4 contains information on gradation, dispersion, and tribological behaviors of nanometric diamond particles in lubricating oils. Chapter 5 is dedicated to basics and applications of thermal engineering. Chapter 6 describes alternate fuels for IC engine. Chapter 7 contains information on robotics (history, trends, and future directions). Chapter 8 is dedicated to computer vision in industrial automation and mobile robots. Chapter 9 describes advanced machining processes. Chapter 10 contains information on comparative assessment and merit appraisal of thermally assisted machining techniques for improving machinability of titanium alloys. Chapter 11 describes smart machining system using preprocessor, postprocessor, and interpolation techniques. Finally, Chap. 12 is dedicated to comparison of non-conventional intelligent algorithms for optimizing sculptured surface CNC tool paths.

The present book can be used as a book for undergraduate engineering course or as a topic on mechanical engineering at introductory professional level. Likewise, this book can serve as a valuable reference for students, mechanical engineers, professionals in mechanical engineering and related industries. The book may be convenient for anyone interested to know about mechanical engineering. The interest of this book is evident for many important institutes and universities as well as industry.

The editor acknowledges Springer for this opportunity and professional support. Finally, I would like to thank all the chapter authors for their availability for this editorial project.

Aveiro, Portugal   J. Paulo Davim
April 2018

Contents

viii

About the Editor

J. Paulo Davim received Ph.D. degree in Mechanical Engineering in 1997, M.Sc. degree in Mechanical Engineering (materials and manufacturing processes) in 1991, Mechanical Engineering degree (5 years) in 1986, from the University of Porto (FEUP), the Aggregate title (Full Habilitation) from the University of Coimbra in 2005, and the D.Sc. from London Metropolitan University in 2013. He is Eur Ing by FEANI, Brussels, and Senior Chartered Engineer by the Portuguese Institution of Engineers with an MBA and Specialist title in Engineering and Industrial Management. Currently, he is a Professor at the Department of Mechanical Engineering of the University of Aveiro, Portugal. He has more than 30 years of teaching and research experience in Manufacturing, Materials and Mechanical Engineering with special emphasis in Machining and Tribology. He has also interest in Management and Industrial Engineering and Higher Education for Sustainability and Engineering Education. He has guided large numbers of postdoc, Ph.D., and master's students as well as coordinated and participated in several research projects. He has received several scientific awards. He has worked as evaluator of projects for international research agencies as well as examiner of Ph. D. thesis for many universities. He is the Editor-in-Chief of several international journals, Guest Editor of journals, books Editor, book Series Editor, and Scientific Advisory for many international journals and conferences. Presently, he is an Editorial Board member of 25 international journals and acts as reviewer for more than 80 prestigious Web of Science journals. In addition, he has also published as editor (and co-editor) more than 100 books and as author (and co-author) more than 10 books, 80 chapters, and 400 articles in journals and conferences (more than 200 articles in journals indexed in Web of Science core collection/h-index 45+/6000+ citations and SCOPUS/h-index 52+/8000+ citations).

Part I
Materials, Design and Manufacturing

Chapter 1
Mechanical Properties of Engineering Materials: Relevance in Design and Manufacturing

Viktor P. Astakhov

Abstract The chapter provides an introduction to mechanical engineering, covering fundamental concepts of mechanical properties of materials and their use in the design and manufacturing. It first explains the notion of mechanical properties of materials and then elaborates on the proper definition of most relevant properties as well as materials testing to obtain these properties. The role of mechanical properties at the design stage in form of the design criterion is explained. The use of material properties to assess equivalent stress and strain in complex loading conditions is revealed. At the manufacturing stage, the notion of additive (material is added to the workpiece), neutral (the volume of the workpiece is preserved), and substantive (the volume of the workpiece is reduced) processes is introduced. The relevant properties of materials in the neutral (forming) and substantive (cutting) processes are considered.

Keywords Engineering materials · Basic mechanical properties
Design and manufacturing · Complex state of stress · Failure criteria

1 Conceptual Introduction

The term "knowledge-based economy" results from a fuller recognition of the role of knowledge and technology in economic growth. Knowledge in technology has always been central to economic development. But only over recent years has its relative importance been recognized, just as that importance is growing. This is because, in today's rapidly changing product marketplace, the critical requirements for product quality, productivity, and reliability have been becoming the most powerful driving force behind any new product design and development.

Products achieve success through a combination of sound technical design and effective manufacturing process to achieve the requirements set by the design. The

V. P. Astakhov (✉)
General Motors Business Unit of PSMi, 1792 Elk Ln, Okemos, MI 48864, USA
e-mail: astakhov@physicist.net

© Springer International Publishing AG, part of Springer Nature 2018
J. P. Davim (ed.), *Introduction to Mechanical Engineering*, Materials Forming,
Machining and Tribology, https://doi.org/10.1007/978-3-319-78488-5_1

amalgam creates product quality, which is the way material and processes are used to provide functionality, usability, and satisfaction in ownership. This chapter aims to explain (not just formally lists as it commonly done in the reference literature) the basic mechanical properties of part materials in terms of their relevance in the design and manufacturing. Although the material of this chapter is kept at the introductory level following the known Einstein's famous saying: "Make it as simple as possible, but not simpler," the chapter presents unique synergetic approach unifying the design and manufacturing stage in terms of material properties. Moreover, the standard designation and proper definition of the terms related to mechanical properties of engineering materials provided in the chapter aim to provide an essential help to beginners in mechanical engineering.

2 Mechanical Properties of Materials Reported in Reference Sources

2.1 Properties Commonly Reported in Reference Sources

Physical properties (e.g., density, thermal conductivity, specific heat, anisotropy, electrical conductivity, magnetic properties, type of bonds) are usually associated with a particular materials type (steel, wood, plastic, oxide ceramic, etc.), whereas mechanical properties are mostly attributed to a particular grade within the chosen material type (e.g., mild steel 1012, tool steel H13, oak wood).

Mechanical properties of engineering materials are obtained from testing. Standard ASTM E6.14406-1 "Terminology Relating to Methods of Mechanical Testing" covers the principal terms relating to methods of mechanical testing of solids. The general definitions are restricted and interpreted, when necessary, to make them particularly applicable and practicable for use in standards requiring or relating to mechanical tests. These definitions are published to encourage uniformity of terminology in product specifications.

The most common and very useful test of mechanical properties of materials is a tension (a.k.a. tensile) test. It is carried out on a tensile machine (Fig. 1a) according to standard ASTM E8/E8M—16a "Standard Test Methods for Tension Testing of Metallic Materials." This standard describes the test specimens design, test procedure, and evaluation of the test results. Tension tests provide information on the strength and ductility of materials under uniaxial tensile stresses. This information may be useful in comparison with materials, alloy development, quality control, and design under certain circumstances.

Compression tests are carried out on the same machine. Standard ASTM E9 covers the apparatus, specimens, and procedure for axial load compression testing of metallic materials at room temperature.

Shear testing is different from tensile testing in that the forces applied are parallel to the upper and lower faces of the object under test. Materials behave differently in

(a)   **(b)**

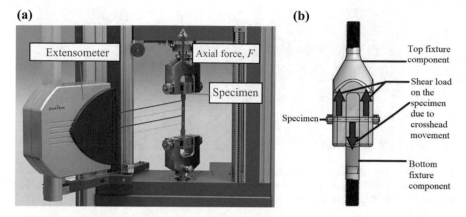

Fig. 1 Basic materials testing: **a** tension and **b** shear

shear than in tension or compression, resulting in different values for strength and stiffness. Usually performed on fasteners, such as bolts, machine screws, and rivets, shear testing applies a lateral shear force to the specimen until failure results. Apart from tension tests, there are a number of various standards on shear testing. One of the most common shear tests is the double shear test of metallic materials. This test can be carried out on the tensile machine using a standard fixture shown in Fig. 1b.

Tensile tests are performed for several reasons [1]. The results of tensile tests are used in selecting materials for engineering applications. Basic tensile properties are normally included in material specifications to be used in the design and manufacturing of products. Tensile properties are often used to predict the behavior of a material under forms of loading other than uniaxial tension (discussed later in this chapter).

The strength of a material often is the primary concern. The strength of interest may be measured in terms of either the stress necessary to cause appreciable plastic deformation or the maximum stress that the material can withstand before fracturing.

These measures of strength are used, with appropriate caution (in the form of safety factors discussed later in this chapter), in engineering design. Also of interest is the material's ductility, which is a measure of how much it can be deformed before it fractures. Rarely is ductility incorporated directly in design; rather, it is a key property in materials manufacturing including forming and cutting as discussed later in this chapter.

In the tension test, a specimen of standard dimension (one of the standard specimens is shown in Fig. 2) is subjected to a continually increasing uniaxial tensile force while simultaneous observations of the elongation of the specimen.

The results of the tension test are stress–strain diagrams. A stress–strain diagram is a diagram in which corresponding values of stress and strain are plotted against each other. Values of stress are usually plotted as ordinates (vertically) and values

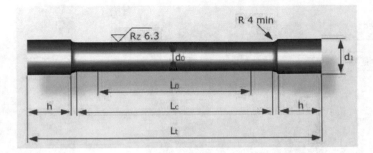

Fig. 2 Geometry of one of the standard specimens: d_0 is the diameter of specimen; d_1 is the diameter of grip ($>1.2d_0$); L_0 is the gage length ($L_0 = 5d_0$); L_c is parallel length ($L_c = L_0 + d_0$) L_t is the total length; h is the height of grip

of strain as abscissas (horizontally). Figure 3b shows a typical diagram for a ductile material. Engineering stress, s, is defined as

$$s = \frac{F}{A_0} \tag{1}$$

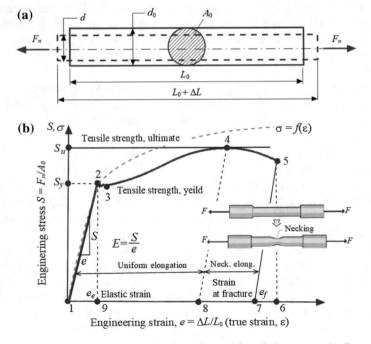

Fig. 3 Standard tension test: **a** deformation in testing and **b** typical stress–strain diagram

where F is the tensile force and A_0 is the initial cross-sectional area of the gage section. Obviously, if the specimen is round then $A_0 = \pi d_0^2/4$.

Engineering strain, or nominal strain, e, is defined as

$$e = \frac{\Delta L}{L_0} \tag{2}$$

where L_0 is the initial gage length and ΔL is the change in gage length $(L - L_0)$.

Multiple detailed descriptions of a typical stress–strain diagram is presented in the literature, e.g., in [2]. The most important aspect relevant in mechanical engineering is segments shown by numbered points in Fig. 3b. When a solid material is subjected to small stresses, the bonds between the atoms are stretched. When the stress is removed, the bonds relax and the material returns to its original shape. This reversible deformation is called elastic deformation represented by segment 1–2 in the diagram. For most materials, segment 1–2 is linear. The slope of this linear segment is called *the elastic modulus or Young's modulus*.

$$E = \frac{s}{e} \tag{3}$$

In the elastic range, the ratio, μ, of the magnitude of the lateral contraction strain to the axial strain is called *Poisson's ratio*.

$$\mu = -\frac{e_y}{e_x} \text{ (in an x-direction test)} \tag{4}$$

At point 2, stresses become high enough to cause planes of atoms slide over one another. This deformation, which is not recovered when the stress is removed, is termed plastic deformation. Note that the term "plastic deformation" does not mean that the deformed material is a plastic (a polymeric material).

Segment 2–3 represents the so-called the transient effects (or yielding instability). For some materials (e.g., low carbon steels and many linear polymers), the stress–strain curves have initial maxima followed by lower stresses as shown in Fig. 3b. After the initial maximum, all the deformation at any instant is occurring within a relatively small region of the specimen with no stress increase. This segment is noticeable only for relatively ductile materials, while for relatively brittle materials a smooth transition of segment 1–2 into segment 3–4 is the case.

Segment 3–4 is called the strain-hardening region. When a metal is stressed beyond its elastic limit, it enters the plastic region (the region in which residual strain remains upon unloading). When the load is increased further (a kind of rearrangement occurs at atom level and the mobility of the dislocation decreases), "dislocation density" increases that in turn makes the metal harder and stronger through the resulting plastic deformation. It means it is more difficult to deform the metal as the strain increases, and hence it is called "strain hardening." This tends to increase the strength of the metal and decrease its ductility. Note that the maximum stress in the diagram is achieved at point 4.

Segment 4–5 is the necking region. Necking is defined by standard ASTM E8/ E8M—16a as the onset of nonuniform or localized plastic deformation, resulting in a localized reduction of cross-sectional area. In this region, relatively large amounts of strain localize disproportionately in a small region of the material. The resulting prominent decrease in local cross-sectional area provides the basis for the name "neck." Note that because engineering stress [see Eq. (3)] is calculated through the initial diameters of the specimen, d_0 (and hence A_0), the stress in the diagram on segment 4–5 decreases although in reality, when the actual diameter of the neck is considered, it increases.

Two more segments shown in the diagram (Fig. 3b) are of interest. The first is segment 1–9 that represents the maximum elastic strain e_e; segment 1–6 that represents the maximum strain at the instant of fracture; segment 6–7 is elastic recovery of the specimen after fracture. Note that line 5–7 is drawn parallel to line 1–2. Although this segment is of small interest in the design, it is of significant importance in part manufacturing; segment 1–7 is known at strain at fracture, i.e., the plastic strain in the specimen as measured as two parts of the fractured specimen are brought together. From the beginning of loading (point 1) to the beginning of necking (point 4), the specimen undergoes uniform elongating whereas starting from point 4 till point 4 it undergoes necking elongation.

The common tensile properties directly obtained from the stress–strain diagram are:

- *Modulus of Elasticity, E* [see Eq. (3)]. Young's modulus, also known as the elastic modulus, is a measure of the *stiffness* of a solid material. Stiffness is understood as resistance of a material to deformation like elongation, twisting, bending, and deflection.
- *Yield Tensile Strength, S_y.* Yield strength is the stress at which a material exhibits a deviation from the proportionality of stress to strain; i.e., the occurrence permanents plastic strain is the case.
- *Ultimate Tensile Strength, S_u.* Ultimate tensile strength (a.k.a. *UTS*) is the maximum stress developed by the material based on the original cross-sectional area.
- *Elongation at fracture or engineering strain, e_f.* It is normally reported in %, that is $e_f \cdot 100\%$.

Some other important properties and terms used in engineering literature are:

Toughness. There are a number of different approaches to toughness characterization and methods of measurements discussed in the literature [3] or set by standards (e.g., ISO 26843:2015 "Metallic materials. Measurement of fracture toughness at impact loading rates using precracked Charpy-type test pieces"; ASTM E23 "Standard Test Methods for Notched Bar Impact Testing of Metallic Materials"). In the author's opinion, the most engineering-sound approach having clear physical meaning is the energy need to fracture of a unit volume of a material so it is measured in J/m^3. It is characterized by the modulus of toughness which is the total energy absorption capabilities of the material to fracture, and thus is given by the total area under the s–e curve such that

$$U_t = \int_0^{e_f} s\,de \approx \frac{(S_y + S_u)}{2} e_f \tag{5}$$

Energy absorbed per volume of material (toughness) is obtained from numerical integration of data in a measured stress–strain experiment. When all else is equal, this value will be the same for the material from one test to another. Multiple tests of the same material will give a statistically representative report (average and standard uncertainty).

Resilience. It is the strain energy stored by body up to elastic limit. In other words, it is the recoverable/elastic strain energy. It is characterized by the modulus of resilience, which is the amount of energy stored in stressing the material to the elastic limit as given by the area under the elastic portion of the *s–e* diagram and can be defined as

$$U_r = \int_0^{e_e} s\,de \approx \frac{S_y e_e}{2} \tag{6}$$

Typical values of U_r and U_t are listed in Table 1 for some common engineering materials.

Hardness. Standard ASTM E6.14406-1 defines hardness as the resistance of a material to deformation, particularly permanent deformation, indentation, or scratching. In reality, it is the resistance of a material to indenter penetration so one should not have any extended interpretation of this characteristic. It is tested for with an indenter hardness machine usually (but not solely) by measuring the size of the indentation after releasing the load.

The most known of the hard materials is diamond. It is so hard; it is usually used as the penetration material (for the Vickers hardness, for example). A typically soft material is aluminum metal or any plastic.

Table 1 Energy properties of materials in tension

Material	Yield strength (MPa)	Ultimate strength (MPa)	Modulus of resilience (kJ/m^3)	Modulus of toughness (kJ/m^3)
SAE 1020 annealed	276	414	186	128,755
SAE 1020 heat treated	427	621	428	91,047
Type 304 stainless	207	586	102	195,199
Ductile cast iron	400	503	462	50,352
Red brass	414	517	828	13,795

Hardness correlates well with scratch-proof ability meaning that harder materials are harder to scratch, i.e., the greater its abrasion resistance. For metallic materials, hardness is uniquely correlated with the ultimate tensile strength of the material as documented in [4]. Therefore, harder materials are also stronger materials and vice versa.

Ductility. This property is defined by standard ASTM E8/E8M—16a as the ability of a material to deform plastically before fracturing. Ductility is usually evaluated by measuring (1) the elongation or reduction of area from a tension test, (2) the depth of cup from a cupping test, (3) the radius or angle of bend from the bend test, or (4) the fatigue ductility from the fatigue ductility test. *Malleability* is the ability to deform plastically under repetitive compressive forces.

Unfortunately, there is no usable theory of ductile behavior. Ductility is measured by dozens of methods, all contradictory, none leading to practical forecasts of a part's behavior. Moreover, the limit is not sharp either. At our present degree of understanding, ductility is a qualitative, subjective property of a material. In general, measurements of ductility are of interest in three ways:

1. To indicate the extent to which a material can be deformed without fracture in metalworking operations such as rolling and extrusion.
2. To indicate to the designer, in a general way, the ability of the metal to flow plastically before fracture. A high ductility indicates that the material is "forgiving" and likely to deform locally without fracture should the designer made a mistake in the stress calculation or the prediction of severe loads.
3. To serve as an indicator of changes in impurity level or processing conditions. Ductility measurements may be specified to assess material quality even though no direct relationship exists between the ductility measurement and performance in service.

A common definition is a material with less than 0.05 (5%) elongation (strain) at fracture is brittle. This includes ceramics, glasses, and some metal alloys. Cast iron would be classified as brittle. For reference, Fig. 4 shows the elongation at fracture of some common engineering materials.

In shear tests (Fig. 1b), the shear force F_s acts tangentially to the top area A as shown in Fig. 5a. The shear stress τ measures the intensity of a reaction to externally applied loading sustained by the material as it maintains equilibrium with this force. This stress is calculated as

$$\tau = \frac{F_s}{A} \tag{7}$$

The material response on the application of force F_s is represented as the shear stress versus shear strain plot shown in Fig. 5b. In full analogy with tension tests, the yield shear strength, τ_y, as the stress at which a material exhibits a deviation from the proportionality of stress to strain, i.e., the occurrence permanents plastic strain is the case, and ultimate shear strength, τ_u, as the maximum stress developed by the material based on the original cross-sectional area are distinguished. Note

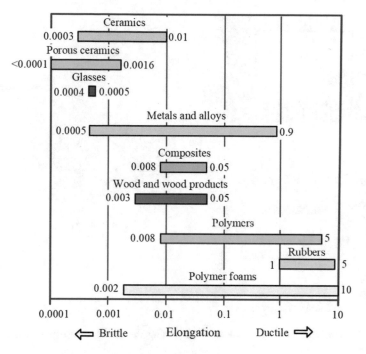

Fig. 4 Elongation at fracture of common engineering materials

that the term "engineering" is used here to explain the occurrence of segment 3–4 in the stress–strain diagram. In testing (schematic of which is shown in Fig. 5c), the ultimate shear strength occurs just before two cracks begin to develop at points A and B. After the cracks start to run toward each other along the shear plane, the initial shearing area represented by shear plane reduces in full analogy with the necking in tension tests.

In this plot, the engineering shear strain refers to the angular distortion that a material suffers in shear. The shear strain is a dimensionless measure of distortion and is defined in Fig. 5a as

$$\gamma = \tan \phi \tag{8}$$

In Eq. (8), ϕ is the angular change in the right angle measured in radians. Within the elastic region, the shear displacement ΔL is small so ϕ is also small. For small enough angles, the tangent of an angle is equal to the angle itself (when measured in radians) so that the engineering shear strain can be approximated as $\gamma \approx \phi$ (rad). Note that this is not nearly the case in metal forming and cutting when large deformations take place.

In full analogy with the normal stress, Hooke's law is valid for the considered case provided the shear stress varies linearly with shear strain and the shear

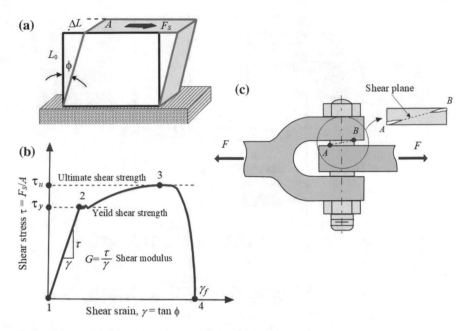

Fig. 5 Standard shear test: **a** deformation in testing, **b** typical stress–strain diagram, **c** test schematics

modulus, G, is the slope of the shear stress–strain (Fig. 5b), and thus it may be determined from the slope of the stress–strain curve or by dividing stress by strain,

$$G = \frac{\tau}{\gamma} \tag{9}$$

Shear properties can also be found from a right circular cylinder loaded in torsion, i.e., in widely performed torsion tests. In solid mechanics, torsion is the twisting of an object due to an applied torque. Torque, like a linear force, will produce both stress and the strain. Torsion causes a twisting stress, called shear stress (τ), and a rotation, called shear strain (γ). A simple model of a torsion test shown in Fig. 6 includes a cylindrical body, which essentially is cantilever beam one end of which is rigidly fixed and torque T is applied to the other (free) end.

In Fig. 6a, a line AB is drawn on the surface parallel to the beam longitudinal axis. When the torque is applied to the free end of the circular beam (as shown in this figure), line AB becomes helical assuming position AB' in which angle of helix is γ_{tw}. The beam twists by an angle θ_{tw}. This angle is a function of the beam length, L, and stiffness represented by shear modulus G. The twist angle starts at 0 at the fixed end of the beam and increases linearly as a function of z-distance from this end. The change of angle, γ_{tw}, is constant along the length.

A small differential element, dz, is sliced from the beam as shown in Fig. 6b. Because the cross sections bounded this element are separated by an infinitesimal

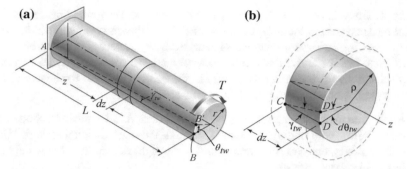

Fig. 6 Model of torsion: **a** A circular beam loaded by torque T and **b** small circular element dz

distance, the difference in their rotations, denoted by the angle $d\theta_{tw}$, is also infinitesimal. As the cross sections undergo the relative rotation $d\theta_{tw}$, straight line CD deforms into the helix CD'. By observing the distortion of the sliced element, it should be recognized that the helix angle γ_{tw} is the shear strain of the element.

Two angles γ_{tw} and $d\theta_{tw}$ must be compatible at the outside edge (arc length D–D'). This gives the relationship

$$\text{Arc Length D} - \text{D}' = \rho d\theta_{tw} = \gamma_{tw}dz \tag{10}$$

from which the shear strain γ_{tw} is

$$\gamma_{tw} = \rho\frac{d\theta_{tw}}{dz} \tag{11}$$

The quantity $d\theta_{tw}/dx$ is the angle of twist per unit length, where θ_{tw} is expressed in radians. The corresponding shear stress (Fig. 7a) is determined from Hooke's law as

$$\tau = G\gamma_{tw} = G\rho\frac{d\theta_{tw}}{dz} \tag{12}$$

Fig. 7 Shear stress due to torsion (**a**) and shear stress distribution (**b**)

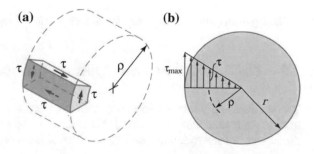

A simple analysis of Eq. (12) reveals that the shear stress varies linearly with the radial distance ρ from the axial of the beam. This variation is shown in Fig. 7b. As can be seen, the maximum shear stress, denoted by τ_{max}, occurs at the surface of the beam. Note that the above derivations assume neither a constant internal torque nor a constant cross section along the length of the beam, i.e., valid for a general case. Two important conclusions follow from this analysis:

- As the maximum shear stress occurs at the surface of a beam/shaft, shaft can be made hollow with minimum compromising of the shaft torsional strength. Using hollow shafts, one can achieve a significant weight reduction—it is widely used in the aerospace industry.
- In the direct torsion test/application, the critical shear stress (yield or ultimate) occurs over the entire cross-sectional area of the specimen/part that leads to failure. In torsion, the critical stress occurs only on the surface that does not lead to failure as the rest of the shaft takes the applied torque.

The shear modulus, G, is related to Young's modulus, E, as

$$G = \frac{E}{2(1+\mu)} \tag{13}$$

As Poisson's ratio, μ (Eq. 4), varies between 0.3 and 0.5 for most materials, the shear modulus is often approximated by $G \sim E/3$.

Two other torsion-related parameters are considered in the technical literature. Torsional rigidity is the ability of an object to resist torsion or in layman's terms "twisting," due to an applied torque *torsional stiffness* or *rotational stiffness* K_T is defined as the ration of the applied torque, T to the angle of twist, θ i.e.,

$$K_T = \frac{T}{\theta} \tag{14}$$

Torsional stiffness can also be expressed in terms of the modulus of rigidity, G (units Pa), torsion constant J (units m^2) known as the polar moment of inertia of the cross section (for a round shaft $J = \pi d^4/32$), and the characteristic length, L (Fig. 6a), as

$$K_T = \frac{JG}{L} \tag{15}$$

The product JG is commonly known as *torsional rigidity*.

2.2   The Concepts of True Stress and Strain

The engineering stress–strain curve does not give a true indication of the deformation characteristics of a metal because it is based entirely on the original dimensions of the specimen, and these dimensions change continuously during the test. Also, ductile metal which is pulled in tension becomes unstable and necks down during the course of the test.

The initial cross-sectional area of the specimen, A_0, does not change noticeably over the elastic region unless rubber-like materials are considered. However, the cross-sectional area of the specimen is decreasing rapidly after the yield point (point 2 in the diagram shown in Fig. 3b). As a result, the load required continuing deformation falls off as the engineering stress based on original area likewise decreases, and this produces the fall-off in the stress–strain curve beyond the point of maximum load (segment 2–3 in Fig. 3b). In reality, the metal continues to strain-harden all the way up to fracture, so that the stress required to produce further deformation should also increase. The true stress, designated as σ, is based on the actual cross-sectional area of the specimen. If it is used, then the stress–strain curve increases continuously up to fracture as shown by the dashed curve $\sigma = f(\varepsilon)$ in Fig. 3b. Obviously that the true stress, σ, is related to engineering stress/elongation, e, as

$$\sigma = s(1+e) \tag{16}$$

The concept of the true strain, designated as ε, seems to be more complicated as presented in many literature sources. In reality, it is rather simple and straightforward when properly explained. Strain is normalized deformation so it can be expressed in the differential form for a bar undergoing tension as $d(strain) = dL/L$ or an infinitesimal deformation dL normalized to length L. Therefore, the total axial strain ε (known as the true strain) is found by integrating this expression from the initial length L_0 to the final length $L_F = L_0 + \Delta L$ (see Fig. 3a), i.e.,

$$\varepsilon = \int_{L_0}^{L_F} \frac{dL}{L} = \ln(L)|_{L_0}^{L_F} = \ln\left(\frac{L_F}{L_0}\right) = \ln\left(\frac{L_0 + \Delta L}{L_0}\right) = \ln\left(1 + \frac{\Delta L}{L_0}\right) = \ln(1+e) \tag{17}$$

How close are the values of e and ε? Figure 8 shows the engineering and true strains for values up to 1. The agreement is quite good for strains of less than 0.1.

Advantages of using the true strain ε compared to engineering strain e are:

1. It is the exact value, not an approximation.
2. Sequential strains can be added: If two strains ε_1 and ε_2 are executed sequentially, the total strain is

$$\varepsilon_1 + \varepsilon_2 = \ln\left(\frac{L_1}{L_0}\right) + \ln\left(\frac{L_2}{L_1}\right) = \ln\left(\frac{L_1}{L_0} \cdot \frac{L_2}{L_1}\right) = \ln\left(\frac{L_2}{L_0}\right) \tag{18}$$

This is not the case with the engineering strain where the total strain is

$$\frac{L_2 - L_0}{L_0} \neq e_1 + e_2 = \frac{L_1 - L_0}{L_0} + \frac{L_2 - L_1}{L_0} \tag{19}$$

Fig. 8 Engineering and true
strains as function of bar
deformation

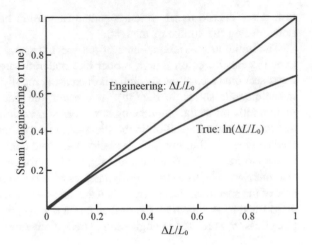

3. It is used to characterize materials that deform by large amounts (considerable fractions of their length up to many times their length). A quick look at the literature shows that true strain has been recently used to characterize materials like polyamide yarn, epoxy, rubber, and cartilage. Moreover, this strain is of prime concern in the analyses and modeling of manufacturing process involved great amounts of plastic deformation of materials.
4. It is geometrically symmetric: That is, if the strain associated with being stretched to n times the original length is ε, then the strain associated with being compressed to $1/n$ the original length is $-\varepsilon$.

3   Materials Properties at the Design Stage

3.1   The Concept of the Critical/Limiting Stress Known as the Design Criterion

At the design stage, the concept designs of the machine and units are developed to meet the intended requirements of machine performance, quality, and reliability. As such, the power inputs and outputs, velocities, limiting weight, operating environment, and many other operating parameters are defined. Once approved, the units are decomposed into parts and a model for each responsible part is constructed to determine the limiting stress considered as the design criterion. Based on the limiting stress, size, and weight limitations as well as the physical properties required for the part's intended performance (e.g., thermal conductivity), the type and grade of the part material are selected by the designer. The problem is in the proper selection of this limiting stress. The practice of the design shows that such a selection is not simple and straightforward.

The first encounter of mechanical engineers with the limiting stress takes place in university/college courses on mechanics of material where stresses and stress distribution in mainly beams loaded by various forces, moments, and torques are calculated and the maximum stress over a certain (often referred to as the critical) section of a beam is found. When the beam is made of a ductile material, this maximum stress is compared to the yield tensile or shear strength (chosen as the design criterion) making sure that this maximum is less or equal to the design criterion. As such, plastic deformation of the beam is considered as its failure mode. If, however, the beam material is brittle, the ultimate strength is selected as the design criterion, and thus, fracture of the beam is considered as its failure mode. The listed mechanical properties of few common materials are provided in the student's textbook that creates impression that these properties are constants for the listed materials, and thus can readily be found when needed.

Being simple and clear for students, such an approach is an oversimplified version of the real design practice where many other factors are to be considered in obtaining the design criterion in each and every design situation.

3.2 Yield Strength for Ductile Materials

Although the notion of ductility is discussed in Sect. 2 and the 5% elongation criterion to distinguish the "brittle–ductile" boundary is introduced, it is not that "sharp" at the design stage. For elongation under 1%, most people would say "brittle"; over 5%, most people would say "ductile"; at 3% like brass many people would say "a bit brittle". The final call should be made in the consideration of the stress–strain diagram of a material in question. If this diagram (Figs. 3b and 5b) shows the distinguishable shear strength and if the plastic deformation after the initial yielding is of an appreciable amount (elongation/strain) which can lead to undesirable change of the part configuration, then this material is considered as ductile.

Although the yield strength is defined by standard ASTM E8/E8 M—16a as the stress at which a material exhibits a deviation from the proportionality of stress to strain, i.e., the occurrence permanents plastic strain is the case, it is not that certain in the stress–strain diagram. In Sect. 2, it is discussed that segment 2–3 represents the so-called the transient effects (or yielding instability) (see Fig. 3b). To avoid this region, the yield strength S_y is commonly defined by the offset method. According to this method, the yield strength is determined at a certain offset Om (Fig. 9) of the original gage length of the specimen. Commonly, the yield strength at 0.2% offset is obtained by drawing through the point of the horizontal axis of abscissa $e = 0.2\%$ (or $e = 0.002$), a line parallel to the initial straight line portion of the stress–strain diagram. The 0.2% of the initial gage length has been chosen by standardization organizations (mainly ASTM) because it is small enough to ensure the accuracy of the yield point but it is sufficiently large to be measured with conventional methods in the laboratory, just with a caliper or micrometer. For brittle materials, such as gray cast iron, the offset is 0.05% of the initial gage length, because the plastic deformation is small. The yield strength obtained using this method is designated as $S_{y0.2}$ and thus actually used in the design calculations.

3.3 Relevance of Material Properties in Complex Loading

The complex loading includes, in the most general case, a three-dimensional state of stress. To understand the stresses involved, let us consider an infinitesimal element in a parallel-sided form with its faces oriented parallel to the three coordinate planes as shown in (a). Each plane will have normal and tangential components of the stress resultants. The tangential or shear stress resultant on each plane can further be represented by two components in the coordinate directions. Although nine components of stress are shown in Fig. 10a, only six of them are independent, namely σ_x, σ_y, σ_z, $\tau_{xy} = \tau_{yx}$, $\tau_{xz} = \tau_{zx}$, and $\tau_{yz} = \tau_{zy}$. To deal with these components, the so-called equivalent tensile stress or von Mises stress, σ_e, is used

$$\sigma_e = \sqrt{\sigma_x^2 + \sigma_y^2 + \sigma_z^2 - \sigma_x\sigma_y - \sigma_y\sigma_z - \sigma_z\sigma_x + \left(\tau_{xy}^2 + \tau_{yz}^2 + \tau_{zx}^2\right)} \qquad (20)$$

In two-dimensional application, i.e., when $\sigma_x = \tau_{xz} = \tau_{yx} = 0$, the equivalent stress is calculated as

$$\sigma_e = \sqrt{\sigma_x^2 + \sigma_y^2 - \sigma_x\sigma_y + 3\tau_{xy}^2} \qquad (21)$$

Fig. 9 Determination of yield strength by the offset method

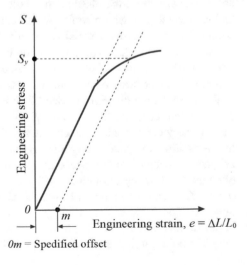

$0m$ = Spedified offset

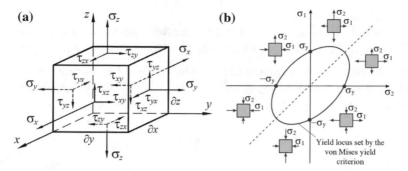

Fig. 10 General state of stress: **a** stress components representation and **b** yield locus set by the von Mises criterion

The von Mises yield criterion [5] (with Hencky's interpretation [6]) states the yielding occurs when the equivalent stress, σ_e, reaches the *yield strength of the material in simple tension*, S_y (σ_y). In other words, using this criterion, one can calculate the design criterion/limiting stress for any general type of loading/state of stress. The corresponding equivalent or von Mises strain is calculated as

$$\varepsilon_e = \frac{\sqrt{2}}{3}\left[\left(\varepsilon_x - \varepsilon_y\right)^2 + \left(\varepsilon_y - \varepsilon_z\right)^2 + \left(\varepsilon_z - \varepsilon_x\right)^2 + 6\left(e_{xy}^2 + e_{yz}^2 + e_{zx}^2\right)\right]^{1/2}. \quad (22)$$

3.4 Safety Factor

Factor of safety (FoS), also known as *safety factor* (SF), is a term describing the structural capacity of a part or system beyond the expected loads or actual loads. Essentially, how much stronger the part or system is than it usually needs to be for an intended load. Safety factors are often calculated using detailed analysis because comprehensive testing is impractical on many engineering projects, such as bridges and buildings, but the structure's ability to carry load must be determined to a reasonable accuracy [7].

Many systems are purposefully built much stronger than needed for normal usage to allow for emergency situations, unexpected loads, misuse, or degradation. Any structure or component can be made to fail if it is subjected to loadings in excess of its strength. Structural integrity is achieved by ensuring that there is an adequate safety margin or reserve factor between strength and loading effects. The margin of safety, or alternatively the safety factor, which is appropriate for a particular application must take into account the following:

- The scatter or uncertainty in the variables which form the input data for load and resistance effects.
- Any uncertainty in the equation used to model failure.

– The consequences of failure.
– The possibility of unknown loadings or mechanisms of failure occurring.
– The possibility of human error causing unforeseen events.

Factors of safety can be incorporated into design calculations in many ways [8]. For most calculation, the following equation is used to obtain the design criterion—the allowable working stress $[s_w]$

$$[s_w] = \frac{S_m}{f_s} \tag{23}$$

where S_m is the strength of the material (the yield strength for ductile materials, $S_{y0.2}$, and ultimate strength, S_u, for brittle materials) and f_s is the factor of safety.

Particular values of the factor of safety and what is actually covered by this factor are regulated by various standards and design recommendation in various industries. Table 2 presents some general recommendations.

3.5 Materials Properties' Outcome of the Design Stage

The output of the design is *a drawing* where all materials specification, properties (e.g., hardness), and quality requirements to be achieved in part/unit/machine manufacturing are recorded. In other words, the drawing is the primary document for part/unit/machine quality. In advanced industries, the drawing may have some reference to certain manufacturing procedures and processes to be used to assure the required part quality. For example, a surface roughness callout can have note indicating "ground," brazing joint can have a note to use a certain brazing procedure. The properties of the part set by the drawing are the prime information used in designing manufacturing process for this part/assembly/structure.

Table 2 General recommendation for values of factors of safety

1.3–1.5	For use with highly reliable materials where loading and environmental conditions are not severe, and where weight is an important consideration
1.5–2	For applications using reliable materials where loading and environmental conditions are not severe
2–2.5	For use with ordinary materials where loading and environmental conditions are not severe
2.5–3	For more brittle materials where loading and environmental conditions are not severe
3–4	For applications in which materials properties are not reliable and where loading and environmental conditions are not severe or where reliable materials are to be used under difficult loading and environmental conditions

4 Materials Properties at the Manufacturing Stage

A manufacturing specialist cannot change the material type/grade chosen by the part/unit/machine designer. His or her responsibility is to manufacture parts efficiently while fully complying with quality requirements set by the drawing.

In part production, three general types of manufacturing processes, which can be conditionally classified as additive, "neutral," and subtractive, are used. Additive manufacturing (AM) describes the technologies that build 3D objects by adding layer-upon-layer of material, whether the material is plastic, metal, concrete or one day it will be human tissue. "Neutral" processes (commonly referred to as forming processes) change the shape of the original workpiece (blank), while the volume of the work material remains unchanged. Subtractive processes change the shape of the original workpiece (blank) by removing a part of its original volume. AM processes are relatively new so the standard on material properties is not yet fully developed. Therefore, we will consider only the relevant properties of materials in "neutral" known as forming and subtractive known as cutting processes showing that considerably different properties of the work materials are involved in these processes.

4.1 Forming Processes

Forming is defined by standard DIN 8580 as manufacturing through three-dimensional or plastic modification of a shape while retaining its mass and material cohesion. In other words, forming is the modification of a shape with controlled geometry. Forming processes are categorized as chipless or non-material removal processes. Forming includes a large set of manufacturing operations in which the material is deformed plastically to take the shape of the die geometry. The tools used for such deformation are called die, punch, etc., depending on the type of process. At the most general level [9], these are divided into bulk metal forming and sheet metal forming as shown in Fig. 11.

Bulk deformation processes are generally characterized by significant deformations and massive shape changes, and the surface-area-to-volume of the work is relatively small. The term bulk describes the workparts that have this low area-to-volume ratio. Starting work shapes for these processes include cylindrical billets and rectangular bars. Figure 12 illustrates the following basic operations in bulk deformation:

- Rolling: In this process, the workpiece in the form of slab or plate is compressed between two rotating rolls in the thickness direction, so that the thickness is reduced. The rotating rolls draw the slab into the gap and compress it. The final product is in the form of a sheet.
- Forging: The workpiece is compressed between two dies containing shaped contours. The die shapes are imparted to the final part. Forging is traditionally a hot working process, but many types of forging are performed cold.

Fig. 11 General
classification of
metal-forming processes

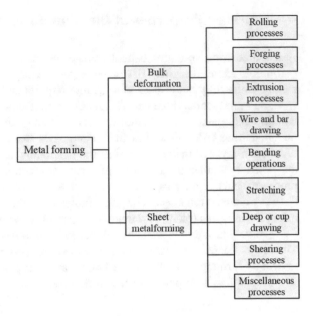

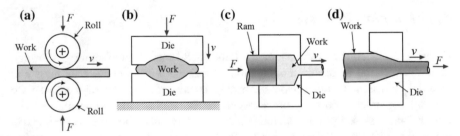

Fig. 12 Basic bulk deformation processes: **a** rolling, **b** forging, **c** extrusion, and **d** drawing. Relative motion in the operations is indicated by *v*; forces are indicated by *F*

- Extrusion: This is a compression process in which the work metal is forced to flow through a die opening, thereby taking the shape of the opening as its own cross section.
- Drawing (often referred to as wire or rod drawing): In this forming process, the diameter of a round wire or bar is reduced by pulling it through a die opening.

Sheet metal forming involves operations performed on metal sheets, strips, and coils. The surface-area-to-volume ratio of the starting metal is relatively high; thus, this ratio is a useful means to distinguish bulk deformation from sheet metal processes. Pressworking is the term often applied to sheet metal operations because the machines used to perform these operations are presses. A part produced in a sheet metal operation is often called a stamping.

Fig. 13 Basic sheet metalworking operations: **a** bending, **b** drawing, **c** shear spinning, and **d** roll bending

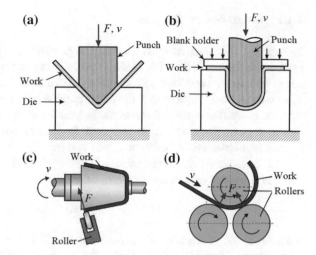

Sheet metal operations are always performed as cold working processes and are usually accomplished using a set of tools called a punch and die. The punch is the positive portion, and the die is the negative portion of the tool set. The basic sheet metal operations are sketched in Fig. 13a, b which are defined as follows:

- Bending: In this, the sheet material is strained by punch to give a bend shape (angle shape) usually in a straight axis.
- Drawing: In sheet metalworking, drawing refers to the forming of a flat metal sheet into a hollow or concave shape, such as a cup, by stretching the metal. A blank holder is used to hold down the blank, while the punch pushes into the sheet metal. To distinguish this operation from bar and wire drawing, the terms cup drawing or deep drawing are often used.

The miscellaneous processes within the sheet metalworking classification include a variety of related shaping processes that do not use punch and die tooling. Examples of these processes are stretch forming, roll bending, spinning, and bending of tube stock. Schematic representations of shear spinning and roll bending are shown in Fig. 13c, d.

4.1.1 Material Properties Involved

When one designs a forming process, the following important parameters are considered:

- Work material characterization, and thus formability.
- Process parameters: plastic deformation characterization, strain and strain rate.
- Microstructural alternations of the work material.
- Forces and energies involved.
- Friction in metal forming.

In the section to follow, only work material characterization is considered.

4.1.2 Stress–Strain Diagram

In forming, the most relevant material properties are represented by the so-called
flow curve shown in Fig. 14, which is actually the stress–strain diagram in the true
stress-true strain coordinate, i.e., obtained from the standard stress–strain diagram
discussed in Sect. 2. Note that the elastic region is not accounted for as the elastic
strain is insignificant for many metals. In forming of other materials with significant
elastic region, the curve does not start from zero but rather from the point of
maximum elastic strain before material yields.

The flow curve for many metals in the region of plastic deformation can be
expressed by the simple power curve relation

$$\sigma = K\varepsilon^n \tag{24}$$

where n is the strain-hardening exponent and K is the strength coefficient. These
parameters are determined as follows. A log–log plot of the flow curve will result in
a straight line. The linear slope of this line is n, and K is the true stress at $\varepsilon = 1.0$ as
shown in Fig. 15. The strain-hardening exponent may have values from $n = 0$
(perfectly plastic solid) to $n = 1$ (perfectly elastic solid). For most metals, n has
values between 0.10 and 0.50; see Table 3.

Only a part of the entire flow curve 1–2 (Fig. 14) is feasible to use in forming to
avoid fracture and fracture-associated defects of the finished parts. That is, why a
certain safety margin (segment 3–5 in Fig. 14) is always assigned so the actual flow
curve is represented by segment 1–4 in Fig. 14. The area under this actual flow
curve (area bounded by segments 1–4–5–1) defines the energy needed for deformation. This energy is used to calculate the process power and forces involved.

Fig. 14 Flow curve

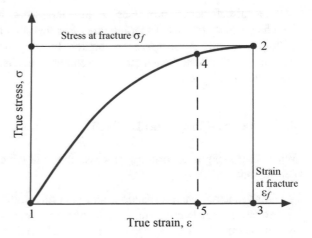

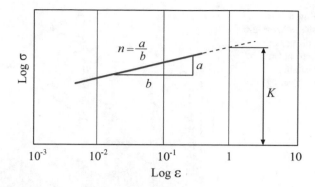

Fig. 15 Log–log plot of true stress-true strain curve to determine strain-hardening exponent n and the strength coefficient K

4.1.3 Accounting for the Process Temperatures, Strain Rates, and the State of Stress

In forming, high temperatures and potentially high strain rates occur particularly in complicated die designs and high-productivity hot (with preheated blanks) forming operations. It is known that the flow stress decreases with temperature while the strain at fracture increases. The opposite is true when the strain rate increases [10]. Figure 16 shows that the influence of temperature and strain rake is highly non-linear. Moreover, different temperature and strain rate may occur in various regions of the blank (workpiece, billet, etc.) that complicates the forming process design in terms of defects (cracks, wrinkles) prevention.

Another materials' aspect of forming operations design is the accounting for the complex state of stress. It was discussed in Sect. 3.3 that when a complex state of stress is the case then the equivalent (von Mises) stress is calculated and the von Mises yield criteria are used to define the beginning of material yielding, i.e., the appearance of irreversible plastic deformation. In mechanical engineering, this criterion is often considered as failure criteria as a part of a mechanism or machine normally cannot function properly if its shape changes due to plastic deformation.

In deforming manufacturing process where severe plastic deformation takes place, the von Mises stress and yield criterion are also fully applicable, thus used in many FEM commercial software packages for designing forming operations. The so-called rule of isotropic hardening is discussed here as it is allocable for most of

Table 3 Values for n and K for metals at room temperature [1]

Materials	Conditions	n	K (MPa)
0.05% carbon steel	Annealed	0.26	530
SAE 4340 steel	Annealed	0.15	641
0.6% carbon steel	Quenched and tempered at 540 °C	0.10	1572
0.6% carbon steel	Quenched and tempered at 705 °C	0.19	1227
Copper	Annealed	0.54	320
70/30 brass	Annealed	0.49	896

Fig. 16 Showing the
influence and the temperature
and strain rate on the flow
stresses and strain at fracture
of a common metallic
material used in forming

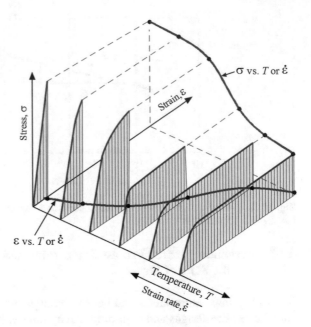

materials used in forming. Therefore, the equivalent stress and strain are used as materials' properties in the design of forming processes

Figure 17 illustrates this rule within two common plane stress states: (a) tension–torsion and (b) principal biaxial stress. The two shown figures connect the rule to simple shear and uniaxial hardening curves *OPQ*, as yield loci expand to contain flow stresses between *P* and *Q*. The single most attractive feature of isotropic hardening over alternative hardening rules is its mathematical simplicity, while the use of this rule can predict plastic strain paths acceptably.

To account for process temperatures, strain rate, and complex state of stress, in the practice of modern CAD of forming processes, a more sophisticated [compared to the flow approximation represented by Eq. (24)] representation of the flow curve known as the Johnson and Cook model [11] is used. Its full form is

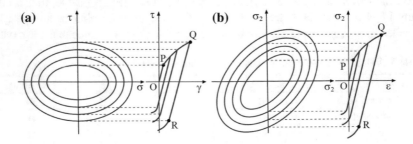

Fig. 17 Rule of isotropic hardening under plane stress: **a** tension–torsion and **b** principal biaxial stress

$$\sigma_e = \left(A + B\varepsilon_{eq}^n \right) \left(1 + C \ln\left(\frac{\dot{\varepsilon}_{eq}}{\dot{\varepsilon}_{eq}^0} \right) \right) \left[1 - \left(\frac{T - T_0}{T_F - T_0} \right)^m \right] \tag{25}$$

where ε_{eq}^n is the equivalent plastic strain [see Eq. (22)], $\dot{\varepsilon}_{eq}$ is the equivalent plastic strain rate, $\dot{\varepsilon}_{eq}^0$ is the reference equivalent plastic strain rate (normally $\dot{\varepsilon}_{eq}^0 = 1 s^{-1}$), T is the temperature, T_0 is the room temperature, T_F is the melting temperature, and A, B, C, n, and m are constants, which depend on the material. These constants are determined through material tests.

In the Johnson and Cook model, the equivalent stress and strains are used to account for any 3D state of stress. The first term of this equation accounts for work material strain hardening, the second term accounts for the process and local strain rates, and the third term accounts for the process and local temperatures.

Deformation in sheet forming is limited by necking, tearing, fracture, or wrinkling that define forming limits (i.e., allowable strains). As a result, only a part of the flow curve 1–2 (shown as 1–4 in Fig. 14) is actually utilized in forming. The science and art of sheet forming is to achieve the required final shape without producing strains that approach any of these limits. Forming limit diagram (FLD) first introduced by Keeler and Backofen [12] and Goodwin [13] has been developed for decades and is widely used in deep-drawing industry as a useful tool for predicting strain limits of sheet forming operations.

A schematic of FLD is illustrated in Fig. 18a. The flow limit curve (FLC) in this diagram can be split into two branches known as the "left branch" and "right branch." The "right branch" of FLC is valid for positive major and minor strains, whereas the "left branch" of FLC is applicable for positive major and negative minor strains. FLC is solely applicable for proportional strain path. Therefore, to construct FLC, different ratios of major and minor strains are chosen in proportional strain paths.

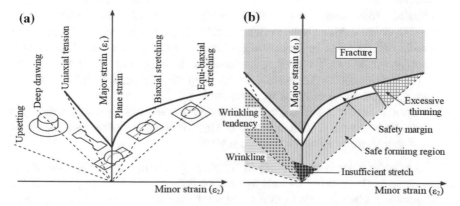

Fig. 18 Forming limit diagram: **a** principle of construction and **b** schematic representation of save the forming regions

While forming, strains in the workpiece should be located below FLC everywhere in order to avoid forming defects. By offsetting FLC (generally 10%), a safety margin is normally introduced. The risk of necking/failure is determined by how close is the maximum strain in the process to that set by the boundary set by FLC. Rejection of the formed part is based not only on occurrence of necking but also involves other defects such as excessive thinning, wrinkling, or insufficient stretch. Therefore, apart from necking, all other failure conditions are normally also evaluated by studying the strain levels involved in the process. Such considerations result in further modifications of FLD as shown in Fig. 18b.

As mentioned above, forming operations are successfully modeled using FEM in manufacturing practice. It is to say that such a modeling is widely used in the designing forming operations and forming tools. Standard ASTM E2218—14 "Standard Test Method for Determining Forming Limit Curves" are used to construct FLDs, which now are included in modeling FEM packages. In the process of design of forming operations and forming tools, strains at various regions of the part being formed are calculated and plotted on the corresponding FLD. The design/optimization of a given forming operations is carried out until all the strains are found below FLC accounting for the above-mentioned safety margins.

4.1.4 Summary of Relevant Materials Properties in Forming Operations

What unites forming processes listed in Fig. 11 in terms of materials properties involved is all of them are accomplished by plastic deformation in order to change the shape of the workpiece/blank into the final desirable shape of the part. This implies the following:

- The stress needed for plastic deformation of the work material must be well above the yield strength of the work material. This sets manufacturing materials-related calculating apart from those used in the design.
- Plasticity (commonly referred to as ductility) as the amount of plastic deformation of the work material before undesirable phenomena occur, for example, cracks, wrinkling, is the prime mechanical characteristic of the work material in forming operations. The greater the plasticity/ductility of the work material, the better its formability. To increase ductility, special measures are often used, for example preheating of the workpiece/blank.
- Fracture is the "principal enemy" in forming processes. As a result, only a part of the flow curve 1–2 (shown as 1–4 in Fig. 14) and many other precautions are used to stay away from the fracture regions.

As the relevant materials properties in forming operations are clearly understood, simulations of the forming process are successfully carried out using commercial FEM packages obtaining vital information on the optimal tool/process design. As a result, tool development and production time have been reduced by about 50% due to the use of simulations in recent years and a further 30% reduction over the next few years appears realistic. The simulation of forming tool has already reached the stage where its result can be fed directly into the press tool digital planning and validation process. Thus today, starting from the design model and through practically all process steps as far as the actual design of the press tool, the production of component can be fully simulated before a first prototype is built [14].

4.2 Cutting Processes

The notion "*cutting*" is not defined by any standard that creates a lot of confusion. To clarify the issue, the author would like to state the following. First, cutting is *physical separation* of a material into two or more parts (a subtractive process). Second, by definition, *fracture is physical separation* of a body into two or more parts. Therefore, *cutting is fracture*. Third, cutting in the manufacturing sense is physical separation of a material into two or more parts carried out by a cutting tool (blade) in a controllable manner. Therefore, in terms of materials properties, those related to fracture should be considered/analyzed in the design of cutting operations including process parameters, tools.

Figure 19 shows a generalized classification of cutting processes/operations. Besides some special cases (e.g., cutting by a knife or splitting by a hatchet), all manufacturing cutting operations are divided into two principal groups, namely shearing operations/processes and metal cutting operations/processes. Both processes aim to physically separate a blank/workpiece into two or more parts so materials should be brought to fracture in both processes. The only difference is how this fracture is achieved; i.e., how much energy (per unit volume of the work material) is spent and how good are the fractured surfaces.

4.2.1 Relevant Material Properties in Shearing Processes

Figure 20 shows some examples of common shearing operations. In punching, the sheared slug is discarded, while in blanking, this slug is the part and the rest is scrap. Die cutting includes perforating or punching a number of holes in a sheet; parting or shearing the sheet into two or more pieces; notching or removing pieces of various shapes from the edges; lancing or leaving the tab without removing any material.

All shearing operations include the deformation and then separation of a material substance in which parallel surfaces are made to slide past one another. In shearing, one layer of a material is made to move on the adjacent layer in a linear direction due to action of two parallel forces F_{sh} located at distance a_{cl} known as the clearance distance as shown in Fig. 21a. A plane over which the sliding occurs is known as the shear plane. A typical example of shearing is cutting with a pair of scissors (Fig. 21b). Scissors are cutting instruments consisting of a pair of metal blades connected in such a way that the blades meet and cut materials placed

Fig. 19 General
classification of cutting
processes/operations

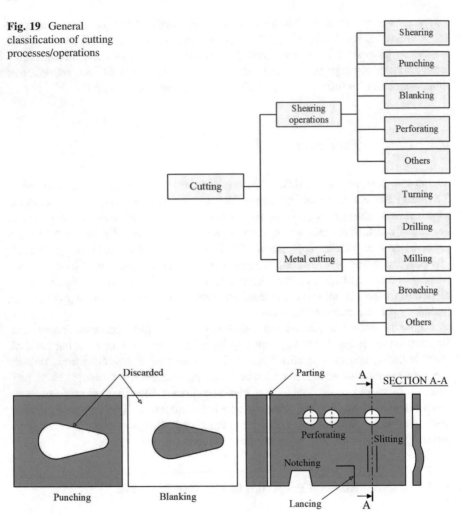

Fig. 20 Examples of common shearing operations

between them when the handles are brought together. Shearing machines (sheet metal shears) are yet another example.

A shearing or scissor-like action is used to cut metal into sheets or strips. Shearing machines and shearing machinery are multipurpose devices used in the cutting of alloys and other sheet metals. Shearing machines are used in steel furniture industries, refrigeration, doorframe manufacturers, automobile industries, and control panel manufacturers. There are many types of shearing machines. Examples include hydraulic, conventional, mechanical, cut-to-length line, and plastic devices.

In these operations, the sheet is cut by subjecting it to shear stress typically between a punch and a die as shown in Fig. 22. Shearing usually starts with the formation of the shear planes and then cracks on both the top and bottom edges of

(a)   **(b)**

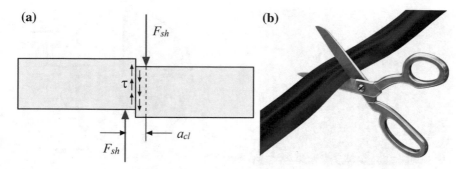

Fig. 21 Principle of shearing operations: **a** force and deformation model and **b** typical example

the workpiece (A and B, and C and D in Fig. 22). These cracks eventually meet each other and separation occurs. The rough fracture surfaces are due to these cracks.

In students' textbooks (e.g., [15]), the punching force is calculated as the product of the shear strength, τ_u, of the work material (see Fig. 5) obtained in the direct shear test (see Fig. 1) and the shearing area, A_{sh}, that is,

$$F_{pn} = A_{sh}S_{sh} = L_{pn}T_w S_{sh} \tag{26}$$

where L_{pn} is the length or perimeter of cut. If the punch is round, then $L_{pn} = \pi d_{pn}$ where d_{pn} is the punch diameter and T_w is the thickness of sheet being sheared.

Although the same formula for calculating the punching force appears in engineering reference books (e.g., [8]), it is pointed out that this force needs to be increased by 20–40%. No valid reasons for such an increase are provided. The problem is that being seemingly simple and straightforward, Eq. (26) is incorrect as it stems from misunderstanding of the essence of the shear stress–strain diagram shown in Fig. 5b. The area under the shear stress–strain curve defines the energy needed for shearing of a unit volume of the work material. Knowing this energy, one can easily calculate the power needed in punching, and then dividing this power over the punching speed (v in Fig. 22), one can obtain the punching force.

As the product $L_{pn}T_w$ is significant, a great force in punching is needed. Even for punching relatively small holes (10 mm), a 50 ton punching press is used whereas 200 ton presses are common for many applications.

4.2.2 Relevant Material Properties in Metal Cutting Processes

Machining is one of the oldest yet still most common manufacturing operations with a wide range of techniques such as turning, milling, drilling, grinding. Regardless of this great variety, the essence of metal cutting can be explained using a simple model shown in Fig. 23. The cutting tool is actually a cutting wedge having the rake (the rake angle γ) and the flank (the clearance angle α) faces that

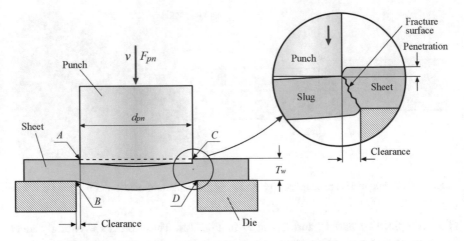

Fig. 22 Schematic illustration of shearing with punch and die

meet to form the cutting edge. As pointed out by Taylor [16], while a metal cutting tool looks like a wedge, its function is far different from that of the wedge. The flank face is never (at least theoretically) allowed to touch the workpiece. Therefore, the presence of the clearance angle α distinguishes a metal cutting tool from other cutting tools with cutting edges. As such, it does not matter what work material is actually cut—metal wood, plastic—the process is still referred to as metal cutting.

When the tool is in contact with the workpiece, the application of the force N leads to the formation of a deformation zone ahead of the cutting edge (point A). The tool moves forward with a cutting speed v. The workpiece first deforms elastically and then plastically. As a result of the plastic deformation, a "bump" forms in close contact with the rake face. Thus, the plastic deformation of the layer

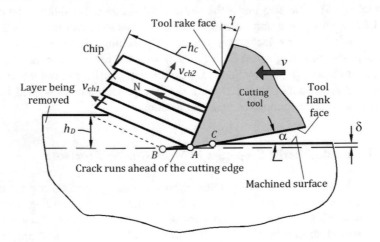

Fig. 23 Model of metal cutting

to be removed and formation of the bump in front of the cutting tool is actually beginning of the metal cutting process as the state of stress ahead of the tool becomes complex including a combination of bending (as the bumps act as a lever due to which the bending stress in the deformation zone is formed), compressive (compression of the layer being removed by the rake face), and shear (due to the friction over the tool–chip interface) stresses. When the combined stress just in front of the cutting edge (point A) reaches a limit (for a given work material), a small crack AB forms ahead of the cutting edge so that a chip fragment of the layer being removed disengages from the rest of the workpiece and starts to slide over a sliding surface formed in the direction of the maximum combined stress with velocity v_{ch1} and simultaneously over the tool rake face with velocity v_{ch2}. The chip fragment continues to slide until the force acting on this fragment from the tool is reduced, when a new portion of the work material enters into the contact with the tool rake face. This new portion attracts part of the cutting force, and thus the stress along the sliding surface diminishes, becoming less than the limiting stress and arresting the sliding. A new fragment of the chip starts to form.

The uniqueness of the metal cutting process (among other cutting operations, e.g., shearing operations and methods, e.g., cutting by a knife, scissors, splitting by a hatchet [17]) in terms of materials properties involved is in the following:

- The cutting process is a cyclic process. Each cycle includes elastic, then plastic deformation and finishes with fracture; i.e., the whole stress–stress diagram (as shown in Figs. 3b and 14) up to fracture is followed.
- In machining of brittle materials, a well visible crack forms in each cycle, whereas in machining of ductile materials, a series of micro-cracks hampered and then healed by a great hydrostatic pressure [18] within each chip element occurs in each cycle. As plastic deformation and fracture take place only within a small chip fragment, the power per unit volume of the cut material is much low in metal cutting compared to other cutting operations and methods.
- The plastic deformation imposed by the tool is concentrated in the chip and measured by the ratio of the chip thickness, h_C, to the uncut chip thickness, h_D [19]. In contrary to other cutting operations, the machine surface is not subjected to significant plastic deformation. Only a few percent the total plastic deformation that is imposed into the machined surface (coldworked layer on this surface) by the tool–workpiece interface AC (Fig. 23) due to the spring back δ of this surface. The essence of this spring back is not unfortunately clear to many. In the author's opinion, it is rather simple as it follows from the stress–strain diagram shown in Fig. 3b. The amount of spring back is actually represented by segment 6–7 so it appears to be the elastic recovery of the work material. Reading this diagram intelligently, one can calculate the value of spring back. Moreover, he or she can conclude that two major materials properties, namely its ultimate strength and elastic module, fully define this spring back as it follows from Fig. 3b.
- This allows achieving the closest tolerance and low surface roughness of the machined surface (approximately two orders high compared to other cutting

operations). That is why metal cutting operations are most widely used in industry to manufacture finished parts made of variety of different work materials.

The energy needed for fracture of the unit volume of the work material is the area under the true stress–strain curve considered up to fracture known as the damage curve [20] shown in Fig. 24. This curve is obtained from the flow curve shown in Fig. 14 by gining attention to, and thus adding the work material degradation and fracture regions. As can be seen, the real damage curve which describes behavior of the work material in fracture has well-defined, and thus measurable, area.

The elastic–plastic undamaged path *abc* is the same as in the flow curve (Fig. 14) followed by the departure of the experimental yield surface from the virtual undamaged yield surface at point *c*. Point *c* can be considered as the hypothetic damage initiation site where the material hardening modulus becomes progressively sensitive to the amount of damage leading to the declination of the material loading capacity. Due to increased damage, the material reaches its ultimate stress capacity at *d* where the hardening modulus becomes zero. This usually occurs in ductile metals when the material loading capacity decreases by 30–70% of its full capacity due to the accumulated damage [21]. The observed fracture initiation site is denoted by point *e,* and finally, the fracture is indicated by point *f*.

The damage curve is approximated as follows: Its segment *ac* is described by Eq. (25), whereas the scalar damage parameter (*D*) [20] is used to describe the material flow past damage initiation (segment *cf*)

$$\sigma_{e-cf} = (1 - D)\sigma_e \tag{27}$$

where σ_e is the hypothetic undamaged stress as predicted by Eq. (25).

The partition of the total energy spent in cutting is represented through the power balance as

$$P_c = P_{pd} + P_{fR} + P_{fF} \tag{28}$$

Fig. 24 Damage curve

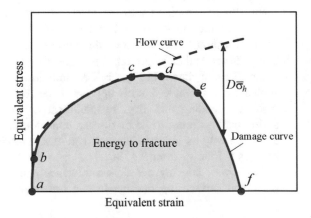

where P_c is the cutting power, i.e., the power required by the cutting system for its existence. It is calculated using the total energy supplied to this system, E_{cs}; P_{pd} is the power spent on the plastic deformation of the layer being removed, calculated through the energy of plastic deformation of the work material to fracture, E_f; P_{fR} is the power spent at the tool–chip interface; P_{fR} is the power spent at the tool–workpiece interface.

The analysis of the energy/power partition in metal cutting represented by Eq. (28) revealed that the power spent on plastic deformation of the work material in its transformation into the chip during machining P_{pd} constitutes up to 75% of the total energy supplied to the cutting system [22]. The greater the ductility of the work material, the higher is P_{pd}. As plastic deformation of the chip serves no useful purpose after machining, up to 75% of the energy supplied to the cutting system is simply wasted. Moreover, only less than 2% of this energy is actually stored in the chip whereas remaining 98% converts into the thermal energy flowing as heat in the chip, tool and workpiece. This causes high temperatures in the machining zone, that, in turn, reduces tool life and restricts the allowable cutting speed, and thus productivity of machining.

For many years of machining history, the cycle time due to actual machining was "a drop in the bucket" compared to the total time part spent in the machine as part loading–unloading, part proper positioning in the chuck (e.g., with an indicator), tool change, part measuring, inspection within the machine, etc. Even a 100% reduction of the actual machining time or increasing tool life by 50% did not contribute noticeably to machining productivity and efficiency.

This "gloomy picture" has been rapidly changing since the beginning of the twenty-first century as global competition forced many manufacturing companies, first of all car manufacturers, to increase efficiency and quality of machining operations. To address these issues, leading tool and machine manufacturers have developed a number of new products as new tool materials and coatings, new cutting inserts and tool designs, new tool holders, powerful precision machines, part fixtures, advanced controllers that provide a wide spectrum of information on cutting processes, and so on. These increase the efficiency of machining operations in industry by increasing working speeds, feed rates, tool life, and reliability. These changes can be called the "silent" machining revolution as they happened in rather short period of time. Implementation of the listed developments led to a stunning result: For the first time in the manufacturing history, the machining operating time became a bottleneck in the part machining cycle time.

To address this challenge of a new era, a closer look at the materials properties involved in cutting should be taken. As the energy of plastic deformation of the material to fracture is greatest by far and thus affects the productivity and efficiency of machining operations, this energy should be analyzed so that some means to reduce this energy can be developed.

The energy of plastic deformation of the material to fracture is calculated as the area under the damage curve [18] as

$$E_f = \int\limits_0^{\varepsilon_f} \sigma_e \mathrm{d}\varepsilon \tag{29}$$

Therefore, a question is if this area and thus energy can be reduced or it is an inherent property of the work material.

It was pointed out earlier that metal cutting involved a three-dimensional state of stress in the deformation zone ahead of the cutting tool. It is known that the material ductile fracture (i.e., the strain at fracture) is affected by the path under which this deformation was developed [23]. The process of fracture is strictly path dependent, and the fracture strain in one process may differ from another. Bai and Wierzbicki [24] showed that the stress triaxiality state parameter (η), which reflects the effect of the mean stress (σ_m), can be used to characterize the state of stress. This dimensionless parameter is calculated as

$$\eta = \frac{\sigma_m}{\sigma_e} \tag{30}$$

and is considered as an important factor in formulating ductile fracture models in the literature [25, 26].

In practical assessments of the influence of the state of stress and in materials fracture tests, another parameter of the stress state, known as the normalized third deviatoric invariant, ξ, that affects a material's ductility and thus affects its fracture strain [27] is used. Parameters ξ and η are correlated as

$$\xi = \frac{27}{2}\eta\left(\eta^2 - \frac{1}{3}\right) \tag{31}$$

The tensile and compression axial symmetry loading states $\xi = 1$ and $\xi = -1$ can be achieved experimentally by using the classical notched or smooth round bar specimen.

According to the authors' opinion, reducing strain to fracture by controlling the state of stress in the deformation zone in machining through cutting tool design and selection of machining regime is the most efficient measure of the metal cutting process optimization because this does not involve additional costs. To explore this option, one must be able to determine the stress-state dependence of the work material, incorporate this dependence in material constitutive models, and then correlate the parameters of the tool geometry and machining regime with this state in order to find their combination for the tool and process optimization.

The last question to be answered is how much the state of stress influences the strain at fracture, and thus the energy of plastic deformation to fracture. To visualize this influence, the fracture locus, which is a curve representing the dependence of the strain at fracture on the state of stress, should be constructed using experimental data [20]. Figure 25 shows fracture locus for steel AISI 1045. As can be seen, varying the stress triaxiality from −0.25 to 0.6, the plastic strain at damage initiation

Fig. 25 Fracture locus for
steel AISI 1045 obtained
using the developed
double-notched specimen

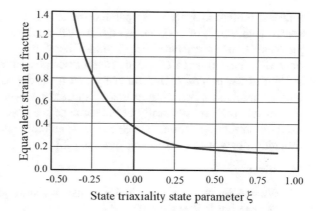

State triaxiality state parameter ξ

decreased from 0.81 to 0.17. In other words, the area under the damage curve for a given work material (Fig. 24), which represents the work of plastic deformation in cutting, can be altered to a wide extent by varying stress triaxiality.

4.2.3 Summary of Relevant Materials Properties in Cutting Operations

What unites forming processes listed in Fig. 19 in terms of materials properties involved is all of them are accomplished by fracture in order to bring the workpiece into the final desirable shape part through cutting chip/slag. This implies the following:

- The stress needed for fracture of the work material must be achieved. This sets manufacturing materials-related calculating apart from those used in the design.
- Contrary to forming processes/operations (see Fig. 11), plasticity (commonly referred to as ductility) as the amount of plastic deformation of the work material before fracture occurs should be brought to its minimum as it causes: (a) de-formed edges of the part after fracture (i.e., poor quality), (b) increased cutting force and temperature that increase tool wear and thus lower tool life. The smaller the plasticity/ductility of the work material, the better its machinability.
- The objective of any cutting process, and thus its criterion of optimization, is achieving fracture with minimum possible energy defined by the area under the shear stress–strain diagram (Fig. 5b) or under the damage curve (Fig. 14). While not much can be done to reduce this energy in shearing operations besides optimizing the clearance between punch and die [17], in metal cutting, on the contrary, this energy can be varied in wide range and thus successfully mini-mized by altering the state of stress in the deformation zone [23].

To obtain the full advantage of the state of stress in meat cutting, one should realize that this state depends on practically all geometrical and design parameters of the cutting tool (mainly the rake, clearance, cutting edge, and inclination angles), tool material (adhesion properties) parameters of the machining regime (the cutting speed, feed, depth of cut), and so on. A cyclic nature of the cutting process complicates the whole picture even further. Therefore, to design/optimize machining operation the influence of the listed parameters on the state of stress should be accounted for properly. Unfortunately, the available software packages for metal cutting simulations do not meet these requirements even to the first approximation. As a result, they are not really of the same help as the commercial FEM packages used in forming operations (see Sect. 4.1.4) although metal cutting process is often thought of as a metal deforming process [28].

Questions

1. What is the most common and very useful test to obtain mechanical properties of materials? Why is it so?
   The most common and very useful test of mechanical properties of materials is a tension test. Tensile tests are performed for several reasons. The results of tensile tests are used in selecting materials for engineering applications. Basic tensile properties are normally included in material specifications to be used in the design and manufacturing of products. Tensile properties are often used to predict the behavior of a material under forms of loading other than uniaxial tension.
2. What is the major outcome of a tension test?
   The major outcome of the tension test is a stress–strain diagram. A stress–strain diagram is a diagram in which corresponding values of stress and strain are plotted against each other. Values of stress are usually plotted as ordinates (vertically) and values of strain as abscissas (horizontally).
3. What is the design criterion?
   It is the allowable working stress obtained by dividing the strength of this material (the yield strength for ductile materials and ultimate strength for brittle materials) by the factor of safety.
4. What is a forming process in the manufacturing sense?
   A forming process is a chipless or non-material removal process through three-dimensional or plastic modification of a shape while retaining its mass and material cohesion.
5. What are two types of forming processes?
   At the most general level, forming processes are divided into bulk metal forming and sheet metal forming.
6. What is formability of a material?
   Formability of a material is defined by its plasticity (ductility) considered as the amount of plastic deformation of the work material before undesirable phenomena occur, for example cracks, wrinkling.
7. What is cutting in the manufacturing sense?
   Cutting in the manufacturing sense is physical separation of a material on two or more parts carried out by a cutting tool (blade) in a controllable manner.

8. What is the most relevant property of the work material in shearing processes?
   The area under the shear stress–strain curve defines the energy needed for shearing of a unit volume of the work material. Knowing this energy, one can easily calculate the power needed in punching, and then dividing this power over the punching speed one can obtain the punching force.
9. What is the prime work material property in metal cutting?
   The energy of plastic deformation of the material to fracture is calculated as the area under the damage curve. This energy defines all other process outcomes as the cutting force, cutting temperature, tool life, quality of the machined surface, etc.
10. What is the technical objective of the cutting process optimization?
    The objective of any cutting process, and thus its criterion of optimization, is achieving fracture with minimum possible energy defined by the area under the damage curve. This energy can be varied in wide range and thus successfully minimized by altering the state of stress in the deformation zone.

Glossary

A stress–strain diagram A diagram in which corresponding values of stress and strain are plotted against each other. Values of stress are usually plotted as ordinates (vertically) and values of strain as abscissas (horizontally).

Cutting Physical separation of a material into two or more parts.

Ductility Ability of a material to deform plastically before fracturing.

Damage curve Curve is obtained from the flow curve by adding the work material degradation and fracture regions.

Factor of safety (FoS) also known as safety factor (SF) A term describing the structural capacity of a part or system beyond the expected loads or actual loads.

Flow curve A curve in the true stress-true strain coordinate describing the material behavior up to fracture.

Forming processes Manufacturing processes done through three-dimensional or plastic modification of a shape while retaining its mass and material cohesion

Fracture locus An experimentally obtained curve representing the dependence of the strain at fracture on the state of stress.

Hardness The resistance of a material to indenter penetration. Measured in MPa or GPa.

Metal cutting (machining) Purposeful fracture of a thin layer on the workpiece by a wedge-shaped cutting tool occurring under a combined stress in the deformation zone in cyclic manner.

Modulus of elasticity The slope of elastic segment of a stress–strain diagram. It is a measure of the stiffness of a solid material. Measured in GPa.

Poisson's ratio The ratio of the magnitude of the lateral contraction strain to the axial strain. Dimensionless.

Resilience The strain energy stored by body up to elastic limit. Measured in J/m^3.

Shear modulus The slope of the elastic segment of a shear stress–strain diagram. Measured in GPa.

Shearing operations Cutting operations the deformation and then separation of a material substance in which parallel surfaces are made to slide past one another.

Stress triaxiality state parameter The ratio of the mean and equivalent stress. Used to characterize the state of stress.

True stress and true strain Stress and strain determined based on the actual cross-sectional area of the specimen.

Toughness The energy need to fracture of a unit volume of a material. Measured in J/m^3.

von Mises stress criterion States that yielding of a material occurs when the equivalent stress reaches the yield strength of the material in simple tension.

Ultimate shear strength The shear maximum stress developed by the material before fracture based on the original cross-sectional area. Measured in MPa.

Ultimate tensile strength (UTS) The maximum normal stress developed by the material before fracture based on the original cross-sectional area. Measured in MPa.

Yield shear strength The shear stress at which a material exhibits a deviation from the proportionality of stress to strain. Measured in MPa.

Yield tensile strength The normal stress at which a material exhibits a deviation from the proportionality of stress to strain, i.e., the occurrence permanents plastic strain is the case. Measured in MPa

References

1. Davis, J. R. (2004). *Tensile testing* (2nd ed.). Materials Park, OH: ASTM.
2. Dieter, G. (1976). *Mechanical metallurgy* (2nd ed.). New York: McGraw-Hill.
3. Zhu, X.-K., & Joyce, J. A. (2012). Review of fracture toughness (G, K, J, CTOD, CTOA) testing and standardization. *Engineering Fracture Mechanics, 85,* 1–46.
4. Isakov, E. (2000). *Mechanical properties of work materials.* Cincinnati, OH: Hanser Gardener Publications.

5. von Mises, R. (1913). Mechanik der festen Körper im plastisch deformablen Zustand. *Göttin Nachr Math Phys, 1,* 582–592.
6. Hencky, H. Z. (1924). Zur theorie plastischer deformationen und hierduch im material hervorgerufenen nebenspannunger. *Z Angerw Math Mech, 4,* 323–334.
7. Burdekin, F. M. (2007). General principles of the use of safety factors in design and assessment. *Engineering Failure Analysis, 14,* 420–433.
8. Oberg, E. (Ed.) (2016). *Machinery's handbook. Technology & Engineering.* Industrial Press: New York.
9. Groover, M. P. (2010). *Fundamentals of modern manufacturing: materials, processes and systems* (4th ed.). Hoboken, NJ: Wiley.
10. Atkins, A. G., & Mai, Y. W. (1985). *Elastic and plastic fracture: metals, polymers. Ceramics, composites, biological materials.* New York: Wiley.
11. Johnson, G. R., & Cook, W. H. (1983). A constructive model and data for metals subjected to large strains, high strain rates and high temperatures. In *Proceedings of the 7th International Symposium on Ballistics,* The Hague.
12. Keeler, S., & Backofen, W. (1963). Plastic instability and fracture in sheets stretched over rigid punches. *ASM Trans Quart, 56*(1), 25–48.
13. Goodwin, G. M. (1968). Application of strain analysis to sheet metal problems in press shop. *Metallurgica Italiana, 60,* 767–774.
14. Roll, K. (2008). Simulation of sheet metal forming—necessary developments in the future. In: *LS-DYNA Anwenderforum* (pp. A-1-59–A-1-68), Baberg.
15. Kalpakjian, S., & Schmid, S. R. (2001). *Manufacturing engineering and technology* (4th ed.). New Jersey: Prentice-Hall.
16. Taylor, F. W. (1907). On the art of cutting metals. *Transactions of ASME, 28,* 70–350.
17. Astakhov, V. P. (2010). *Geometry of single-point turning tools and drills: Fundamentals and practical applications.* London: Springer.
18. Astakhov, V. P. (1998/1999). *Metal cutting mechanics.* Boca Raton, USA: CRC Press.
19. Astakhov, V. P. (2006). *Tribology of metal cutting.* London: Elsevier.
20. Abushawashi, Y., Xiao, X., & Astakhov, V. P. (2013). A novel approach for determining material constitutive parameters for a wide range of triaxiality under plane strain loading conditions. *International Journal of Mechanical Sciences, 74,* 133–142.
21. Zhang, Y. C., Mabrouki, T., Nelias, D., & Gong, Y. D. (2011). Chip formation in orthogonal cutting considering interface limiting shear stress and damage evolution based on fracture energy approach. *Finite Elements in Analysis and Design, 47,* 860–863.
22. Astakhov, V. P., & Xiao, X. (2008). A methodology for practical cutting force evaluation based on the energy spent in the cutting system. *Machining Science and Technology, 12,* 325–347.
23. Astakhov, V.P., & Xiao, X. (2016). The principle of minimum strain energy to fracture of the work material and its application in modern cutting technologies. In J. P. Davim (Ed.), *Metal cutting technology* (pp. 1–35). Berlin: De Gruyter.
24. Bai, Y., & Wierzbicki, T. (2008). A new model of metal plasticity and fracture with pressure and Lode dependence. *International Journal of Plasticity, 24*(6), 1071–1096.
25. Wierzbicki, T. B., Lee, Y. W., Bao, Y., & Bai, Y. (2005). Calibration and evaluation of seven fracture models. *International Journal of Mechanical Science, 47,* 719–743.
26. Bai, Y., Teng, X., & Wierzbicki, T. (2009). On the application of stress triaxiality formula for plane strain fracture testing. *Journal of Engineering Materials and Technology, 131,* 021002.
27. Teng, X., & Wierzbicki, T. (2006). Evaluation of six fracture models in high velocity perforation. *Engineering Fracture Mechanics, 73*(12), 1653–1678.
28. Shaw, M. C. (2004). *Metal cutting principles* (2nd ed.). Oxford: Oxford University Press.

Chapter 2
Analysis and Material Selection of a Continuously Variable Transmission (CVT) for a Bicycle Drivetrain

Ewan M. Berge and A. Pramanik

Abstract There has been much research and deliberation on what the optimum cadence for a cyclist is to achieve maximum efficiency and hence maximum performance. Research to date has shown that the ideal pedal rate varies based on the particular cycling task, such as course geography and duration. Several components of a CVT bicycle were analysed using SolidWorks finite element analysis to determine the von Mises stress. These results were used to determine and examine candidate materials for the components of the drivetrain by using Granta Design software. Technical feasibility study was carried out on the drivetrain where a v-belt was subjected to a pretension load which is uniform around the belt before pedalling. As the cyclist pedals, one side of the belt tightens whilst the other side of the belt slackens. However, the sum of the tension in the belt never changes and hence the load on the pulley. For a professional cyclist starting their sprint, 300 Nm applied torque at the highest gear ratio, the required pretension to prevent slippage is roughly 1500 N. The simulation yielded an initial maximum von Mises stress of 1250 MPa in the corner of the spindle. This stress was reduced to 1059 MPa by adding 0.7 mm radius to the corners of the cut-outs. Cadence is the rate at which a cyclist is pedalling/turning the pedals. Higher cadence in the order of (100–120) rpm would improve sprint cycling over shorter distances where the muscle fatigue is reduced and the cycling power output is increased. A well-maintained chain drive system with a derailleur setup can achieve an efficiency of 98.6%, whereas a well-maintained v-belt drive 97%. The reduced efficiency of a v-belt CVT drivetrain plus design constraints will be a hindrance to its adoption into professional cycling racing; however, a market for the recreation cyclist exists. The design will require further iterations to reduce the number of parts, improve integration with a standard bicycle, increase efficiency and will also require an extensive testing/commissioning phase. Results from the FEA indicated that materials with yield strengths exceeding 500 MPa would be required. Due to the

E. M. Berge · A. Pramanik (✉)
Department of Mechanical Engineering, Curtin University, Perth, WA 6845, Australia
e-mail: alokesh.pramanik@curtin.edu.au

© Springer International Publishing AG, part of Springer Nature 2018
J. P. Davim (ed.), *Introduction to Mechanical Engineering*, Materials Forming,
Machining and Tribology, https://doi.org/10.1007/978-3-319-78488-5_2

nature of loading, materials with properties such as high specific stiffness and strength would be necessary.

Keywords Bicycle drivetrain · Continuously variable transmission Design · Material selection

1 Introduction

Bicycles are enjoyed by people from all backgrounds and ages throughout the world. The common chain and sprocket drivetrain developed in the late nineteenth century for bicycles in principle have remained largely unchanged. In contrast, a car's transmission has seen many advances, from manual, automatic and most recently a continuously variable transmission (CVT). Unlike a multiple gear chain drive, a CVT offers a smooth transition through a defined gear range by providing an infinite number of gears. The first commercially successful, true CVT for a bicycle was manufactured by Fallbrook Technologies Inc., which became publicly available in 2006. The NuVinci CVT comprises a set of balls positioned between an input disc and an output disc. The balls are free to rotate and tilt on their own axis around a central hub. Torque is transmitted from the input disc to the output disc through the balls via a layer a viscous fluid. The balls are held in place by the clamping force of the discs onto the balls [1].

The speed ratio of the system is determined by the angle of the ball axis compared to that of the central transmission axis. When the balls are tilted so that the perpendicular distance between the input disc edge and the ball axis is greater than the perpendicular distance between the output disc edge and the ball axis, i.e. $r_i > r_o$, the output disc will rotate slower than the input disc. When $r_i < r_o$, the output disc will rotate slower than the input thus creating an overdrive speed ratio. When the ball and transmission axis are collinear, then the input speed and output speed are the same. The system comprises of two units, a front pulley and a rear pulley which work in tension to maintain the same belt length and thus correct pretension in the system. As the name suggests, the drivetrain is able to deliver an unlimited number of gear ratios between the upper and lower limits.

This study looks at the material selection and analysis of a 3D prototype CVT drivetrain on a bicycle. The main objective was to refine the current design and further develop it mechanically so a physical prototype could be manufactured to aid in further research and development. The CVT drivetrain consists of two variable diameter pulleys connected via a v-belt. The principle operation is similar to that of a convention pulley-style CVT allowing infinite gear ratios over a predefined range at a constant input speed. The study also looks at the feasibility of applying CVT technology to a bicycle. The use of a CVT on a bicycle is not new, with the NuVinci CVT having seen success in this area. This success has been limited to non-performance-orientated cyclists, who are after simplicity and ease of maintenance. The major factor that has prevented CVT adoption in professional cycling racing is due to weight, size and inefficiency. A typical chain and derailleur system can have a drivetrain efficiency above 98%. A typical v-belt design that uses friction to transmit power is around 95%. Third party analysis conducted on the NuVinci CVT indicated efficiencies below 90%. These less-efficient drivetrains do not appeal to the performance-oriented cyclists who are looking at converting the most power to the rear wheel with the least amount of energy. In addition to the study of efficiency, physical factors and limitations to the design are addressed. These design constraints including pedal width, aerodynamics, weight, maintenance, crank length, aesthetics and durability are discussed. It is important that these design constraints are met in order to maintain the cyclist's comfort and ease of use. The study indicated that the current design would not meet the requirements of a performance-oriented cyclist; however, the use of the transmission on e-bike could have merit.

2   Background

The bicycle is one of the most identifiable creations of mechanical equipment worldwide due to its availability and mass production. The bicycle is a common occurrence in most countries of the world being used as the primary mode of transport in developing countries and for predominately recreational use throughout the developed world. Advances in mechanical design in today's market of the bicycle are more focused on a niche set of consumers. These advances include components such as the braking system, material selection and drivetrain components. This part will focus on the development and design of a drivetrain system for professional cyclists. A general history of the evolution of the bicycle drivetrain follows.

During the late 1830s and early 1840s, the first mechanically propelled two-wheel bicycles started to appear and were known as a treadle bicycle as they were powered by a treadle instead of a crank [2]. The rear wheel was driven using a

Fig. 1 Treadle (left) and Ariel High-Wheeler Bicycle (right), Woodforde [2: 18 and 42]

set of front-mounted treadles connected by rods to a rear crank as shown in Fig. 1. This design is similar to the transmission of a steam locomotive.

The 1860s saw the move from powering the rear wheel to the front wheel. The bicycle was coined the "boneshaker" and was constructed of a metal frame which was lighter than its wooden counterparts. The wrought iron frame variants were assembled in a diamond shape similar to that of today's bicycles. Rotary cranks and pedals were mounted directly onto the front wheel hub which made pedalling the bicycle at higher speeds much easier. Later improvements included rubber tyres and ball bearings [3]. The high-wheeler bicycle, also commonly referred to as the "penny farthing" due to the difference between the front and rear wheel diameters, came to shape during the 1870s and early 1880s. It featured a high wire-spoke tensioned wheel equipped with cranks and pedals connected to the front hub. The larger front wheel compared to its predecessor created a smoother ride which allowed for increased speeds; however, it was very dangerous due to the rider's high centre of gravity and proximity to the front wheel [3]. The bicycles were equipped with solid rubber tyres, cast iron frames and incorporated plain bearings for the steering, pedals and wheels. The speed ratio was obviously determined by the height of the wheel, which was limited somewhat to the cyclist's height.

Further refinements to the high wheeler were focused on making the bicycle safer and improving its stability by reducing the front wheel diameter. This allowed the seat to be lower to the ground which in turn lowered the cyclist's centre of gravity. This principle was applied to the Xtraordinary bicycle, featuring levers connected to cranks at the hub allowing for a smaller diameter wheel as shown in Fig. 2.

In 1879, the Facile bicycle emerged which incorporated two gears in mesh. A central sun gear was driven by a smaller planetary gear. The cyclist would push down on the pedals alternatively in a stepping motion that in turn drove the sun gear via rods as shown in Fig. 2. By 1884, due to advancements in chain pins, the first chain-driven bicycle named the Kangaroo was invented. The front wheel was driven by a set of cranks which sat below the centre of the wheel. These cranks

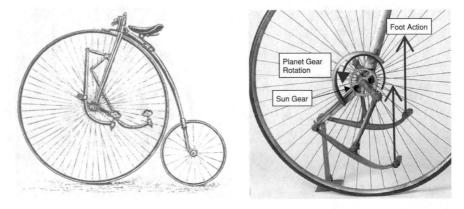

Fig. 2 Xtraordinary 1878 (left) and Geared Facile 1879 (right) bicycles, Woodforde [2: 59], Streeter [31]

were connected by a sprocket which transferred power to the centre of the hub by chains either side of the wheel as shown in Fig. 3 [2].

By 1885, today's common bicycle had taken shape. These bicycles were referred to as the "safety bike", due to the lowered seat height in conjunction to that of its predecessors. The Rover bicycle combined many of the idiosyncrasies we associate with modern bicycles. These included direct steering, equal diameter wheels, chain and sprocket drive to the rear wheel and the familiar diamond-shaped frame for stiffness. It contained all the basic components that form the style of the present-day bicycle.

During the early twentieth century, bicycle manufacturers looked at ways to produce bicycles with multiple gears. The concept was to equip the rear wheel with different sized sprockets either side of the hub. To achieve the other gear ratio, a cyclist would have to dismount the bicycle, flip the wheel and reconnect the chain. In the 1930s, a mechanical system for changing gears was developed in France and introduced to professional cycling racing in 1937 [4].

The derailleur allowed the shifting of the chain onto different sized rear sprockets, known as a "cassette" by a mechanical cable. Figure 4 is a pictorial representation of a current derailleur setup. It incorporates a front shifter that is typically used to change between two or three front sprockets known as a chainring and a cassette with ten sprockets. Advancements of the derailleur have focused on improvements in shifting precision, changing under load, ease of shifting and a move towards electronic control as opposed to mechanical control.

A modern professional road bike with two chainrings and a ten-speed cassette theoretically has twenty gear ratios. In reality, many of these ratios are overlapping so the number of usable gears would be much less. For this setup, the range between the lowest and highest gears is 428% meaning a cyclist would be able to travel 4.28 times faster assuming that the pedalling rate known as cadence remains the same. This equates to a mean step between each of the gears being 12.9% as shown in Table 1 along with other bicycle drivetrains.

In the professional world of competitive cycling, a cyclist aspires to maximise their efficiency. The maximum efficiency is determined in part by the combination

Fig. 3 Kangaroo drivetrain 1884 (left) and Premier Safety 1885 bicycle (right), Vintage Bicycle [32], Woodforde [2]

Fig. 4 Derailleur mechanism for changing sprockets sizes, Brown [33]

Table 1 Drivetrain transmission types showing ranges and mean step

Range (%)	Transmission type	Usable gears	Mean step (%)
180	Three-speed hub gear	3	34.2
380	NuVinci CVT	Continuous	None
428	Road 2 chainring derailleur setup (2 × 10, 50 − 34 × 11 − 32)	13	12.9
526	Rohloff Speed hub 14-speed hub gear	14	13.6
599	Treadlie CVT	Continuous	None
698	Touring 3 chainring derailleur setup (3 × 10, 48 − 34 − 20 × 11 − 32)	15	14.9

of cadence and resistance that will produce the most power with the least amount of stress on the body. Factors that contribute to this stress include oxygen consumption, heart rate level and muscle fatigue. Studies suggest that the optimum efficiency was around 90 rpm [5]. Since the mean steps in gear ratios for derailleur drivetrains are limited to 12% as shown in Table 1 there will be situations where the optimal cadence will not be possible. A continuously variable transmission (CVT) will allow for a continuous gear range between the lowest and highest gears. This will allow a professional cyclist to ride at the most efficient cadence and a recreational cyclist to ride at the most comfortable cadence.

A typical cone CVT system includes a driven and drive pulley, with cones at each pulley that are connected by a steel belt that is in the shape of a wedge. The cones move closer together or further apart which in turn increases or decreases the effective diameter at the point the belt rotates [6]. Since the belt length must remain constant, when one of the pulley diameters changes, the other pulley changes its diameter accordingly. A method used to move the cones is by a centrifugal force generated from weights inside the cone. Torque is transmitted via friction from the belt to the cone as shown in Fig. 5. CVT's are commonplace on scooters, small tractors, snowmobiles and are increasing in popularity amongst cars. CVT use as a bicycle transmission is limited. One particular CVT system named the NuVinci CVT has been adopted by some bicycle manufacturers.

The first commercially successful, true CVT for a bicycle was manufactured by Fallbrook Technologies Inc., which became publicly available in 2006. The NuVinci CVT comprises a set of balls positioned between an input disc and an output disc as evident in Fig. 6. The balls are free to rotate and tilt on their own axis around a central hub. Torque is transmitted from the input disc to the output disc through the balls via a layer a viscous fluid. The balls are held in place by the clamping force of the discs onto the balls [1]. The speed ratio of the system is determined by the angle of the ball axis compared to that of the central transmission axis. As shown in Fig. 6, when the balls are tilted so that the perpendicular distance between the input disc edge and the ball axis is greater than the perpendicular distance between the output disc edge and the ball axis, i.e. $r_i > r_o$, the output disc will rotate slower than the input disc. When $r_i < r_o$, the output disc will rotate slower than the input thus creating an overdrive speed ratio. When the ball and transmission axis are collinear, then the input speed and output speed are the same [1]. The NuVinci CVT is incorporated into the rear hub of a bicycle. A rear sprocket is connected to the input disc which in turn is driven by a chain connected to a front sprocket. The setup has the appearance of a typical, single-speed bicycle. The NuVinci N380 CVT released in 2015 is capable of providing a continuous gear width of 380%, 0.5 underdrive to 1.9 overdrive [7]. This is less in comparison with that of a typical road racing setup which has a gearing ratio of between 420 and 450%.

The main hurdle for not just the NuVinci CVT but all CVT systems is the efficiency of the overall transmission. Efficiency details have not been released by the manufacturer, however; independent efficiency tests resulted in measurements of 78–86% [8]. A well-maintained chain drive system with a derailleur setup can

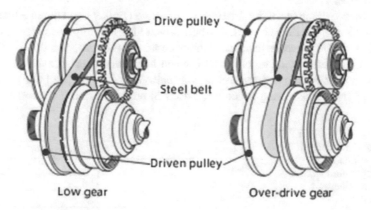

Fig. 5 Continuously variable transmission (CVT) of a car [34]

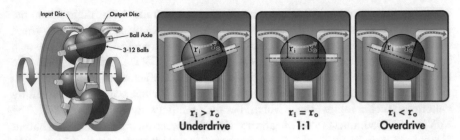

Fig. 6 Schematic of NuVinci CVT principle [1]

achieve an efficiency of 98.6% [9]. Other factors such as scalability and weight have kept CVTs from widespread adoption.

The major benefit with a CVT is that it allows the engine to constantly run at the most efficient power band, which in turn reduces fuel consumption. Additionally, a CVT can often be manufactured lighter than a manual or automatic transmission. A major drawback to CVTs is the amount of torque that can be applied to the transmission. This is restricted by the belt which requires friction to transfer the torque and is susceptible to slippage under high torques. As such, no automobile with an eight-cylinder engine has been manufactured with a CVT. A belt-driven CVT offers an overall drivetrain efficiency of 88% [10]. This is a lot lower than that of a manual or automatic transmission and, however, is offset by the fact the engine can run at the most optimal speed thus reducing fuel consumption (Table 2).

V-belt drive efficiency loss is caused by many factors. Notched style v-belts have shown to be run cooler and last longer than standard unnotched belts. They are generally about 2% more efficient. The reason behind this is due to the reduction in compressive stresses during belt bending and flexing. The notches allow the belt to bend without clogging and provide better engagement against the pulley sheaves. This helps to discourage any energy-consuming slippage. This is clearly evident in Fig. 7 that the energy lost through heat from an unnotched v-belt compared to a notched v-belt.

The sheave diameter of the driven and non-driven pulley has a direct effect on the drive efficiency. Studies show that v-belt drive efficiency improves steadily as the diameters of the sheaves increase. Under certain conditions, the drive efficiency could be increased by as much as 10% when larger diameter sheaves are utilised. Factors such as torque and tension play a major role. The largest pulleys possible should be used considering space limitations in order to achieve the most efficient

Table 2 Efficiencies of various transmissions (https://www.physicsforums.com/threads/cvt-efficiency.304935/)

Transmission type	Efficiency (%)
Manual	97
Automatic	86
CVT belt	88
CVT toroidal	93

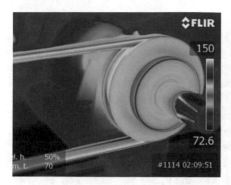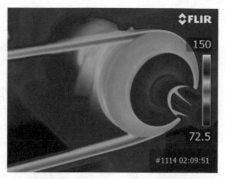

Fig. 7 Comparison between heat dissipation from a notched (right) and unnotched (left) v-belts [11]

transmission. The benefits of using larger diameter pulleys include less tension due to more surface contact with the pulley. This translates to reducing loading on the bearings. Further, the v-belt bends less, therefore generating less heat resulting in improved efficiency. The wedging action of the v-belt into the pulley groove is improved allowing for increased tension. Less bending stresses in the belt prolong the longevity with regard to failure [11]. Alignment of v-belts refers to the positioning of the pulleys in relation to one another. Losses of up to 2% can result in parallel or offset misalignment. In the case, the v-belt favours one side of the pulley resulting in uneven loading causing heat build-up and loss of efficiency. Another consequence of this misalignment is the unnecessary and additional wear on the pulleys. Improper tensioning of the v-belts is the most energy-consuming mistake. Energy losses of up to 20% may be experienced due to over-tensioned or under-tensioned belt. Under-tensioning accelerates the wear of the belt and the pulley due to slippage. This causes the belt to harden or "heat age" which resists bending and flexing this reducing efficiency. It is important the regular maintenance is carried out on the transmission. The tension is the system can reduce significantly over an hour of operation. This is not so much due to elongation in the belt but more to do with the belt heating up, which causes the belt to fit better into the pulley.

Cadence is the rate at which a cyclist is pedalling/turning the pedals. It is the number of revolutions of the crank per minute (rpm). The wheels in turn spin proportional to the cadence and, however, change depending on the gearing setup. It is an important factor, which influences the power output, energy burnt and development of fatigue during cycling [12]. Research has also suggested that the best cadence is associated with neuromuscular fatigue, blood flow, comfort and oxygen consumption. It has also been suggested that the efficiency of a cyclist is dictated by the individual's muscle composition.

There has been much research and deliberation on what the optimum cadence for a cyclist is to achieve maximum efficiency and hence maximum performance. Research to date has shown that the ideal pedal rate varies based on the particular cycling task, such as course geography and duration. Higher cadence in the order of

100–120 rpm would improve sprint cycling over shorter distances. In these situations, muscle fatigue is reduced and the cycling power output is increased [12]. For cycling events, up to four hours in length, a reduced cadence of 90–100 rpm would be more beneficial. For an increased length in cycling time, the optimal cadence falls. For endurance racing staged over multiple days, relativity low cadences of 70–90 rpm improve cycling economy and lower metabolic consumption [12]. This is also true for high climbs where relativity lower cadences are better suited at 70 rpm. Lucia et al. [12] stated that lower cadences are known to increase the pedal forces necessary to maintain a given power output. The ideal cadence is also related to the workload of the cyclist. Extremely low cadences of 50–60 rpm appear to be most efficient at low power outputs of less than 200 W. However, at lower cadences, more force is required per pedal stroke to maintain the power output which results in additional muscle fibre recruitment at a higher metabolic expenditure. The optimal cadence also increases linearly with increased output. With increased workloads of approximately 350 W, higher cadences of 80–100 rpm are more suitable.

With advancements in technology and design, bicycles are being manufactured with an ever-increasing number of gears. This advancement is mainly due to the invention of the derailleur, which allowed for multiple sprockets on the rear hub and front crank arm assembly letting for a wide selection of gears. A typical road bicycle setup has an 11-speed cassette with two chainrings resulting in a possible twenty-two gears; however, not all are a distinct ratio. The gearing arrangement shown in Fig. 8 is for a typical road bicycle as described above. The two chainrings are 52 and 36 teeth represented by the horizontal lines. The eleven-speed cassette ranges from 11 to 28 teeth as shown by the inverted triangles. The number above the triangle represents the speed in km/h based on a tyre size of 27.5″ and a cadence of 90 rpm for the top chart and 96 rpm for the bottom chart as shown in Fig. 8. Additionally, the top bar represents the metres travelled for one complete revolution of the cranks. Also included is the mean step displayed as a percentage between two adjacent gears. The range of gears between the low and high is 368% as shown in the bottom corner. Not only is the range of gears is important, the mean step between gears is vital to ensure no awkward or uncomfortable gear changes. A moderate change in gears is considered as being less than 15%. A professional cyclist would aim to have step in gears less than 10% to ensure relative constant leg speed [13]. The largest mean step throughout the gear range is 13% as in Fig. 8. This is between the 52:17 and the 52:15 ratios. At 90 rpm, this results in a jump in road speed of 4.9 km/h from 37 to 41.9 km/h. In order for a cyclist to ride at a speed directly between these gears say at 39.5 km/h, an increase in cadence from 90 to 96 rpm will be required as shown in Fig. 8. Therefore, the cyclist can effectively travel at any given speed between the upper and lower gear ratios if the cadence is varied by only ±6 rpm.

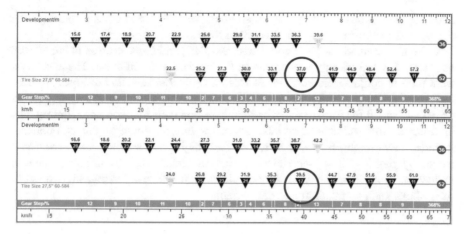

Fig. 8 Typical gearing of a two chainring setup on a road bike. Top figure at 90 rpm, bottom figure at 96 rpm, Feeken [35]

3   Analysis

The design of the CVT drivetrain was analysed using SolidWorks, which is a 3D modelling software package. The main focus was to determine the external forces and torques acting on the system so an FEA simulation could be conducted. These external loads were to be justified by acquiring cycling data and performing engineering calculations from first principles. The forces present in the system included the pretension from the v-belt and torque applied at the pedals from a cyclist. The model was firstly separated into individual components to simplify the analysis, and then the applicable forces/torques were applied and analysed using SolidWorks inbuilt FEA simulation package. The analysis was to be initially limited to a static FEA. The key information recorded from this analysis included deflection/elongation and von Mises stress values. Additionally, results from this analysis were verified using different solving methods such as solid bodies and shell elements. The FEA study suggested that materials with elastic modulus greater than that of steel will be required to achieve an adequate safety factor for the spindle. The crank arm FEA study was achieved by using shell elements. This allowed the shell thickness to be optimised in order to reduce stress whilst limiting its weight.

Following this analysis, an appropriate selection criterion was established based on the dependent variables and a suitable factor of safety was applied. The components were prepared for FEA by simplifying and cleaning the geometry and by suppressing non-structural features and components. Boundary conditions were defined to the model such as loads, fixtures and contact sets. Succeeding this, an appropriate mesh was applied to the components and iterations were made to ensure correct aspect ratios. A simulation was run prioritising in resolving problems, meshing issues, identifying singularities, checking general behaviour, adjustment of boundary conditions if required, examining von Mises stresses/displacements, checking the deformed shape and reaction forces.

Once the simulations were completed and the von Mises stresses obtained, the next step was to make improvements to the model based on these. These stresses were used in conjunction with Granta Design a material selection database, to help select candidate materials for the CVT components. Constraints and objectives were determined for parts of the drivetrain based on their function and requirements in order to find suitable materials. Important criteria amongst the components included low weight, high stiffness, low cost, high yield strength and wear resistance. By plotting these material properties on graphs and by utilising material indices, suitable materials were found. These included low density, high strength steel alloys such as titanium and aluminium.

3.1 V-Belt Theory

V-belt drives transfer torque entirely through friction where maximum belt tension ratio can be derived. Assume that a section of the belt has an angle of $d\phi$ at the centre. The length of the segment of belt is then

$$dl = rd\phi \tag{1}$$

The forces acting on the belt segment are shown in Fig. 9. There is a normal force acting on the belt segment due to the rotation and a friction force which is given by

$$f = \mu dl \tag{2}$$

where μ is the coefficient of friction between the belt and the pulley. The centrifugal force (CF) due to the motion of the belt acting on the belt segment is

$$CF = \frac{1}{r}m(rd\phi)v^2 \tag{3}$$

Fig. 9 Free body diagram of a belt segment

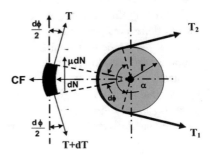

$$CF = mv^2 d\phi \tag{4}$$

where v is the peripheral velocity and m is the mass of the belt per unit length.

$$m = bt\rho \tag{5}$$

where b is the width, t is the belt thickness and ρ is the belt density. For the system to be in equilibrium, the sum of all forces must equal zero.

$$\sum F_t = 0 \quad T \cos\frac{d\phi}{2} - (T + dT)\cos\frac{d\phi}{2} + \mu dN = 0 \tag{6}$$

$$\sum F_n = 0 \quad mv^2 d\phi + dN + T\sin\frac{d\phi}{2} - (T + dT)\left(\sin\frac{d\phi}{2}\right) = 0. \tag{7}$$

For a small angle of $d\phi$

$$\cos\frac{d\phi}{2} \cong 1, \quad \sin\frac{d\phi}{2} \cong \frac{d\phi}{2}. \tag{8}$$

Therefore, Eq. 7 is reduced to

$$dN = \frac{dT}{\mu}. \tag{9}$$

From Eqs. 8 and 9,

$$mv^2 d\phi + \frac{dT}{\mu} - Td\phi = 0 \tag{10}$$

$$\frac{dT}{T - mv^2} = \mu d\phi. \tag{11}$$

Considering the entire angle of wrap,

$$\int_{T_1}^{T_2} \frac{dT}{T - mv^2} = \int_0^\alpha \mu d\phi. \tag{12}$$

Thus, the equation for determining the relationship between the belt tensions is

$$\frac{T_1 - mv^2}{T_2 - mv^2} = e^{\mu\alpha}. \tag{13}$$

Assuming negligible belt mass and velocity and taking into account the V-belt angle Eq. 13 reduces to,

$$\frac{T_1}{T_2} = e^{\mu\alpha/\sin\frac{\theta}{2}}. \tag{14}$$

where θ is the v-belt angle.

A v-belt is subject to a pretension which is uniform around the belt before pedalling. As the cyclist pedals, one side of the belt tightens whilst the other side of the belt slackens. However, the sum of the tension in the belt never changes and hence the load on the frame. This is described by Eq. 15.

$$2\text{Tension}_{\text{pretension}} = \text{Tension}_{\text{tight}} + \text{Tension}_{\text{slack}}. \tag{15}$$

By solving Eq. 15 for the two extreme gear ratios with a variety of coefficients of friction, the pretension required to transmit the given torques can be investigated.

As indicated in Eq. 16, the ratio of tension is an exponential function of the coefficient of friction. Figures 10 and 11 show that as the coefficient of friction is increased, the corresponding decrease in the required pretension diminishes. A coefficient of friction of 0.6 is a reasonable starting point as higher coefficients of friction may become more difficult to achieve, and it would result in only a slight decrease in required pretension. Figure 8 shows that for a professional cyclist starting their sprint (300 Nm applied torque) at the highest gear ratio, the required pretension to prevent slippage is roughly 1500 N. Under this pretension, up to 175 Nm can be applied without belt slippage occurring at the lowest gear ratio, with an applied torque at the driving gear as shown in Fig. 11. This is considered reasonable as it is highly unlikely that 300 Nm of torque would be applied at the lowest gear ratio.

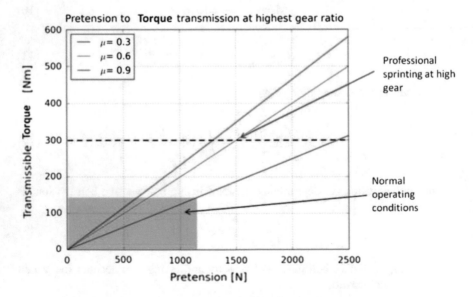

Fig. 10 Pretension to torque transmission at highest gear ratio

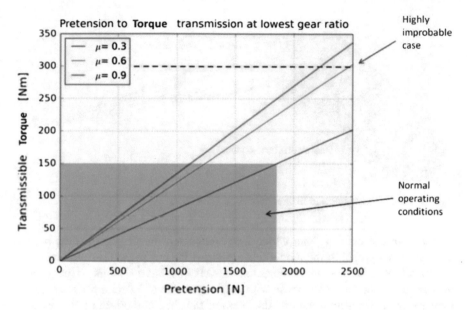

Fig. 11 Pretension to torque transmission at lowest gear ratio

A high pretension also affects other aspects of the bicycle design. The tight side tension in the v-belt when peddling is approximately equivalent to the tension in a chain or timing belt-driven bicycle given the same applied torque. Hence, the peak load on the frame and bearings is equivalent. However, the static pretension, and hence the static load on the frame and bearings, is much higher when using the v-belt considered in the present work.

3.2   FEA of Spindle

The spindle shown in Fig. 12 is supported by two bearings and fits inside the bottom bracket of the bicycle. Dimensionally the outside diameter is restricted at 32 mm, which is a standard size to suit most bicycles. Torque is applied to the spindle from the force exerted by a cyclist on the pedals. A force of 1820 N is applied at a distance of 170 mm (typical crank length) from the centre of the spindle resulting in a torque of 310 Nm. This value is based on independent studies into the maximum power a professional cyclist can generate in a sprint.

Firstly, the model was simplified by removing redundant geometry such as the splines which connect the cranks to the spindle. All radii were removed to achieve a more consistent mesh. Multiple holes were removed from an internal bracket. The outside surface of the spindle was split into multiple surfaces to allow the appropriate application of loads and constraints. A fixed geometry constraint was applied

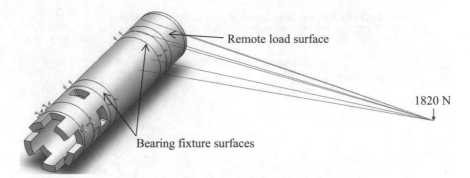

Fig. 12 Loads, constraints and fixtures applied to the spindle

on the opposite crank spline, which fixed movement in all planes. Two bearing fixtures were applied to the model that replicates the bearings. These fixtures were set to allow axial and circumferential movements but no radial expansion or movement. A standard solid mesh with an element size of 1.4 mm was used for the FEA study. This ensured that the thickness of the shell comprised of at least three elements as shown in Fig. 13. An aspect ratio <3 was achieved for 99.7% of all elements, and also no Jacobian elements were present.

The study was simulated yielding an initial maximum von Mises stress of 1250 MPa. This maximum was shown to be on the edges of the square cut-outs as shown in Fig. 12. As a result, a 0.7 mm radius was applied to the corners of the cut-outs and the simulation was rerun. This reduced the maximum von Mises stress

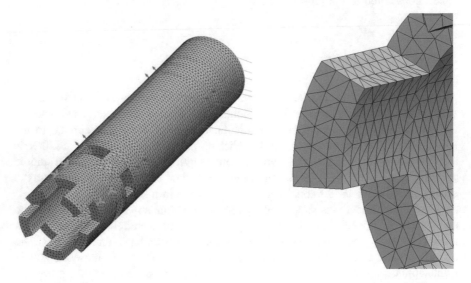

Fig. 13 Meshing applied to the spindle

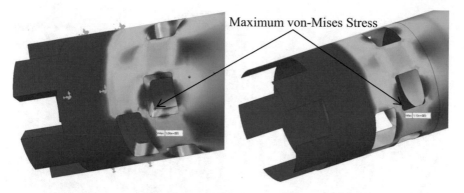

Fig. 14 Comparison of results using solid body (left) and shell elements (right)

to 1059 MPa. As a comparison, the model was also analysed as a shell, improving
solving time of which the results can be seen on the right of Fig. 14. The maximum
von Mises stress was 1112 MPa with the location coinciding with the location from
the solid body analysis, which in turn provided a level of verification for the
analysis. By clipping the results to show only the areas where Von Misses stresses
were greater than 500 MPa, it was evident that stresses were localised and concentrated around the radii of the cut-outs as shown in Fig. 15. The location of these
stresses was also matching.

3.3   FEA of Crank

The crank as pictured in Fig. 16 is dimensionally the same length as a standard
crank on bicycles being 170 mm from the centre of the pedal to centre of the
spindle. This allows for commonality and provides the cyclist with the option of
changing in the future. An eccentric load of 1820 N has been applied at the location
of the pedal being 60 mm offset. This will result in a design torque of 300 Nm. The
crank arm has been given a fixed constraint on the internal surface to replicate the
connection onto the spindle.

The original thickness of the crank was not uniform which would lead to the
manufacturing issues. To achieve a consistent mesh, the crank was converted from
a solid body into a surface. This would allow the thickness of the crank to be easily
changed in order to optimise von Mises stresses and displacements. A standard
solid mesh with an element size of 2.8 mm was used for the FEA study. An aspect
ratio <3 was achieved for 98.7% of all elements. The results of the FEA study can
be seen in Fig. 17. A maximum von Mises stress of 211 MPa was present on the
outer surface where the crank clamps onto the spindle. Initially, the clamping force
was ignored; however, due to the location of maximum stresses, this may require
further investigation.

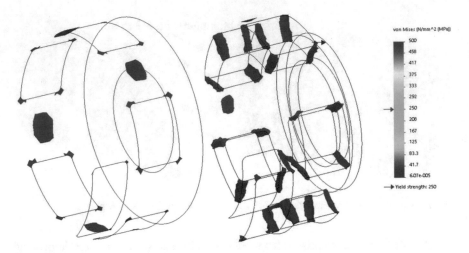

Fig. 15 Comparison between shell (left) and solid body (right) of von Mises stress above 500 MPa

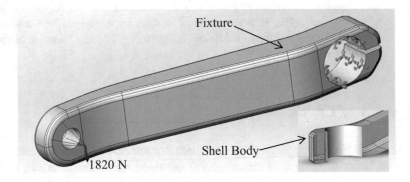

Fig. 16 Loads, constraints and fixtures applied to the crank arm

Although the von Mises stresses were in a reasonable range that would allow the use of many materials, the amount of deflection indicated otherwise. Figure 18 shows the deflected state of the crank with an elastic modulus of 105 GPa, representing a typical titanium alloy. The maximum deflection occurred at the point of the applied load resulting in 0.61 mm. This was deemed to be not acceptable as the deflection was greater than 1/500 industry standard. It would be recommended to increase the thickness of the shell if this material is considered as a suitable candidate for the application.

Table 3 is a summary of the FEA study on the crank and spindle. In summary, the maximum von Mises stress for the spindle and crank were 1059 and 211 MPa, respectively. It is noted once one of the deflections is known the other deflections can be calculated by using ratios. Likewise, if the applied load was reduced to

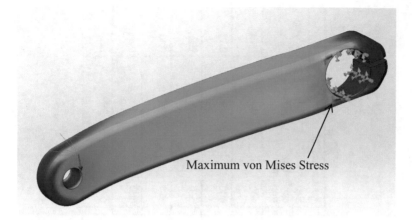

Fig. 17 FEA results of crank showing maximum von Mises stress for a thickness of 3 mm

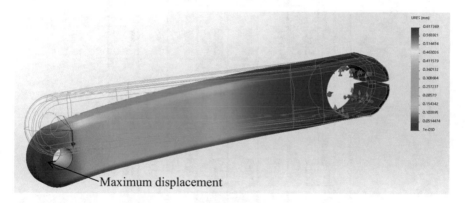

Fig. 18 Deflection results for a titanium alloy with 105 GPa elastic modulus

600 N, the resulting stresses and displacements could be determined mathematically without requiring additional simulations.

4   Material Selection

The successful design is closely related to the correct selection of materials for each of the components. With thousands of materials to choose from, a tool to assist with the material selection would be required. Granta Design CES Selector provides a collection of software tools and materials property data to support systematic materials selection. CES Selector allows a designer to combine considerations of properties such as environmental, engineering and economic factors. This is

Table 3 FEA results summary

FEA results											
				VM stress		Displacement		Actual loading			
	Material type	E (GPa)	Force/torque applied in FEA (N)	Max (MPa)	Location	Max (mm)	Location	Force/torque applied in FEA (N)	VM$_{Max}$ (MPa)	Dis$_{PMax}$	
Spindle	Steel	200	1820	1059	Cut out corners	0.134	End of spindle	600	349	0.04	
	Aluminium	69	1820	1059	Cut out corners	0.403	End of spindle	600	349	0.13	
	Titanium	105	1820	1059	Cut out corners	0.262	End of spindle	600	349	0.09	
Crank	Steel	200	1820	211	Edge of clamp	0.321	Pedal	600	70	0.11	
	Aluminium	69	1820	211	Edge of clamp	0.943	Pedal	600	70	0.31	
	Titanium	105	1820	211	Edge of clamp	0.617	Pedal	600	70	0.20	

achieved through the use of a systematic rational selection process that involves trade-offs between cost, engineering, performance and eco-behaviour [14].

Metals exhibit characteristics such as stiffness or rigidity which allows them to resist deformation under an applied load. They also display relatively high elastic moduli due to this stiffness. In a pure form, metals are soft and easily deformed. The tensile strength of a metal can be increased by alloying with other elements. Strength can also be increased through heat treatment processes. During this process, the steel remains ductile allowing them to be formed by deformation processes. Before a metal fails, it displays elongation known as yielding before it fractures. More than other material groups' metals are least resistant to corrosion and also suffer from fatigue due to their ductility. Ceramics like metals have a high elastic modulus and, however, are brittle. The strength of ceramics in compression is far greater than that in compression. Ceramics can withstand fifteen times more force in compression than tension before brittle failure. Due to not displaying, any ductility ceramics are prone to stress fractures due to holes or upon a clamping force. Brittle materials cannot distribute a load evenly over its shape and thus require a substantial volume of material directly under the load. The main features of ceramics include stiffness, hardness, resistance to abrasion and corrosion as well as maintaining strength under high temperatures. Glasses are non-crystalline (amorphous) solids. The common glasses include the soda-lime and borosilicate glasses used in applications such as ovenware and bottles. Due to the lack of a crystalline structure which suppresses plastic deformation glasses are hard, brittle and prone to stress concentrations. Polymers have a low elastic modulus which is approximately fifty times lower than that of metals, however that can be almost as strong as metals. This means that their deformation under a load can be large. The mechanical properties of polymers can vary greatly with temperature. A polymer may be flexible at room temperature or brittle at 4 °C and few have strength over 200 °C. Polymers are mainly a crystalline structure with some amorphous or a combination of both. The strength to density ratio can be as high as steels. Benefits of polymers include resistance to corrosion, low coefficients of friction and easy to form. Elastomers are long-chain polymers that exist above their glass transition temperature. The polymer chains remain intact from covalent bonds above the glass transition temperature and by van der Waals and hydrogen bonds below this temperature. As such elastomers exhibit unique characteristics such as a very low elastic modulus, as well as an increase in strength as temperatures rise and a vast elastic region. Hybrids are a combination of two or materials in a prearranged shape and scale. These materials aim is to combine the advantages of other materials whilst avoiding their disadvantages. Hybrids include fibre and particulate composites, sandwich structures, lattice structures, foams, cables and laminates. They also included naturally occurring materials such as wood, leaf and bone. Advanced hybrids include a polymer matrix reinforced with fibres such as glass, carbon. The benefits of hybrids are that they are light, stiff, strong and tough. The disadvantages are the cost and formability.

4.1 Properties

Many materials have to conform to standards and codes, i.e. for those that will come in contact with food. Others will be required to conform to environmental considerations. For a product to be successful in the marketplace, it must be economically viable and meet the design considerations in terms of performance, consumer appeal and cost, for example. This performance is directly related to the chosen material. Table 4 represents a list of basic material properties to consider when choosing materials.

These vital properties are inbuilt into the material libraries of CES Selector. The density, ρ (kg/m^3), is the mass per unit volume. This is an important selection criteria since the weight of the unit needs to be minimised. Performance for a professional cyclist can be greatly affected by weight. Even a less performance-orientated cyclist wishes to minimise weight in order to reduce the required energy. The price, C_m ($/kg), is an important consideration. For a bicycle drivetrain component to be marketable and sell, its cost must be relative to existing equipment. The elastic modulus, E (GPa), is the slope of the initial linear elastic region on a stress–strain curve. It describes the response of a material to tensile or compression loading. The curve is used to measure the yield and tensile strength of a material. The yield strength, σ_y (MPa), for metals is the stress at which the stress–strain curve deviates by a strain of 0.2% from the linear elastic line as shown in Fig. 19. For polymers, the yield strength is defined when the stress–strain curve becomes nonlinear. The tensile strength, σ_{ts} (MPa), is the nominal stress at which a round bar of material loaded in tension begins to separate. For brittle materials such as ceramics and glasses, the tensile strength is the same as the yield strength. For ductile metals as shown in Fig. 19, the tensile strength is a lot greater than the yield strength.

Cyclic loading can cause cracks to develop and to grow in a material, culminating in fatigue failure. For many materials an endurance limit, σ_e (MPa), exists which is illustrated in Fig. 20. The endurance limit can be defined as withstanding a very large number of cycles 10^7, without fracture. The spindle undergoes a typical

Fig. 19 Stress–strain curve for a metal showing elastic modulus, yield and ultimate strength, Ashby [14]

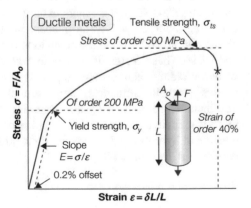

Fig. 20 Endurance limit of a material. *Source* Ashby [14]

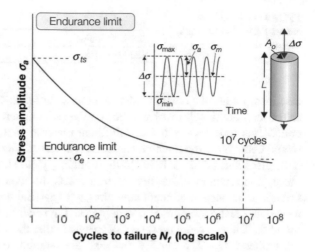

cyclic motion. The spindle would have to withstand approximately 1700 h at a cadence of 100 rpm to have an infinite life.

Wear is known as the loss of material when surfaces slide past each other. Wear can be categorised as adhesive, abrasive, fretting and erosive. Additionally, wear can be due to surface fatigue or corrosion. When solids slide, the volume of material lost from one surface, per unit distance slide, is called the wear rate, W (m^2). The wear resistance of the surface is characterised by the Archard wear constant, K_A, (1/MPa) defined by the equation,

$$\frac{W}{A} = K_A P \tag{16}$$

where A is the area of the slider surface, P is the normal force acting on the adjacent surface and K_A is a constant based on the sliding couple.

4.2 Material Selection Strategy

There are four main steps involved in the process of selecting materials for a product. This includes firstly translation, which involves translating the design requirements into functions, constraints, objectives and free variables. Secondly, materials that do not meet the above requirements must be carefully eliminated. Thirdly, the remaining materials must be ranked based on the objective that will make the material perform the best. Finally, research into the top-ranked materials must be performed to ensure suitability. Material selection will be performed on (i) end disc, (ii) crank arm, (iii) shifter ring and (iv) spindle.

The end discs house the complete assembly comprising of the shifter rings, shifters and guide rings which are then mounted on the spindle. The right-hand end

Table 4 Objective for material selection of crank and end discs

Maximise	Minimise
Elastic modulus	Density
Yield strength	Cost

disc incorporates the crank, whilst the inner or left-side disc enclosures the unit. The selection process will involve choosing the same material for the end discs and the crank. It is unnecessary to have different material candidates for these items as it would complicate the function and add complexity. The constraints imposed on the ends discs and crank include elastic modulus, yield strength, density and cost. Of these, it is imperative that the material meets the required yield strength to avoid failure. It is also non-negotiable that the material must have a certain elastic modulus. Elastic modulus is a variable used in order to determine the stiffness of a member. The end discs must be stiff to ensure that the v-belt maintains contact with the shifter rings. The density and cost are negotiable constraints. Both of these constraints help with the marketability of the product, however, are not critical to the function of the system and thus can be considered free variables. Table 4 highlights constraints which shall be maximised and minimised.

Elastic modulus versus density chart was plotted as shown in Fig. 21. To reduce the number of candidates, limits were applied in line with Table 4. The density of materials was restricted to less than 6000 kg/m^3 with elastic modulus filtering out materials less than 10 GPa. This still resulted in 1200 candidates. Furthermore, material indices were plotted on the graph. These index, $\frac{E^{1/2}}{\rho}$, lines prescribe the light and stiff beam materials. The top-ranking materials from all material groups using this index included carbon fibres, polyethylene fibres, diamond, Kevlar and beryllium as shown in Fig. 21. The best-ranked materials upon material class are shown in Table 5.

The strength versus density plot as shown in Fig. 22 is used to find materials that prescribe a low weight and high strength beam material. Figure 23 represents materials that call for stiffness and strength at a minimum weight. The tensile strength and elastic modulus are divided by the density. These properties measure the mechanical efficiency of the material, meaning the minimum weight to support a force. A high specific strength gives a material stiffness, whilst a high specific strength prevents failure. Again, the density was restricted to materials less than 6000 kg/m^3 with a tensile strength requiring at least 300 MPa to filter out brittle materials. FEA results showed that the von Mises stress was 211 MPa in the crank thus 300 MPa was chosen as a minimum. Additionally, the cost per kilogram of material was restricted to $200 AUD to lower candidates and to meet objectives. The best performing materials were a range of carbon fibres including a metal matrix composite (MMC), bamboo as a natural material, titanium and aluminium alloys. The best performing polymer was PEKK a thermoplastic with 40% carbon fibre.

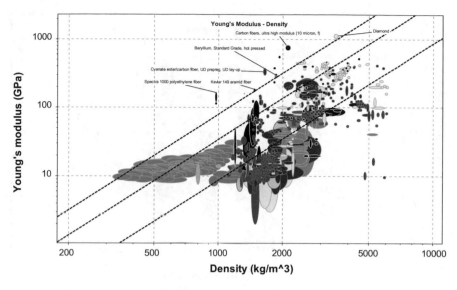

Fig. 21 Elastic modulus E plotted against density ρ. The guidelines of constant $E^{1/2}/\rho$ allow selection of materials when high stiffness and low weight are prescribed

Table 5 Short list of candidate materials for the cranks and end discs

Metals/ alloys	Hybrids	Polymers	Ceramics/ glasses	Fibres
Beryllium	Cyanate ester/carbon fibre	PEKK	Diamond	Carbon fibre
Magnesium	PEEK/carbon fibre	PA66	Boron carbide	Polyethylene fibre
Aluminium	Wood	PEEK	Silicon carbide	Kevlar
Titanium		SPS		

The shifter rings need to meet the same objectives as the end discs and the crank. Additionally, they need to be wear resistant due to the frictional contact between them and the v-belt known as adhesive wear. This wear is non-avoidable since it is required to transmit the torque. Friction is required in the direction of the v-belts motion; however, radially it should be minimised to reduce the force required to change the belts diameter. Material candidates described below are possible chooses; however, methods of improving the surface are required. Kevlar, as indicated in Table 5 by itself, is a candidate material. Kevlar the main material being aramid fibres is also a high abrasion resistant material in body armour. When used in conjunction with carbon and glass fibres, it would be suitable for the shifter rings. Common materials used on v-pulleys in the industry include aluminium, brass, cast iron, plastic, stainless steel and zinc alloys; however, not all these meet the objectives for design.

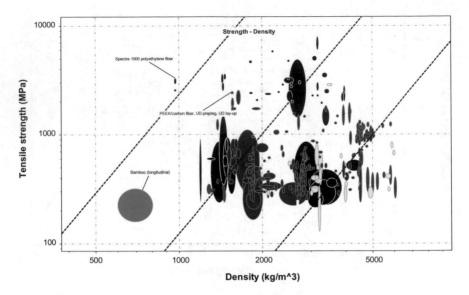

Fig. 22 Strength, σ_f is plotted against density, ρ. The guidelines of constant $\sigma_f^{2/3}/\rho$ allow selection of materials when minimum weight and high strength beam are prescribed

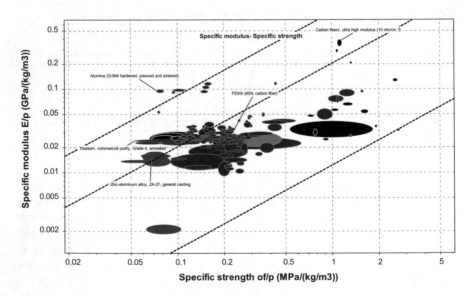

Fig. 23 Specific modulus, E/ρ, is plotted against specific strength, σ_f/ρ

4.3 Material Candidates

Titanium is an abundant metal found in the earth's crust; however, extraction of the metal from the oxide is difficult. This makes titanium expensive compared to other light metal alloys such as aluminium and magnesium. Regardless, titanium is a fast-growing material based on some outstanding mechanical properties. Titanium alloys have a high melting point (1677 °C) and are light, resistant to corrosion from most chemicals protected by a film of oxide on its surface. Titanium alloys are exceptionally strong for their weight and can be used in high-temperature applications. An example is compressor blades on an aircraft turbine. They also exhibit low thermal and electrical conductivity and low expansion coefficients.

Titanium alloys such as Ti-3A1-2.5V has been used on modern sports equipment including golf clubs, tennis rackets and bicycle frames. With carbon fibre composites becoming more advanced, titanium alloys fell out of favour. However, titanium cranks can still be purchased today. Although expensive, titanium alloys are an option for the cranks and end discs. Aluminium alloys have been used in the bicycle industry for many years. Aluminium alloy 7075 composing of zinc, magnesium and copper plus other trace elements is a common material used in the bicycle industry. As shown in Fig. 23, it has a high specific strength and modulus. It is strong compared with carbon steel, has good fatigue strength and is machinable. Its low cost and low weight contribute to making it a candidate for the use of the spindle and shifter rings. Carbon fibre-reinforced polymer (CFRP) is a very strong and lightweight fibre-reinforced plastic that comprises of carbon fibres. CFRP is more expensive than steel alloy options but is used extensively on bicycle components due to their high strength to weight ratio. They are also very stiff, making them good for sports equipment plus usable in aerospace and the automotive industries. Compared to aluminium or titanium, CFRP is lighter whilst maintaining the same strength. Carbon fibre composites are workable meaning they can be made into sections that will maximise stiffness in one direction by the orientation of the weave. Hence, CFRP can be manufactured to maximise the aerodynamics of a cyclist which is important in professional cycling events. Currently, CFRP dominates modern performance-orientated bicycle designs. It can be found on suspension linkages, pedals, handlebars, derailleur parts and brake components. As such, CFRP is a viable choice of material for many parts namely the cranks, spindle and shifter.

4.4 Design Considerations

The design of CVT needs to consider the existing hardware and setup of the common chain drivetrain. This is important to ensure that the new design can integrate into a standard bicycle with the least alterations possible. It is also important that the pulley is designed for manufacture (DFM). This means that the pulley is designed in such way that it can be produced easily. DFM is the practice of designing products with manufacturing in mind, so they can:

- Be designed in the least time with the least development cost,
- Make the quickest and smoothest transition into production,
- Be assembled and tested with the minimum cost in the minimum amount of time,
- Have the desired levels of quality and reliability, and
- Satisfy customers' needs and compete well in the marketplace.

The horizontal distance between the outside of the pedal connection from one crank to the other is known as the tread width or "Q" factor. This lateral distance determines the distance between the feet during a pedal stroke. There is no standardisation of tread distance in bicycle manufacturing. The tread distance is mainly determined by the width of the bottom bracket and the number of chainrings. A touring mountain bike generally has three chainrings compared to two on a road bike and hence would have a greater tread distance to accommodate the extra chainring. The tread width also plays an important role in cycling performance, injury prevention and comfort [15]. The ideal tread width varies based on the individual's biomechanical characteristics. In order to pedal efficiently, the tread width should mimic that of the cyclist's natural stance. Previous research has shown that performing squats at alternating stance widths affected the electromyography (EMG) activity [16]. Also, Donelan et al. [17] found that the metabolic cost of walking was reduced with narrower steps, compared to that of a wider footprint.

Having said these, it is thought that reducing the tread width on a bicycle will result in lower oxygen intake for a given power output thereby increasing the gross mechanical efficiency (GME). GME was calculated using the ratio of the work accomplished in kcal/min to energy expended in kcal/min. Disley and Li [15] examined whether changes in the tread width had any benefit on cycling efficiency and muscular activation. A total of twenty-four professional cyclists were examined on stationary test bike that involved pedalling for five minutes at 90 rpm using tread widths of 90, 120, 150 and 180 mm. Although there was no substantial evidence suggesting more or less muscle activity over the varying tread widths, there was a significant difference of GME as shown in Fig. 24.

The results show that for tread widths below 120 mm, there was an increase in GME approximately 0.3% compared to the tread widths 150 mm and above. This increase results in an additional power output at submaximal levels of approximately 1.5–2.0% [15]. Professional cyclists competing in events such as the Tour de France over three weeks require cyclists to perform at submaximal levels for up to six hours a day. For an 180 km stage with a cyclist working close to the submaximal level with a power output of 180–280 W, this would correlate to around a 3–5 W saving resulting in a potentially finishing the race 2–3 min earlier [18].

A typical road bike has a tread width of approximately 150 mm, whereas a mountain bikes are larger at around 170 mm [19]. Recent new designs in bottom

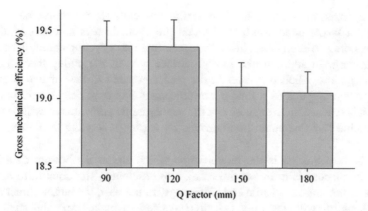

Fig. 24 Changes in gross mechanical efficiency (GME) with different tread widths [15]

bracket design notably the BB30 allow the road bikes to have narrow tread width. It has been noted that some Olympic and World Champion cyclists have ridden custom-made bicycles with tread widths less than 130 mm in order to improve their performance and in doing so have won Olympic and World Championship medals and time trial stages of the Tour de France [15]. From the tests carried out by Disley and Li [15], the cyclists also reported that the 180 mm tread width felt uncomfortable and not natural. In order for a cyclist to perform well, it is important that the design of the bicycle encompasses good ergonomics. Additionally, the tread width affects the manoeuvrability of the bicycle. When a cyclist is banking their bike to navigate a corner, a wider tread width causes the pedals to come in contact with the ground at a lesser angle. This is known as a pedal strike which can cause a cyclist to crash at high speeds. Bicycle manufacturers look to design bottom brackets (BB) as low as possible as this allows for better handling. It allows the seat to be lower to the ground, in turn, lowering the rider's centre of gravity. This provides more responsive rear end steering and sharper cornering [20].

Consumers are expecting more visually on how a product looks more than ever before. In the past, people were satisfied if the product did the intended job and were not so concerned about its visual appeal. The modern consumer is looking for well thought out designs that showcase an understanding of their values and lifestyle. If a product falls short in this respect, the consumer will be disappointed and will start to look elsewhere [21]. The aesthetic appearance will be more important to the daily commuter than the performance-orientated cyclist. However, it is important to understand the benefits that the cyclists want from the product and to understand what is required to achieve these desirable features. In essence, the product will need to bring joy to the end-user [21].

During individual time trials in cycling racing, or in triathlons where a cyclist races on their own, it is important to be as aerodynamic as possible. In these events, a few seconds advantage can result in a podium position. Manufactures and cyclists go to extreme lengths to reduce drag. This includes helmet design, bicycle frame design, cyclists' body position, wheel design and clothing to name a few. It can be seen that a larger tread width will widen the cyclist's legs and feet resulting in additional drag. Although not ideal for a performance, a cycling commuter could be less forgiving. In addition, the frontal surface area of the pulley that houses the shifter rings and v-belt is greater in the area compared to that of two chainrings. Although no computational fluid dynamic simulations have been conducted on the pulley, it is reasonable to suggest that the aerodynamic performance will be weaker. It is possible that this could be improved by adding a cowling to the front of the pulley.

The bottom bracket is the tubular pipe which sits at the bottom of a bicycle frame. Its purpose is to house the crank arm assembly. The crank arm assembly consists of the cranks, spindle and chainrings. In the past, the bottom bracket came fairly standard, with the only decisions needed to make were the shell width, spindle length and in rare occasions, the thread type, being either English or Italian threading [22]. In today's market, there are many different sized bottom brackets each with varying configurations from bicycle manufacturers that claim benefits over the others. These benefits may include at-home serviceability, weight reduction, pedal stiffness, narrower "q" factor, reduced creaking, longevity and reduced manufacturing costs. The conventionally threaded bottom bracket dates back to more than twenty years. The idea is to allow the use of a much larger spindle by moving the bearings to the outside of the shell. The standard diameter spindle for this setup is 24 mm. The major benefit of this design is that it is compatible with the largest range of cranks that isolate the bearing from the frameset and avoid creaking. The consequence for this system is that it is heavier than a press-fit arrangement [22]. In 2007, Trek introduced the BB90 system which involved the manufacturer pressing bare bearing directly into the bottom bracket. To save weight, the bearings are pressed into precision-machined carbon seats with the bottom bracket. This replaced the common arrangement in which bearings were pressed into aluminium cups which were then screwed into the frame. The benefits of the BB90 include maintaining a wide range of compatibility with conventional setups and create a wider bottom bracket without affecting the "q" factor [23]. Disadvantages include not being able to use cup-mounted chain guides and requiring special tooling for serviceability. Similar to the BB90 system, the BF86 standard is used by popular brands such as Giant and Scott. The bearing spacing is identical to a conventional threaded bottom bracket setup with the exception that the bearings are fitted into cups before being press-fitted into the bottom bracket. The tolerances are more relaxed compared to the BB90 system which allows them to easier adapt to alloy frames. The press-fit cups improve the pedalling stiffness by up to 10% on the traditional threaded bottom brackets [22]. The benefits and negatives are the same as the BB90 system. In 2006, Cannondale released the BB30 system as an open standard. As with the BB90 system, the bearing cartridges press directly into a tight tolerance shell. The difference is that the spindle and shell diameters are increased and not the shell width. This results in a lighter weight system due to not requiring separate cups that house the bearings. Another difference is that the BB30 system has a larger 30 mm diameter spindle which is fitted into a bottom bracket that is just 68 mm. This compares to 86 and 90 mm for the BB86 and BB90 systems, respectively. This shorter width allows for more heel clearance against the frame of the bicycle and depending on the crank offset makes it possible to have a narrower "q" factor. In fact, the BB30 allows for the narrowest "q" factor on the market. Additional advantages include reduced weight and increased stiffness [22]. The increased stiffness is due to the increased diameter of the spindle which has a greater radius of gyration compared to a 24 mm spindle.

As with most products, a product that requires as little maintenance as possible will appeal to consumers. A typical derailleur chain drivetrain requires regular cleaning of road built-up grime. This is as simple as using a brush to loosen debris, applying degreaser and wiping clean. It is also important to clean and lubrication the chain. Research has shown that an inadequate lubricated chain can cause a loss in efficiency [24]. Lubrication of the cable/housing is important to ensure smooth shifting. Additional maintenance would include calibration of the front and rear derailleurs to ensure a smooth transition when shifting between gears. As expected, all manufactured parts do not have an infinite life expectancy. A typical derailleur chain drivetrain requires replacement of chains, cassettes and chainrings at regular intervals. A chain comprises of many links made up from a pin, outer and inner plates a bushing and a roller. The typical road bike can have in excess of one hundred links. The spacing between each chain link is known as the pitch. Depending on the brand of chain, it may require replacing every 5000 km due to stretch. This sketch occurs from the bushings rubbing on the chain pins. Over a period of time, the inner diameter of the bushings increases and hence pins reduce in diameter. A bicycle chain is generally said to be worn when the growth exceeds 1% of the original length [25]. It is critical that a worn chain is replaced in order to reduce wear on the cassette and chainrings. A worn chain can easily wear out a brand new cassette within 1000 km. A chain is very good at making the sprockets match the pitch of the chain; therefore, it is important to change the chain before its worn out to ensure the longevity of the cassette and chainrings [25]. Even so, the cassette will require replacing every second or third chain replacement, 10,000–15,000 km [26].

As with a chain drivetrain, a v-belt will stretch over time. In automobile applications and some industrial applications, the sketch of the belt is controlled by the use of a tensioner. Tensioners can be automatic or manual depending on the equipment and required maintenance schedule. A chain derailleur system takes up the slack chain by the use of a spring mechanism built into the body of the rear derailleur. These belts are made of several flexible high strength steel bands that are held together by bow-tie-shaped pieces of metal. Theses bow-tie-shaped pieces or elements work together with the metal bands to transmit torque. This occurs because of the friction between the bands and the elements [27]. This causes a combined push–pull action in the belt that allows torque transmission.

Bicycle cranks are made to varying lengths to suit different sized riders and different types of cycling. The length of the crank is measured from the centre of the spindle to the centre of the pedal. The typical and most common crank length on a bicycle is 170 mm; however, manufacturers offer sizes anywhere from 165 to

180 mm usually in 2.5 mm increments. Speciality cranks and adjustable cranks can also be purchased outside of this range [28]. The general idea of the perfect crank length has been centred on the length of the cyclist's leg. A longer leg would lead to a longer crank, whereas a shorter leg would require a shorter crank. Research in this area has produced numerous considerations for crank length selection. Considerations include the inseam length to the leg length. The inseam length is a measurement from the ground to the crotch, whereas the leg length is a measure from the ground to the femoral head. This ratio is not always proportional to one another; therefore, one would expect these lengths to be used in the calculation of the optimal crank length. Also, the leg proportions between the upper and lower leg should be considered. A cyclist with a long tibia bone will experience higher knee lift which can cause the thigh to interfere with their rig cage especially on longer cranks and low handlebars. A longer foot relative to the leg will enable the cyclist to reach longer pedals and may be a factor in determining the crank length. Other considerations may include the flexibility of the cyclist in terms of their hamstrings [28]. Research done by Martin and Spirduso [29] investigated the effects of cycle crank length on maximum cycling power, cadence, pedal speed and crank length to leg ratio. Crank lengths of 120, 145, 170, 195 and 220 mm were analysed. The results on maximum power to crank length from their work are shown in Fig. 25. It can be seen that 145 mm cranks resulted in the maximum power output under test conditions. The optimal crank length was 20% of the leg length or 41% of the tibia length [29]. The conclusion to this work suggested that even though maximum cycling power was affected by crank length, using standard 170 mm cranks should not dramatically affect maximum cycling power in cyclists.

The majority of new bicycle equipment on offer targets weight reduction by using advanced materials and hollow components. For a performance-orientated cyclist, 100 grams' reduction can result in seconds or even minutes' advantage depending on the event. That being said, in 2000 the world governing body for sports cycling put a limit on the weight of bicycles in tournaments to be greater than 6.8 kg [30]. This was mainly due to safety reasons because of the rapid development of bicycle componentry towards the end of the century. Due to this current restriction, it is possible for an amateur cyclist to be riding a bicycle that is lighter than that of the professionals. In major cycling tournaments, weights are added to the bicycle to ensure compliance or heavier equipment is used. Table 6 represents a

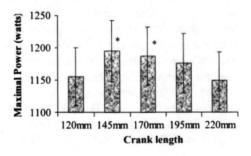

Fig. 25 Maximum power versus crank length [29]

Table 6 Breakdown of bicycle component weights by the manufacturer [36]

2014	Campagnolo	Shimano	SRAM Red
	Super record	Dura Ace 9000	BB30
Crank set	585	636	605
BB cups	45	67	105
Shift lever	330	363	280
Rear derailleur	155	160	143
Front derailleur	72	67	76
Chain	239	248	250
Cassette	177	163	155
Brake callipers	272	294	254
Group weighting	1875	1998	1868

breakdown of bicycle component weights by three popular and high-end equipments. Currently, since bicycles are easily manufactured under the weight restrictions of the UCI, additional weight may be tolerated. A benefit is that the weight is being added quite low on the bicycle which reduces the centre of the bicycle of gravity. This, in turn, will assist with the stability and cornering capabilities.

5 Conclusion

There has been a small success with some drivetrains such as the internal hub gears and the NuVinci CVT; the chain derailleur system has been dominated the market since its inception in the late 1930s.

The spindle and cranks had to bare a 300 Nm torque in a performance-orientated bicycle. This torque results from the FEA analysis suggested the spindle would require yield strengths exceeding 500 MPa. The von Mises stresses in the spindle could be reduced by increasing the shell thickness; however, thickness needed to be minimised due to the requirement of housing an internal electric motor. Results from the FEA on the crank indicated a maximum von Mises stress of 211 MPa. Maximum deflection occurred at the pedal is approximately 0.6 mm which was deemed acceptable.

The material selection process involved the use of an advanced material database within CES Selector software. The goal was to determine materials for the initial physical prototype. By determining the desirable conditions of the material and their objectives, graphs were plotted along with material indices to highlight and systematically determine the optimal materials. Materials such as aluminium and titanium alloys performed well due to their low weight-to-strength ratio. Composite materials that incorporated carbon fibres such as CFRP showed to be the most suitable due to their high stiffness and low cost, as well as the low weight and a high tensile strength common amongst low weight steel alloys.

A technical feasibility study was performed on the drivetrain. Some of the key areas explored included the "q" factor, aesthetics, aerodynamics, bottom bracket design, maintenance, durability, crank length and weight. Compared to a standard chain derailleur drivetrain, the CVT system was greater in volume, which in this case resulted in significantly higher overall drivetrain weight. A larger 'q' factor and poorer aerodynamics will lead to further inefficacies. The current design can only fit a BB30 style bottom bracket which limits its implementation on popularly branded bicycles.

The overall drivetrain efficiency of a standard chain derailleur system can have high as 98.6%. The major problem with CVT systems compared to the currently preferred derailleur chain setup is that they are less efficient at transferring torque to the rear wheel. This is due to heat dissipation from the friction of the v-belt onto the shifter rings. Another factor that is responsible for the overall drivetrain efficiency is the diameter of the rear pulley. The rear pulley is not included in the scope of this project of work; however, the relatively small diameter reduces the drivetrain efficiency. The CVT will need to overcome many hurdles to be accepted and marketable as an alternative drivetrain. The drivetrain has the potential to exploit the rapidly growing e-bike industry as many of the issues facing the current design would be alleviated.

Questions

1. What is the significance of a bike?

   Answer: The bicycle is one of the most identifiable creations of mechanical equipment worldwide due to its availability and mass production. The bicycle is a common occurrence in most countries of the world being used as the primary mode of transport in developing countries and for predominately recreational use throughout the developed world. Advances in mechanical design in today's market of the bicycle are more focused on a niche set of consumers. These advances include components such as the braking system, material selection and drivetrain components.

2. What is the function of derailleur in a bike?

   Answer: The derailleur allowed the shifting of the chain onto different sized rear sprockets, known as a "cassette" by a mechanical cable. It incorporates a front shifter that is typically used to change between two or three front sprockets known as a chainring and a cassette with ten sprockets. Advancements of the derailleur have focused on improvements in shifting precision, changing under load, ease of shifting and a move towards electronic control as opposed to mechanical control.

3. How to define maximum efficiency of a bike?

   Answer: In the professional world of competitive cycling, a cyclist aspires to maximise their efficiency. The maximum efficiency is determined in part by the combination of cadence and resistance that will produce the most power with the least amount of stress on the body. Factors that contribute to this stress include oxygen consumption, heart rate level and muscle fatigue.

A continuously variable transmission (CVT) will allow for a continuous gear range between the lowest and highest gears. This will allow a professional cyclist to ride at the most efficient cadence and a recreational cyclist to ride at the most comfortable cadence.

4. How is the cadence related to the cycling activity?

   Answer: The ideal pedal rate varies based on the particular cycling task, such as course geography and duration. Higher cadence in the order of 100–120 rpm would improve sprint cycling over shorter distances. In these situations, muscle fatigue is reduced and the cycling power output is increased. For cycling events, up to four hours in length, a reduced cadence of 90–100 rpm would be more beneficial. For an increased length in cycling time, the optimal cadence falls. For endurance racing staged over multiple days, relativity low cadences of 70–90 rpm improve cycling economy and lower metabolic consumption. This is also true for high climbs where relativity lower cadences are better suited at 70 rpm. The lower cadences are known to increase the pedal forces necessary to maintain a given power output. The ideal cadence is also related to the workload of the cyclist. Extremely low cadences of 50–60 rpm appear to be most efficient at low power outputs of less than 200 W. However, at lower cadences, more force is required per pedal stroke to maintain the power output which results in additional muscle fibre recruitment at a higher metabolic expenditure. The optimal cadence also increases linearly with increased output. With increased workloads of approximately 350 W, higher cadences of 80–100 rpm are more suitable.

5. How to select materials for an application?

   Answer: There are four main steps involved in the process of selecting materials for a product. This includes firstly translation, which involves translating the design requirements into functions, constraints, objectives and free variables. Secondly, materials that do not meet the above requirements must be carefully eliminated. Thirdly, the remaining materials must be ranked based on the objective that will make the material perform the best. Finally, research into the top-ranked materials must be performed to ensure suitability.

6. What is the significance of tread width?

   Answer: The tread width also plays an important role in cycling performance, injury prevention and comfort. The ideal tread width varies based on the individual's biomechanical characteristics. In order to pedal efficiently, the tread width should mimic that of the cyclist's natural stance. Previous research has shown that performing squats at alternating stance widths affected the electromyography (EMG) activity. The metabolic cost of walking is reduced with narrower steps, compared to that of a wider footprint.

7. What is the significance of weight of a bicycle?

   Answer: The majority of new bicycle equipment on offer targets weight reduction by using advanced materials and hollow components. For a performance-orientated cyclist, 100 g reduction can result in seconds or even minutes' advantage depending on the event. That being said, in 2000 the world governing body for sports cycling put a limit on the weight of bicycles in tournaments to be greater than 6.8 kg. This was mainly due to safety reasons because of the rapid development of bicycle componentry towards the end of the century. Due to this current restriction, it is possible for an amateur cyclist to be riding a bicycle that is lighter than that of the professionals. In major cycling tournaments, weights are added to the bicycle to ensure compliance or heavier equipment is used. A benefit is that the weight is being added quite low on the bicycle which reduces the centre of the bicycle of gravity. This, in turn, will assist with the stability and cornering capabilities.

8. Please indicate whether the following statements are True or false

    (1) Continuously variable transmission (CVT) offers a smooth transition through a defined gear range by providing an infinite number of gears.
    (2) The derailleur allowed the shifting of the chain onto different sized rear sprockets, known as a "cassette" by a mechanical cable.
    (3) A modern professional road bike with two chainrings and a ten-speed cassette theoretically have twenty gear ratios.

    Answer: All are true

9. Please indicate whether the following statements are True or false

    (1) The maximum efficiency is determined in part by the combination of cadence and resistance that will produce the most power with the least amount of stress on the body.
    (2) The press-fit cups improve the pedalling stiffness by up to 10% on the traditional threaded bottom brackets.
    (3) CVT use as a bicycle transmission is limited though NuVinci CVT has been adopted by some bicycle manufacturers.

    Answer: All are true

10. Please indicate whether the following statements are True or false

    (1) A well-maintained chain drive system with a derailleur setup can achieve an efficiency of 98.6% in comparison with 78–86% of NuVinci CVT.
    (2) The spindle and cranks had to bare a 300 Nm torque in a performance-orientated bicycle.
    (3) The tight side tension in the v-belt when peddling is approximately equivalent to the tension in a chain or timing belt-driven bicycle given the same applied torque.
    (4) The CVT will need to overcome many hurdles to be accepted and marketable as an alternative drivetrain.

    Answer: All are true.

Glossary

Cadence Cadence is the rate at which a cyclist is pedalling/turning the pedals. It is the number of revolutions of the crank per minute (rpm). The wheels in turn spin proportional to the cadence and, however, change depending on the gearing setup. It is an important factor, which influences the power output, energy burnt and development of fatigue during cycling.

Cassette Cassette is the set of multiple sprockets that attaches to the hub on the rear wheel. It works with a rear derailleur to provide multiple gear ratios to the rider. Cassettes are a variety and newer development of cogsets.

Continuously variable transmission (CVT) A continuously variable transmission (CVT) is also known as a single-speed, step-less, pulley and twist-and-go transmission. This is an automatic transmission where a continuous range of effective gear ratios can be achieved seamlessly. CVT offers higher functional power, improved fuel efficiency and a smoother driving practice than a traditional automatic transmission.

Sprockets A sprocket is a profiled wheel with teeth/cogs around the outer edge that fit/mesh into the holes in a chain or a length of film or tape or other perforated or indented material to move it round.

References

1. Nuvinci® Technology. (2016). *Fallbrook intellectual property company LLC*. http://www.fallbrooktech.com/nuvinci-technology.
2. Woodforde, J. (1970). *The story of the bicycle*. London: Routledge & K. Paul.
3. Herlihy, D. V. (2004). *Bicycle: The history/David V. Herlihy*. New Haven, CT: Yale University Press.
4. 1937 Tour De France. (2016). McGann Publishing LLC. http://bikeraceinfo.com/tdf/tdf1937.html.
5. Beer, J. (2008). *Technique: Cadence matters*. Immediate Media Company Limited. http://www.bikeradar.com/au/gear/article/technique-cadence-matters-16394/.
6. Haj-Assaad, S. (2012). *Should you buy a car with a CVT transmission?* VerticalScope Inc. http://www.autoguide.com/auto-news/2012/05/should-you-buy-a-car-with-a-cvt-transmission.html.
7. Sedhom, M. (2015). *Nuvinci Launches New N380™ and N330™ CVT Hubs*. EbikeReviews.com.au. http://www.ebikereviews.com.au/news/nuvinci-launches-new-n380-and-n330-cvt-hubs-for-ebikes/.
8. Oehler, A. (2014). Nuvinci and other—efficiency measurements on Igh—Part 2. fahrradzukunft. https://fahrradzukunft.de/17/wirkungsgradmessungen-an-nabenschaltungen-2/.
9. Sneiderman, P. (1999). *Pedal power probe shows bicycles waste little energy*. Johns Hopkins University. http://pages.jh.edu/~news_info/news/home99/aug99/bike.html.
10. Heath, R. P. G. (2014). Seamless AMT offers efficient alternative to Cvt.Zeroshift.com. http://www.zeroshift.com/pdf/Seamless%20AMT%20Offers%20Efficient%20Alternative%20To%20CVT.pdf. Accessed 29 May, 2017.

11. Recapturing the Lost Efficiency of a V-Belt Drive. (2015). Regal Beloit Corporation. http://www.regalpts.com/PowerTransmissionSolutions/Brochures/Form_9950E.pdf. Accessed 01 June, 2017.
12. Lucia, A., Hoyos, J., & Chicharro, J. L. (2001). Preferred pedalling cadence in professional cycling. *Medicine and Science in Sports and Exercise, 33*(8), 1361–1366.
13. Kifer, K. (1999). *Cycling cadence and bicycle gearing.* https://web.archive.org/web/20120204100228/http://www.kenkifer.com/bikepages/touring/gears.htm. Accessed 28 June, 2017.
14. Ashby, M. (2010). *Materials selection in mechanical design* (4th ed).
15. Disley, B. X., & Li, F.-X. (2014). The effect of Q factor on gross mechanical efficiency and muscular activation in cycling. *Scandinavian Journal of Medicine and Science in Sports, 24* (1), 117–121. https://doi.org/10.1111/j.1600-0838.2012.01479.x.
16. McCaw, S. T., & Melrose, D. R. (1999). Stance width and bar load effects on leg muscle activity during the parallel squat. *Medicine and Science in Sports and Exercise, 31*(3), 428–436. https://doi.org/10.1097/00005768-199903000-00012.
17. Donelan, J. M., Kram, R., & Kuo, A. D. (2001). Mechanical and metabolic determinants of the preferred step width in human walking. *Proceedings of the Royal Society B-Biological Sciences, 268*(1480), 1985–1992. https://doi.org/10.1098/rspb.2001.1761.
18. Laursen, P. B., Rhodes, E. C., Langill, R. H., McKenzie, D. C., & Taunton, J. E. (2002). Relationship of exercise test variables to cycling performance in an Ironman Triathlon. *European Journal of Applied Physiology, 87*(4–5), 433–440. https://doi.org/10.1007/s00421-002-0659-4.
19. Lamy, M. (2015). *What is 'Q Factor', and does it make a difference?* Time Inc. (UK) Ltd Sport & Leisure Network. http://www.cyclingweekly.com/news/latest-news/what-is-q-factor-and-does-it-make-a-difference-187403. Accessed 22 May, 2017.
20. Kavanagh, A. (2016). *How to use bike geometry charts and what they mean.* BikeExchange Inc. https://www.bikeexchange.com.au/article/privacy-and-security. Accessed 23 June, 17.
21. Jordan, P. W. (2003). *Designing pleasurable products: An introduction to the new human factors, designing pleasurable products: An introduction to the new human factors.* Hoboken: Taylor and Francis.
22. Huang, J. (2017). *Complete guide to bottom brackets.* Immediate Media Company Ltd. http://www.bikeradar.com/au/gear/article/complete-guide-to-bottom-brackets-36660/. Accessed 24 May, 17.
23. Sohner, B. (2010). *Bottom brackets 101.* https://www.bikerumor.com/2010/02/17/bottom-bracket-tech-breakdown/. Accessed 25 May, 2017.
24. Huang, J. (2012). *The effects of temperature on chain lubricant and friction.* Immediate Media Company Ltd. http://www.bikeradar.com/au/gear/article/the-effects-of-temperature-on-chain-lubricant-and-friction-35806/. Accessed 29 May, 2017.
25. Rome, D. (2016). *How to know when it's time to replace your bicycle chain.* Immediate Media Company Limited. http://www.bikeradar.com/au/road/gear/article/bicycle-chain-wear-explained-46015/. Accessed 25 May, 2017.
26. Stone, J. (2014). *How many miles will a road bike's chain and cassette last?* John Stone Fitness LLC. http://www.johnstonefitness.com/2014/10/07/how-many-miles-will-a-road-bikes-chain-and-cassette-last/. Accessed 26 May, 2017.
27. Srivastava, N., & Haque, I. (2009). A review on belt and chain Continuously Variable Transmissions (CVT): Dynamics and control. *Mechanism and Machine Theory, 44*(1), 19–41. https://doi.org/10.1016/j.mechmachtheory.2008.06.007.
28. Hogg, S. (2011). *Crank length—which one?* Steve Hogg—All rights Reserved. https://www.stevehoggbikefitting.com/bikefit/2011/06/crank-length-which-one/. Accessed 27 May, 2017.
29. Martin, J. C., & Spirduso, W. W. (2001). Determinants of maximal cycling power: Crank length, pedaling rate and pedal speed. *European Journal of Applied Physiology, 84*(5), 413–418. https://doi.org/10.1007/s004210100400.

30. Arthur, D. (2015). *Could the Uci Scrap the 6.8 kg Weight Limit in 2016?* Farrelly Atkinson (F-At) Limited. http://road.cc/content/tech-news/173097-could-uci-scrap-68kg-weight-limit-2016. Accessed 27 May, 2017.
31. Streeter, C. (2014). *Model of a facile bicycle*. Fine Art America—Buy Art Online. http://fineartamerica.com/featured/model-of-a-facile-bicycle-clive-streeter-dorling-kindersley-science-museum-london.html.
32. Vintage Bicycle. (2014). WordPress.com. https://vintagebicycle.wordpress.com/page/2/.
33. Brown, S. (2008). *Gear theory for bicyclists*. http://www.sheldonbrown.com/gear-theory.html.
34. CVT. (2016). *Mitsubishi Motors Corporation*. http://www.mitsubishi-motors.com/en/spirit/technology/library/cvt.html.
35. Feeken, D. (2017). *Gear calculator*. Dirk Feeken. http://www.gear-calculator.com/. Accessed 28 May, 2017.
36. Component Weights (2017). *Total cycling & export technologies*. http://www.totalcycling.com/en/au/Component-Weights/cc-8.aspx. Accessed 27 May, 2017.

Chapter 3
Coin Minting

Paulo Alexandrino, Paulo J. Leitão, Luis M. Alves, Chris V. Nielsen
and Paulo A. F. Martins

Abstract This chapter addresses ongoing developments in the engineering design
of coin minting processes by application of the finite element method. The presentation draws from a brief overview on the fundamentals of the quasi-static and
dynamic finite element formulations based on implicit and explicit solution procedures to the application of the quasi-static finite element flow formulation to the
numerical simulation of coin minting. Validation of the results from numerical
simulations involved independent determination of the stress–strain curve of the
disk material by means of stack compression tests, verification of the force values
provided by the industrial coin minting press, confirmation of the estimates of the
progressive filling of the relief coin features, and comparison of the numerical and
experimental force versus die stroke evolutions for coins with different diameters
and relief profiles produced by the Portuguese Mint. Results show that finite element analysis can be successfully applied to predict material flow and filling of the
intricate relief coin features, to estimate the required coin minting forces, and to
improve the design of the dies before fabrication.

P. Alexandrino · P. J. Leitão
INCM, Imprensa Nacional Casa da Moeda, Av. António José de Almeida,
1000-001 Lisbon, Portugal
e-mail: paulo.alexandrino@incm.pt

P. J. Leitão
e-mail: paulo.leitao@incm.pt

L. M. Alves · P. A. F. Martins (✉)
IDMEC, Instituto Superior Técnico, Universidade de Lisboa, Av. Rovisco Pais,
1049-001 Lisbon, Portugal
e-mail: pmartins@tecnico.ulisboa.pt

L. M. Alves
e-mail: luisalves@tecnico.ulisboa.pt

C. V. Nielsen
Department of Mechanical Engineering, Technical University of Denmark,
2800 Kongens Lyngby, Denmark
e-mail: cvni@mek.dtu.dk

© Springer International Publishing AG, part of Springer Nature 2018
J. P. Davim (ed.), *Introduction to Mechanical Engineering*, Materials Forming,
Machining and Tribology, https://doi.org/10.1007/978-3-319-78488-5_3

Keywords Coin minting · Finite element method · Experimentation

1 Introduction

Coin minting is a net-shape metal forming processes in which a disk is compressed between dies while it is being retained and positioned by a collar (side ring). The process is often considered as a challenge in combining art and science and technology of metal forming due to the objective of producing a well-defined imprint of a sculptor's piece of work in the opposite faces of a coin by application of very high contact pressures (Fig. 1).

The research in science and technology of coin minting can be classified into three main periods that extended over the past five decades. The first period (1960–1980) draws on the pioneering investigations of Bocharov et al. [1] and Bay and Wanheim [2] to the development of a theoretical framework for understanding the mechanics of coin minting by Kiran and Shaw [3]. The first two investigations were focused on the required stress to form a single isolated groove [1] and on the material displaced by adjacent asperities [2] and gave rise to the characterization of coin minting as a three-stage forming process involving indentation, gross upsetting, and interaction between adjacent relief coin features [3].

The work of Kiran and Shaw [3] considered that the surfaces of coins could be approximated by simple sawtooth relief profiles characterized by a pitch and an asperity included angle under plane strain deformation conditions. In fact, most of the investigations performed in this period were limited to simple triangular and semicircular relief coin features and were exclusively based on analytical models built upon the slab, slip-line, and upper bound methods.

The second period (1980–2000) enhanced previous knowledge on the deformation mechanics of coin minting by giving special attention to the analysis of material flow and calculation of pressure. Delamare and Montmitonnet [4] made use of the upper bound method to characterize plastic deformation inside a coin and to calculate the energy, and the shape of a disk is to produce a coin with a central circular design and an outer annular legend. Brekelmans et al. [5] compared the predictions of material flow and coining pressure for producing a conical relieve feature in the center of a coin obtained by application of the upper bound method and of the finite element method built upon a quasi-static formulation with elastoplastic constitutive equations. This work is considered as the first application of finite elements in coin minting.

Subsequent applications of the finite element method under axisymmetric conditions by Barata Marques and Martins [6] and by Leitão et al. [7] focused on the joining of bimetallic ring coins and on the development of a multistage coin minting process to produce bimetallic foil coins. The research was performed in collaboration with the Portuguese Mint, and the multistage coin minting technology was protected by an international patent [8].

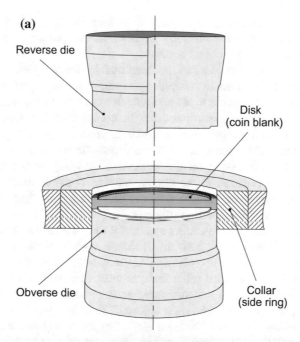

(a)

Reverse die

Disk
(coin blank)

Obverse die

Collar
(side ring)

(b)

Fig. 1 Coin minting process. **a** Schematic representation and notation and **b** photograph of the dies utilized for producing the collection coin dedicated to the age of iron and glass in Europe that was utilized in the investigation

The first three-dimensional application of the finite element method in coining was reported by Choi et al. [9] who developed a computer program based on a quasi-static formulation and rigid–plastic constitutive equations. The program was applied to the piercing and coining of a hole at the center of a disk, but no comparison was given between the numerical estimates and the actual material flow. The second period (1980–2000) ended with the investigation of Ike and

Plancak [10] on the plastic flow to form coin microfeatures, which included a comparison between the experimentally required contact pressure and that calculated by means of the upper bound method.

The third period (2000–until now) has been mainly driven by the availability of three-dimensional finite element computer programs and high-performance computers. Nowadays, it is possible to set up finite element models that account for the actual nonlinear material properties and for the complex contact features that are typical of the coin minting processes in order to produce accurate predictions of plastic flow, strain, stress, damage, and force versus die stroke evolution.

The fundamental discussion on the numerical modeling of the coin minting process is nowadays centered in the choice between dynamic finite element formulations based on explicit solution schemes and quasi-static finite element formulations based on implicit solution schemes.

In 2007, Buffa et al. [11] utilized a commercial three-dimensional finite element computer program based on a quasi-static formulation with rigid–plastic constitutive equations to simulate a coin minting process, where simple three-dimensional circular, o-shaped, and t-shaped relief features were engraved in a coin. The disk was discretized by means of a non-structured mesh of tetrahedral elements, and the dies and collar were assumed to be rigid objects. The accuracy and reliability of the numerical predictions were not validated against experimental results.

In 2008, Xu et al. [12] reported the development of a special purpose finite element computer program based on a dynamic explicit formulation with elastoplastic constitutive equations to simulate the coin minting process. The discretization of the disks was performed by means of hexahedral elements, and the dies and collar were considered rigid and discretized by means of spatial triangular and quadrilateral elements. The program was utilized to optimize the edge shape of the disks.

In 2009, Guo [13] utilized a commercial three-dimensional finite element computer program based on a dynamic explicit formulation to simulate coin minting with serrations on the inner surface of the collar. The dies were considered rigid and discretized by means of triangular elements, but the overall level of detail and complexity of its surfaces were simplified by removing letters and replacing the relief features of the figures by stepped cavities with 0.1 mm depth. The numerical predictions of the coin minting force obtained under these simplifying assumptions were compared against experimental results. However, the above-mentioned simplifications created no conditions to compare finite element predicted and experimentally observed filling of the actual relief coin features.

Later in 2012, Zhong et al. [14] utilized the above-mentioned special purpose finite element computer program [12] to predict the formation of tiny scratches caused by radial friction in the flat regions of the coin adjacent to the edge, in collaboration with the Shenyang Mint of China. The accuracy and reliability of the computer program were assessed by means of qualitative observations related to material flow. Quantitative comparisons between finite element predictions and experimental results were limited.

More recently, Li et al. [15] further developed the special purpose dynamic explicit finite element computer program [12] by introducing locking-free hexagonal elements and an adaptive refinement strategy aimed at increasing the number of elements near the fine relief coin features. The overall strategy was applied in the numerical simulation of a simple coin and of a 'key-shaped' commemorative coin made from 99.9% Ag. Experimental validation of the force with displacement was provided for the 'key-shaped' commemorative coin.

Also in 2016, Shirasaka [16] revealed that Japan Mint is currently using the finite element method to modify relief coin features in order to increase die life. The work reports an increase of die life up to 2.5 times when the relief height is reduced to 60% and confirms that Japan Mint is committed on the research of the utilization of numerical simulation for coin minting. Similar approaches are presently being undertaken or considered by other mints around the world.

This year, for example, Alexandrino et al. [17] revealed details of the ongoing work at the Portuguese Mint for adapting a quasi-static implicit finite element computer program to the numerical simulation of the coin minting process. The publication also addresses the validation of finite element predictions by means of experimentation with disks and coins retrieved from industrial production.

This chapter presents the application of numerical simulation based on the finite element method to the design and fabrication of coins. The aim and objective are to discuss how numerical simulation can help artists who create the modulated reliefs of the coins, and technicians and engineers who fabricate the dies and industrialize the process, to minimizing defects, optimize shapes and reduce tryouts.

2   Finite Element Modeling

The analytical estimate of the coin minting force F is based on the determination of the average contact pressure p_{avg},

$$F = p_{avg}A \quad p_{avg} = Q_f \sigma_R \tag{1}$$

where A is the area of the coin, σ_R is the tensile strength of the coin material, and Q_f is a pressure multiplying factor. Typical values of Q_f are in the range of 4–5 in order to account for the complexity of the relief profiles and the combined influence of flow stress $\bar{\sigma}$, friction μ, diameter D, and thickness t of the coins.

The latter combined influence can be obtained from the slab method analysis of free upsetting of a cylinder between flat parallel platens,

$$p_{avg} \cong \bar{\sigma} \left(1 + \frac{\mu D}{3t}\right) \tag{2}$$

However, when a detailed evolution of force with die stroke is required and the progressive filling of the relief coin features needs to be analyzed, more sophisticated calculation approaches based on the finite element method have to be considered. Two main finite element formulations can be chosen for this purpose:

(i) the quasi-static finite element formulations based on implicit time integration and
(ii) the dynamic finite element formulations based on explicit time integration.

This section starts by presenting a brief overview of both finite element formulations and ends by disclosing the modeling conditions utilized in the numerical simulation of the collection coins produced by the Portuguese Mint. Emphasis is also given to the procedure utilized for transforming the three-dimensional clay models prepared by the sculptor into the triangular spatial meshes that were utilized in the discretization of the dies.

2.1 Theoretical Background

The quasi-static finite element formulations are based on the static governing equilibrium equations in current (say, deformed) configuration (in the absence of body forces),

$$\frac{\partial \sigma_{ij}}{\partial x_j} = 0 \tag{3}$$

and are written in the integral form as follows:

$$\int_V \sigma_{ij} \frac{\partial(\delta u_i)}{\partial x_j} \, dV - \int_{S_T} t_i \, \delta u_i \, dS = 0 \tag{4}$$

In the above equations σ_{ij} denotes the stress tensor, x_j are the coordinates, $t_i = \sigma_{ij} n_j$ are the tractions applied on the boundary S_T with prescribed tractions having the unit normal vector n_j, and u_i are the velocities (or, instead, the displacements). The duality between velocities and displacements is due to the fact that finite element formulations can be set up in the rate form or in the rate independent form.

Equation (4) is commonly referred as the 'weak variational form' of Eq. (3) because the static governing equilibrium equations are now only satisfied in weaker continuity requirements. Its computer implementation requires solving the following matrix set of nonlinear equations,

$$\mathbf{K}^t \mathbf{u}^t = \mathbf{F}^t \tag{5}$$

which expresses the equilibrium condition at the instant of time t. The symbol $\mathbf{K}$ denotes the stiffness matrix, and $\mathbf{F}$ is the generalized force vector resulting from the loads, pressure, and friction stresses applied on the boundary. The basic feature of $\mathbf{K}$ is its nonlinearity with $\mathbf{u}$.

The main advantage of the quasi-static finite element formulations over the dynamic finite element formulations is related to the solution scheme which is commonly based on implicit instead of explicit procedures. As a result of this, quasi-static finite element formulations check equilibrium at each increment of time by means of iterative procedures meant to minimize the residual force vector $\mathbf{R}(\mathbf{u})$ to within a specified tolerance η (the symbol i denotes the current iteration number),

$$\mathbf{R}_i^t = \mathbf{K}_{i-1}^t \, \mathbf{u}_i^t - \mathbf{F}^t < \eta \tag{6}$$

However, quasi-static finite element formulations require the solution of large systems of equations in each iteration and, therefore, require higher computation times and higher memory requirements than dynamic finite element formulations based on explicit solution schemes. These are based on the dynamic equilibrium equations in current configuration (in the absence of body forces),

$$\frac{\partial \sigma_{ij}}{\partial x_j} - \rho \, \ddot{u}_i = 0 \tag{7}$$

In the above equation, ρ denotes the material density and $\ddot{u}_i$ is the acceleration. The matrix form of Eq. (7) for computer implementation is expressed as follows:

$$\mathbf{M}^t \, \ddot{u}^t + \mathbf{F}_{int}^t = \mathbf{F}^t \tag{8}$$

where the symbol $\mathbf{M}$ denotes the mass matrix, $\mathbf{F}_{int} = \mathbf{K}\mathbf{u}$ is the vector of internal forces resulting from the stiffness of the metal forming system, and $\mathbf{F}$ is the generalized force vector.

The above-mentioned advantages in computation and memory requirements of the dynamic finite element formulations result from diagonalization of the mass matrix $\mathbf{M}$ (lumped mass matrix). In fact, its inversion is trivial and the overall solution can be performed independently and very fast for each degree of freedom. However, because there is no check of equilibrium after each increment of time, the increments of time need to be very small, and the overall number of steps to conclude the simulations is, therefore, very large.

Other drawbacks resulting from the utilization of dynamic finite element formulations with explicit solution schemes that are not addressed here are related to the utilization of scaling parameters for mass, velocity, and damping. These scaling parameters are commonly utilized to artificially increase the time increments and may give rise to inaccurate solutions for the deformation, prediction of forming defects, and distribution of the major field variables within the workpiece [18].

From what was said above, it is easy to understand the reason why some investigations in coin minting made use of quasi-static finite element formulations (e.g., references [9, 11, 17]), while other made use of dynamic finite element formulations (e.g., [12–15]). The same applies in the numerical simulation of other metal forming processes.

The in-house finite element computer program that is being developed by the authors in collaboration with the Portuguese Mint is based on the three-dimensional finite element computer program I-form, which has been validated against experimental results of forming processes since the end of the 1980s [19]. I-form is based on the quasi-static finite element flow formulation which may be derived from Eq. (4) as follows:

$$\Pi = \int\limits_V \bar{\sigma}\,\dot{\bar{\varepsilon}}\,\mathrm{d}V + \frac{1}{2}K\int\limits_V \dot{\varepsilon}_v^2\,\mathrm{d}V - \int\limits_{S_T} t_i u_i\,\mathrm{d}S + \int\limits_{S_f}\left(\int\limits_0^{|u_r|}\tau_f\mathrm{d}u_r\right)\mathrm{d}S \qquad (9)$$

In the above functional, $\bar{\sigma}$ is the effective stress, $\dot{\bar{\varepsilon}}$ is the effective strain rate, $\dot{\varepsilon}_V$ is the volumetric strain rate, K is a large positive constant imposing the incompressibility of volume V, t_i and u_i are the surface tractions and velocities on surface S_T, τ_f and u_r are the friction shear stress and the relative velocity on the contact interface S_f between the disk and the dies and collar. Friction is modeled through the utilization of the law of constant friction $\tau_f = mk$.

A detailed description of the computer implementation of the finite element flow formulation is given in Nielsen et al. [20].

2.2 Simulation Conditions in Coin Minting

Figure 2a, b show the original clay models (in a scale 7:1) prepared by the sculptor and the discretization of the corresponding obverse and reverse dies by means of spatial triangular contact–friction elements.

The original clay models are much larger than the dies in order to allow facial details, hair, and hands to be sculptured more lifelike and also to facilitate the inclusion of folds and other realistic features in cloths, flowers, and other motifs commonly available in coins. In fact, all these details would be very difficult to accomplish if the sculptor theme had to be directly expressed in the actual size of the coins.

The scaling down of the clay models to the actual shape and size of the dies is performed digitally and involves three-dimensional scanning and correction of the relief profiles in order to add appropriate small convex curvatures on the obverse and reverse surfaces of the dies. Letters, numbers, and the national coat of arms are digitally added to each die.

The digital representation of each die surface is performed by means of a triangular tessellation and then saved in an stereolithography (STL) file format and converted into a workable finite element input data file. The conversion of the STL file into a workable finite element input data file containing the geometry of the dies is performed automatically by the finite element computer program in order to save time and effort and to prevent the loss of details of the die profiles due to the utilization of intermediate preprocessing systems.

The collar (not shown in Fig. 2) was also discretized by means of spatial triangular contact–friction elements. Two types of collar profiles are used: flat and serrated.

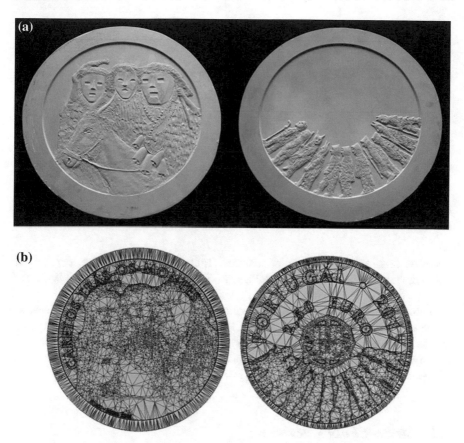

Fig. 2 Discretization of the die surfaces. **a** Original clay models supplied by the sculptor (scale 7:1) and **b** discretization of the die surfaces by means of spatial triangular elements after scanning and scaling down the clay models to the actual coin size

The disks are discretized by a structured mesh of hexahedral elements with five layers of elements across thickness as shown in Fig. 3.

In the proposed investigation, the convergence tolerance η of the residual force vector was specified as 10^{-3}, and the overall numerical simulation was accomplished through a succession of 85 displacement increments each of one modeling approximately 0.15% of the initial disk thickness.

3   Experimentation

The industrial case studies utilized in the investigation were the collection coins dedicated to Portuguese architecture (Fig. 4a), to the age of iron and glass in Europe (Fig. 4b) and to Portuguese Ethnography (Fig. 4c). The selected case studies

Fig. 3 Finite element simulation of the coin minting process. **a** Initial mesh and **b** numerically predicted geometry at the end of the coin minting process

combine different types of relief features. Table 1 summarizes the material and geometry of the disks that were utilized in the investigation performed with each coin.

This section describes the experimental work for determining the stress–strain curve of the disk materials by means of stack compression tests, the verification of the force values provided by the industrial knuckle-joint press where tests were performed and the experimental work plan for assessing finite element predictions of the coin minting process.

3.1 Stress–Strain Curves

The stress–strain curves of the Silver–Copper alloy (Ag500) and of the Copper–Nickel alloy (Cu75Ni25) were determined by means of the stack compression test. This test was chosen instead of the conventional tensile test due to its capability to characterize material stress response to larger values of strain without necking [21].

The stack compression test was performed on multilayer cylinder specimens that were assembled by piling up four circular disks with 15-mm-diameter machined out of the coin blanks utilized in production. The tests were carried out at room temperature on a hydraulic testing machine (Instron SATEC 1200 kN) with a crosshead speed equal to 10 mm/min. The photograph included in Fig. 5 shows a multilayer cylinder test specimen of Ag500 before and after compression.

The experimentally determined stress–strain curves of Ag500 and Cu75Ni25 and their approximation by means of the Ludwik–Hollomon's strain hardening model are shown in Fig. 5,

(a)

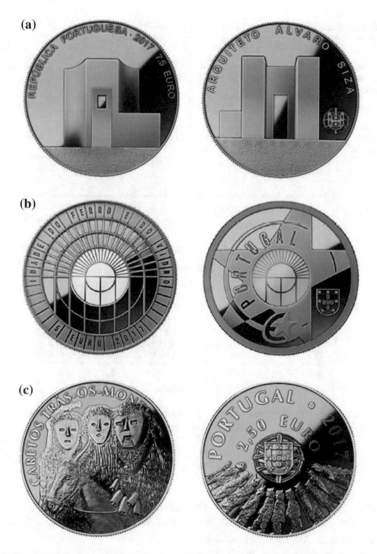

(b)

(c)

Fig. 4 Industrial test cases utilized in numerical modeling and experimentation of the coin minting process. **a** Ag500 collection coin dedicated to Portuguese architecture, **b** Cu75Ni25 collection coin dedicated to the age of iron and glass in Europe, and **c** Cu75Ni25 collection dedicated to Portuguese Ethnography

$$\text{Ag500} : \sigma = 585.5 \; \varepsilon^{0.31} (\text{MPa}) \tag{10}$$

$$\text{Cu75Ni25} : \sigma = 741.2 \; \varepsilon^{0.41} (\text{MPa}). \tag{11}$$

Table 1 Material and geometry of the disks utilized in the investigation

	Material	Diameter (mm)	Thickness (mm)	Edge thickness (mm)
Portuguese architecture	Silver–Copper alloy (Ag500)	32.65 ± 0.05	1.65	2.04 ± 0.1
Age of iron and glass	Copper–Nickel alloy (Cu75Ni25)	29.65 ± 0.05	2.24	2.79 ± 0.1
Portuguese ethnography	Copper–Nickel alloy (Cu75Ni25)	27.65 ± 0.05	1.84	2.28 ± 0.1

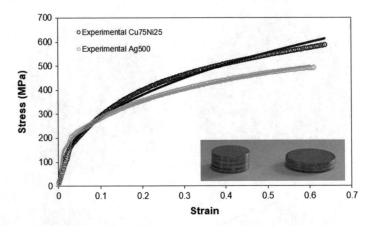

Fig. 5 Stress–strain curves for the Silver–Copper (Ag500) and the Copper–Nickel (Cu75Ni25) disk materials. Solid lines are the Ludwik–Hollomon's approximation

3.2 Experimental Work Plan

The coin minting experiments were performed in a Schuler (MRV300) knuckle-joint press available at the Portuguese Mint. The experimental work plan was structured in two complementary parts.

Firstly, the verification of the force measurements read in the control panel of the industrial press against those obtained in a hydraulic testing machine (Instron SATEC 1200 kN). This was accomplished by compressing five disks between flat parallel dies in the industrial press for different amounts of stroke and by compressing two other disks between the same flat parallel dies up to the maximum amount of stroke in a hydraulic testing machine. Table 2 provides information on the experimental work plan utilized for verifying the forces provided by the industrial press, and the enclosed photograph shows the dies and a disk before and after compression.

The disks utilized in these verification tests had 27.6 ± 0.05 mm diameter and 1.65 mm thickness and were machined out of the Ag500 coin blanks utilized in production. The surfaces of the flat dies had an average roughness $R_a = 0.319$ μm

Table 2 Flat dies, disk geometries, and compression strokes utilized in the verification of the forces measured in the industrial coin minting press

Disk	Equipment	Die roughness (μm)	Diameter (mm)	Thickness (mm)	Compression stroke (mm)
1	Schuler coin minting press	0.319	27.61	1.639	0.100
2			27.60	1.643	0.144
3			27.62	1.653	0.227
4			27.63	1.642	0.331
5			27.63	1.644	0.370
6	Instron testing machine		27.61	1.646	0.380
7			27.63	1.656	0.380
8		<0.01	27.63	1.646	0.380

in order to allow measuring compression forces up to 800 kN, because in case surfaces were polished with a quality similar to that utilized in coin minting dies the values of force would be much smaller, as it will be shown in Sect. 4 (Results and Discussion).

Secondly, the assessment of the accuracy and reliability of the finite element predictions against observations and measurements obtained from coin minting experiments with the three aforementioned types of coins. The corresponding experiments were designed to accomplish two main objectives: (i) validation of material flow during progressive filling of the relief coin features and (ii) validation of the force versus die stroke evolution. Both objectives were accomplished by means of industrial coin minting tests performed for different die strokes. In each test, the coins were intentionally struck one single time in order to prevent possible defects resulting from multiple strikes. At least, two repetitions were made for each testing condition.

4 Results and Discussion

This sections start by assessing the forces given by the industrial press with those obtained in the hydraulic testing machine and follows with validation of finite element predictions of the coin minting process with experimental observations and measurements. The final part includes an example showing how finite element predictions can be utilized to optimize the design of the dies before its fabrication.

4.1 Industrial Versus Laboratory Measured Forces

The black solid circular markers in Fig. 6 are the maximum forces read in the control panel of the industrial knuckle-joint press for the five compression disks labeled as '1' to '5' in Table 2. The gray open diamond markers correspond to the average force versus stroke evolution for the two disks labeled as '6' and '7' in Table 2 obtained from continuous upset compression in the hydraulic testing machine. As seen, the overall agreement is good and enables the utilization of the forces measured by the industrial knuckle-joint press with minor corrections.

Finite element analysis of the upset compression of the disks also allows concluding that the rough surface of the flat dies constrains material outward flow with friction shear stresses corresponding to a friction factor $m \cong 0.35$ (refer to the black dashed line in Fig. 6). This type of surface finishing was needed to ensure that upset compression forces were in the range of coin minting forces because flat parallel dies with typical surface finishing of that utilized in coin minting dies would only be able to provide forces up to 300 kN (refer to the black open rectangular markers).

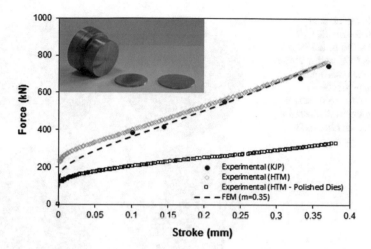

Fig. 6 Force versus stroke results for the upset compression of disks between flat parallel dies performed in the knuckle-joint press (KJP) and in the hydraulic testing machine (HTM). Finite element (FEM) results are included for reference purposes

4.2   Filling of the Relief Coin Features

The numerical simulation of the coin minting process allows analyzing the filling of the relief coin features before fabricating the dies and producing the coins. One of the variables that can be used for this purpose is the finite element predicted distribution of z-stress $\sigma_z = -p$ (where p is the applied contact pressure) in the obverse and reverse of the coins at different percentages of the total die stroke.

This is shown in Figs. 7 and 8 for the collection coins corresponding to Portuguese architecture and to the age of iron and glass in Europe, respectively.

The dark red color in both figures corresponds to pressures close to zero and, therefore, to regions of the coins that are not in contact with the dies. In contrast, the other colors evolving from light red to dark blue correspond to increasing values of the applied pressure and, therefore, to regions of the coins that are in contact with the dies. In addition to this, the enclosed photographs in Figs. 7 and 8 allow distinguishing the actual regions of the obverse and reverse of the coins that are in contact with the dies (refer to the shiny surfaces) from those that are still free. The agreement with the finite element predicted distribution of z-stress for the same amount of die stroke is very good.

Other conclusions that may be drawn from the analysis of the z-stress are (i) the detection of excessive lack of symmetry during coin minting and (ii) the identification of the regions of the coin where contact starts. Ideally, contact should develop as much symmetric as possible in order to minimize rigid body movements and should initiate at the center of the coin and progressively evolve toward its rimmed edge.

Fig. 7 Finite element
predicted distribution of
z-stress (MPa) at **a** 23%,
b 57%, **c** 73%, and **d** 95% of
the total die stroke with
photographs of the
corresponding coin samples
(collection coin dedicated to
Portuguese architecture)

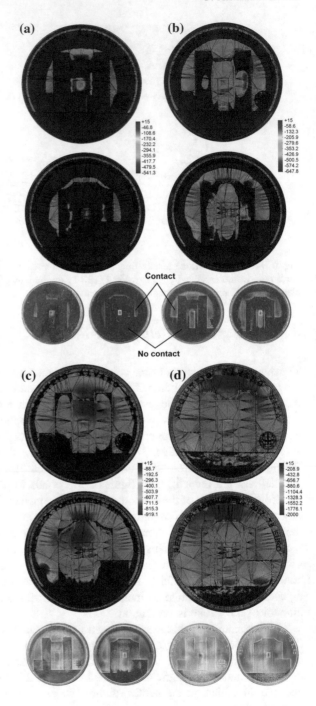

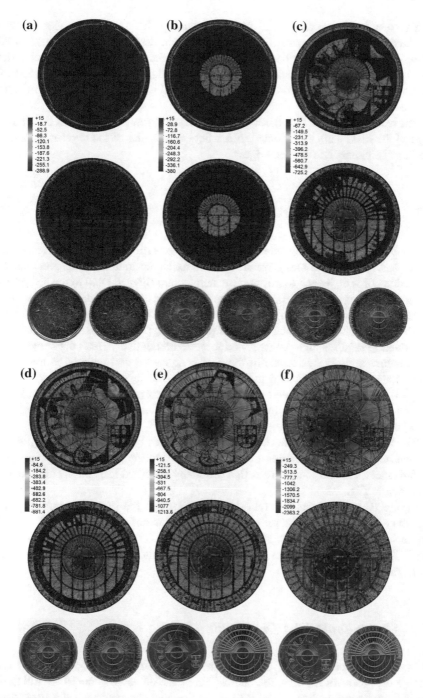

Fig. 8 Finite element predicted distribution of z-stress (MPa) at **a** 17%, **b** 35%, **c** 76%, **d** 85%, **e** 94%, and **f** 100% of the total die stroke with photographs of the corresponding coin samples (collection coin dedicated to the age of iron and glass in Europe)

The results shown in Fig. 7 allow concluding that contact develops with a substantial lack of symmetry in the collection coin dedicated to Portuguese architecture. The top regions of both the obverse and reverse dies get in contact earlier than the bottom regions because the coin contains significant profiles to be filled in the bottom regions. Later, in Sect. 4.4, authors will explain how lack of symmetry can be compensated.

The results shown in Fig. 8 allow concluding that contact initiates at the rimmed edge of the disk. This is attributed to the flat edge profile of the reverse die and to the large thickness of the disk edge.

4.3   Force Versus Die Stroke Evolution

Figure 9 presents the finite element predicted evolution of the force versus die stroke and the actual forces measured in the industrial knuckle-joint press for the collection coins dedicated to Portuguese architecture (Fig. 9a) and to the age of iron and glass in Europe (Fig. 9b). The forces in the industrial knuckle-joint press were measured for different values of die stroke and are plotted as open circular markers.

The force versus die stroke evolution pattern is typical of close die forging, but coin minting requires high compression forces with relatively small die strokes. The force is low until every surface of the disk is squeezed and material is trapped to start filling the remaining and more intricate details in the dies (refer to 'P1' in Fig. 9b). As the dies continue to close, the force increases sharply to 'P2' at which the relief coin features are completely filled with sharp corners, well-defined features, and a brilliant surface finish.

The last positions 'P3' are obtained for die strokes beyond needed and give rise to localized plastic deformation of the tooling and to the production of coins with final diameters slightly larger than that originally specified (30 mm). This situation is shown in Fig. 10.

Figure 10 also reveals a good agreement between the finite element predicted and the experimental evolution of the outside diameter of the disks with the die stroke. The final difference to 30 mm (refer to the continuous and the dashed black lines) is due to the tolerance bandwidth of the finite element contact algorithm (0.03 mm per radius).

4.4   Correction of the Die Profile

Coin minting processes with lack of contact symmetry can be optimized by correcting the profile of the dies. In practical terms, the procedure consists of tilting the obverse and reverse dies in order to guarantee that the center of die pressure is located on the vertical symmetry line of the die.

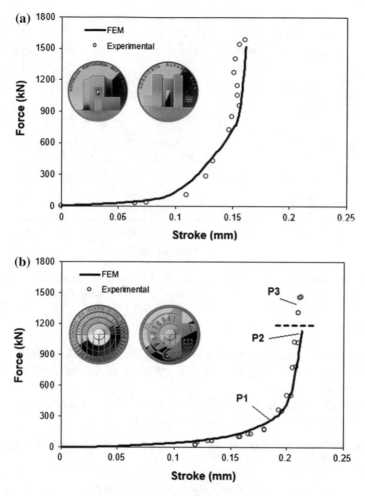

Fig. 9 Experimental and finite element predicted evolution of the force with die stroke for the collection coins dedicated to **a** Portuguese architecture and to **b** the age of iron and glass in Europe

The procedure for correcting the die profile is performed for a die stroke corresponding (or very close) to the end of the coin minting process and involves the following tasks to be carried out for each die:

(a) Calculate the point of application (x_r, y_r) of the resultant vertical force F_r for the original die profile;
(b) Determine the maximum z-distance between the disk surface and the die profile after some compression of the disk;
(c) Perform a first correction of the original die profile based on the maximum z-distance that was determined in (b) and determine the corresponding new point of application (x_r^1, y_r^1) of the resultant vertical force F_r^1;

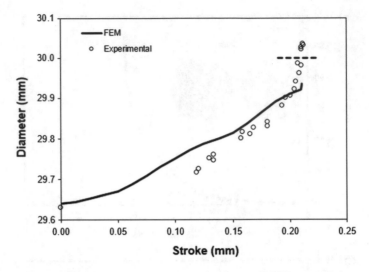

Fig. 10 Experimental and finite element predicted evolution of the outside diameter with die stroke for the collection coins dedicated to the age of iron and glass in Europe

(d) Interpolate the second correction of the original die profile in order to ensure that the new point of application (x_r^2, y_r^2) of the resultant vertical force F_r^2 move toward the center $(0, 0)$;

(e) Repeat (c) and (d) using additional correction values until $(x_r^m, y_r^m) \cong (0, 0)$, where m is the number of necessary iterations.

The point of application (x_r^i, y_r^i) of the resultant vertical force F_r^i after correction i is determined from

$$x_r^i = \frac{\sum_{j=1}^n F_j^i x_j^i}{\sum_{j=1}^n F_j^i} \quad y_r^i = \frac{\sum_{j=1}^n F_j^i y_j^i}{\sum_{j=1}^n F_j^i} \tag{12}$$

where F_j^i is the vertical force of a nodal point j with coordinates (x_j^i, y_j^i, z_j^i) located on the surface of the disk in contact with the die after correction i of the original die profile, and n is the total number of nodal points located on the surface of the disk in contact with the die.

The resultant vertical force F_r^i after correction i is determined from

$$F_r^i = \sum_{j=1}^n F_j^i \tag{13}$$

Regarding the determination of the maximum z-distance between the disk surface and the die profile required by the above-mentioned procedure (refer to task b), it may be claimed that it could be directly obtained from the z-differences in height for each die profile. However, it is better to determine this distance after some amount of die stroke because material flow arising from the initial contact (especially in the rimmed edge) will lead to a better estimate.

In what concerns the collection coin dedicated to Portuguese ethnography, the z-distance between the disk surfaces and the dies was determined for a die stroke corresponding to location 'P' in Fig. 11.

Figure 12 shows the corresponding contact/no-contact distribution plotted as '1—dark red' and '0—dark blue' (Fig. 12a), and the distribution of the distances between the disk surface and the obverse and reverse dies (Fig. 12b).

The maximum z-distances to the obverse and reverse dies for a die stroke corresponding to location 'P' in Fig. 11 were identified as 0.16 mm and 0.14 mm (Fig. 12b). So, by taking into account that the tolerance bandwidth of the finite element contact algorithm was set to 0.04 mm, and it was decided to tilt the obverse and reverse dies by 0.12 mm and 0.10 mm exclusively in the y-direction as a first attempt to correct the die profile. The tilting procedure is schematically shown in the cross-sectional details of the dies that are included in Fig. 13.

Figure 14 shows the results of the above-mentioned procedure for correcting the die profile after two correction attempts performed in the y-direction. As shown in Fig. 14a, the point of application (x_r, y_r) of the resultant vertical force F_r reduces the misalignment in the y-direction from 0.9 to 0.3 mm after performing the second correction attempt of the die profile.

The value of the second correction of the reverse die profile was obtained by interpolation. Figure 14b shows the procedure and the resulting value of 0.05 mm. A similar procedure applied to the obverse die profile (not shown in the figure) gave an interpolated correction value of 0.06 mm.

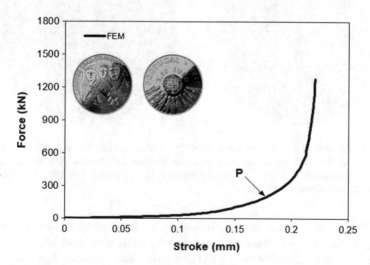

Fig. 11 Finite element predicted evolution of the force with die stroke for the collection coin dedicated to Portuguese ethnography

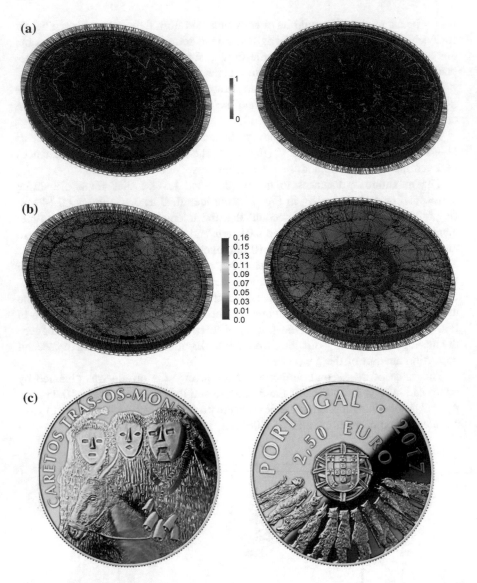

Fig. 12 Finite element estimates of the **a** contact/no-contact distribution and of **b** the z-distances (mm) between the surfaces of the disk and the obverse and reverse dies for a die stroke corresponding to point 'P' in Fig. 11. The collection coin dedicated to Portuguese ethnography is shown in (**c**) for reference purposes

Additional corrections and interpolations in both x- and y-directions allow moving the point of application (x_r, y_r) of the resultant vertical force F_r resultant vertical force to a value close to (0, 0) (absence of misalignment, Fig. 14a). In fact,

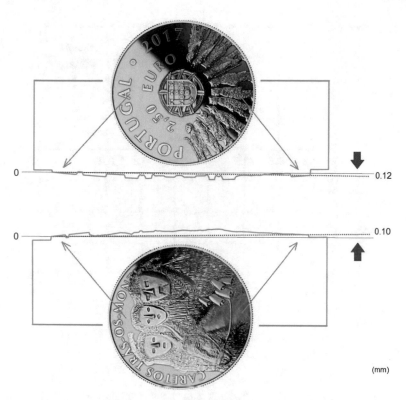

Fig. 13 Cross section of the reverse and obverse dies showing the first correction in the *y*-direction of the coin. Values in (mm)

the bending moment *M* applied in both dies is reduced from (in the original profiles),

$$M = M_x \vec{e}_x + M_y \vec{e}_y = 1008\ \vec{e}_x + 204\ \vec{e}_y\ \text{Nm} \tag{14}$$

into (in the corrected profiles),

$$M = M_x \vec{e}_x + M_y \vec{e}_y = 132\ \vec{e}_x + 24\ \vec{e}_y\ \text{Nm} \tag{15}$$

As a result of this, the bending moment applied in the dies with corrected profiles is 10 times smaller than the bending moment applied in the original dies resulting from the clay models supplied by the sculptor.

To conclude, Fig. 15 presents a comparison of the finite element predicted distribution of z-stress for the original and final corrected die profiles for a die stroke corresponding to 1200 kN (close to the end of coin minting). As seen, the corrected die profile resulting from the above-mentioned procedure improves the overall symmetry of the coin minting process.

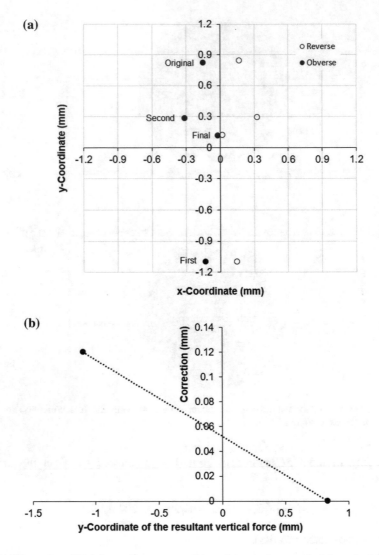

Fig. 14 Correction of the die profile. **a** x, y position of the resultant vertical force in the reverse and obverse dies after two profile correction attempts in the y-direction. **b** Schematic representation of the interpolation procedure that was utilized to determine the second profile correction of the reverse die in the y-direction

The reduction of the bending moment M applied in both dies with corrected profiles extends their service life and facilitates the filling of the relief coin features. The pressure peaks are also lower in case of the corrected die profile of Fig. 15b. In fact, the maximum z-stress is reduced from -3100 MPa (in the original die profiles, Fig. 15a) to -2840 MPa (in the corrected die profiles, Fig. 15b).

(a)

(b)

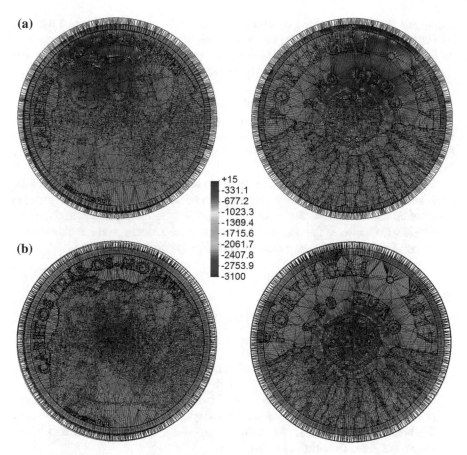

Fig. 15 Finite element predicted distribution of z-stress (MPa) at the end of the coining process for the **a** original die and for **b** corrected die profile (collection coin dedicated to Portuguese ethnography)

5 Conclusions

The first part of this chapter presents a state-of-the-art review of the application of analytical and numerical methods to the coin minting process. Special attention was paid to recent developments in the utilization of finite element computer programs based on quasi-static formulations with implicit solution schemes and dynamic formulations with explicit solution schemes.

Then, by centering the presentation on the assessment of the accuracy and reliability of finite element predictions in coin minting, the chapter entails a comprehensive description of the ongoing developments made by the authors in collaboration with Portuguese Mint. These comprise the independent determination of the stress–strain curves by means of stack compression tests on multilayer cylinder specimens that are assembled by piling up circular disks machined out of the coin blanks utilized in production, the verification of the coin minting forces measured in an industrial knuckle-joint press, and the comparison between finite element predictions and experimental results and observations of the coin minting forces and of the filling of the relief coin features.

Finite element modeling was carried out in a special purpose software based on the in-house three-dimensional computer program I-form, and experimentation made use of three collection coins dedicated to dedicated to Portuguese architecture (Ag500), to the age of iron and glass in Europe (Cu75Ni25), and to Portuguese Ethnography (Cu75Ni25).

The overall results, with special emphasis to those related to the correction of die curvatures, demonstrate that finite elements can be utilized to foster the relationship between artists who create the modulated reliefs of the coins and technicians and engineers who fabricate the dies and industrialize the process with the objective of minimizing defects, optimizing shapes, and reducing tryouts.

Questions

1. Which are the main components of a coin minting system?
   Answer: The obverse and reverse dies, the collar (also known as side ring) and the disk (also known as coin blank).
2. Why is the stack compression test adequate to determine the stress–strain curve of the materials utilized in coin minting?
   Answer: Until recently, Mints fabricated their own coin blanks by cutting them out from metal sheets. Because, nowadays, coin blanks are supplied by third-party manufacturers; it is more difficult for a Mint to cut out tensile test specimens from the original metal sheets.
   Under these circumstances, it is easier for a Mint to assemble compression test specimens by piling up disks cut out from the coin blanks and to determine the stress–strain curve of the materials by means of stack compression tests.
   And, most important, stack compression tests also allow determining the stress–strain curve of the materials for values of strain above those commonly reached in tension tests due to the avoidance of necking.
3. Consider the production of a coin with 30 mm diameter and 1.6 mm thickness made from a Copper–Nickel alloy Cu75Ni25. The tensile strength of the material $\sigma_R = 359$ MPa.

   (a) Calculate the average contact pressure p_{avg} during the initial stage of deformation when there is no contact between the disk and the collar (side ring). Assume the diameter of the disk at this stage to be approximately equal to 29.8 mm and the friction coefficient $\mu = 0.15$
   Answer : $p_{avg} = 693.3$ MPa
   (b) Calculate the average contact pressure p_{avg} at the end of the coin minting process. Assume $Q_f = 5$.

Answer : $p_{avg} = 1795\,MPa$
(c) Calculate the coin minting force at the end of stroke.
     Answer : $F = 1269\,kN$.

4. Consider the production of the above-mentioned coin from a disk with 29.8 mm
   diameter and 2 mm thickness. Will the increase of disk thickness facilitate coin
   minting during the initial stage of deformation?
   Answer: Yes, because Eq. (2) will now provide an average contact pressure
   $p_{avg} = 626.5$ MPa, which is 9.6% lower than that obtained with a disk of
   1.6 mm thickness.
5. Is the force–displacement evolution in coin minting similar to that of close die
   forging?
   Answer: Yes. The typical force–displacement evolution in coin minting shown
   in Figs. 9 to 11 has a pattern similar to that of close die forging but requires high
   compression forces with relatively smaller die strokes. In fact, compression
   forces reach values up to 1200 kN for die strokes below 0.25 mm.
6. Explain the importance of correcting the die profile in order to achieve an
   evolution of contact with die stroke as much symmetric as possible?
   Answer: Contact pressure builds from the interaction between the dies and the
   surface of the disk. Correction by tilting of the obverse and reverse profile of the
   dies is very important for centring the contact pressure, and the resultant force
   applied by the dies in order to facilitate the filling of the relief coin features and
   in order to extend the overall die life. In fact, by centring the contact pressure,
   there will be no regions of the disk with the same thickness experiencing much
   higher pressures than others.
7. Which are the main advantages resulting from the finite element modeling of the
   coin minting process?
   Answer: The finite element method takes into account the practical nonlinear-
   ities of the material properties as well as the complex contact conditions that are
   typical of coin minting to produce accurate predictions of displacements, strains,
   and stresses inside the coins. The utilization of finite elements in coin minting is
   intended to:

- Optimize existing processes by cost and quality;
- Develop new products in shorter time;
- Increase process knowledge;
- Assist marketing efficiently.

Glossary

Coin minting A net-shape metal forming processes in which a disk (coin blank) is
   compressed between two dies while it is being retained and positioned by a

collar (side ring) with the objective of producing well-defined imprints of the dies in its opposite faces.

Finite element method The governing equations of engineering problems solved by the finite element method are typically formulated by partial differential equations in their original form. These are rewritten into a weak form, such that domain integration can be utilized to satisfy the governing equations in an average sense. The domain integration is solved numerically and approximated by a summation over a number of elements utilized for discretizing the domain.

Force–displacement curve A graphical representation showing the evolution of the compression force with the die stroke.

Knuckle-joint press A special press design very appropriate for coin minting in which the connecting rod driven by the eccentric actuates a knuckle-joint system in order to provide a short stroke length and high load capacity.

Stack compression test An alternative experimental procedure for evaluating the stress–strain curve of raw materials. The test makes use of circular disks that are cut out of the coin blanks and stacked to form a cylindrical specimen with an aspect ratio in the range of solid cylinders commonly employed in the conventional compression test.

Stress–strain curve A graphical representation showing the stress response of a material and the corresponding deformation (strain).

Tool system The main components of a coin minting tool system are the obverse and reverse dies and the collar (also known as side ring).

Acknowledgements Luis Alves and Paulo Martins would like to acknowledge the support provided by the Portuguese Mint (Imprensa Nacional Casa da Moeda) and IDMEC under LAETA-UID/EMS/50022/2013.

The authors would also like to acknowledge the technical assistance of Elisabete Novais and Nuno Caetano from Imprensa Nacional Casa da Moeda.

The support of Dr. Silvia Garcia and Dr. Alcides Gama from Imprensa Nacional Casa da Moeda is also acknowledged.

References

1. Bucharov, Y., Kobayashi, S., & Thomsen, E. G. (1962). The mechanics of the coining process. *Journal of Engineering for Industry—Transactions ASME, 84,* 491–501.
2. Bay, N., & Wanheim, T. (1976). Real area of contact between a rough tool and a smooth workpiece at high normal pressures. *Wear, 38,* 225–234.
3. Kiran, C., & Shaw, M. C. (1983). Coining. *Annals of CIRP, 32,* 151–154.
4. Delamare, F., & Montmitonnet, P. (1984). Mechanical analysis of coin striking: Application to the study of byzantine gold solidi minted in Constantinople and Carthage. *Journal of Mechanical Working Technology, 10,* 253–271.

5. Brekelmans, W. A. M., Mulders, L. H. G., & Ramaekers, J. A. H. (1988). The coining process: Analytical simulations evaluated. *Annals of CIRP, 37,* 235–238.
6. Barata Marques, M. J. M., & Martins, P. A. F. (1991). A study of bi-metal coins by the finite element method. *Journal of Materials Processing Technology, 26,* 337–348.
7. Leitão, P. J., Teixeira, A. C., Rodrigues, J. M. C., & Martins, P. A. F. (1997). Development of an industrial process for minting a new type of bimetallic coin. *Journal of Materials Processing Technology, 70,* 178–184.
8. Teixeira, A. C., Leitão, P. J., & Martins, P. A. F. (1999). *A multi-metallic foil technology for minting medals and coins with two or three colors.* EP-97901854, European Patent Office.
9. Choi, H. H., Lee, J. H., Bijun, S. K., & Kang, B. S. (1997). Development of a three-dimensional finite-element program for metal forming and its application to precision coining. *Journal of Materials Processing Technology, 72,* 396–402.
10. Ike, H., & Plancak, M. (1998). Coining process as a means of controlling surface microgeometry. *Journal of Materials Processing Technology, 80–81,* 101–107.
11. Buffa, G., Fratini, L., & Micari, F. (2007). The relevance of the preform design in coining processes of cupronickel alloy. In *AIP Conference Proceedings of NUMIFORM 2007, Materials Processing and Design: Modelling, Simulation and Applications* (Vol. 908, pp. 1005–1010).
12. Xu, J. P., Liu, Y. Q., Li, S. Q., & Wu, S. C. (2008). Fast analysis system for embossing process simulation of commemorative coin–CoinForm. *Computer Modeling in Engineering & Sciences, 38,* 201–215.
13. Guo, K. (2009). *Development of design tools for coining process using FEM* (Ph.D. Dissertation). Ontario, Canada: Carleton University.
14. Zhong, W., Liu, Y., Hu, Y., Li, S., & Lai, M. (2012). Research on the mechanism of flash line defect in coining. *International Journal Advanced Manufacturing Technology, 63,* 939–953.
15. Li, Q., Zhong, W., Liu, Y., & Zhang, Z. (2017). A new locking-free hexahedral element with adaptive subdivision for explicit coining simulation. *International Journal of Mechanical Science.* https://doi.org/10.1016/j.ijmecsci.2017.04.017.
16. Shirasaka, K. (2016). *Application of finite element method for coining process.* In 29th Mint Directors Conference, Bangkok, Thailand.
17. Alexandrino, P., Leitão, P. J., Alves, L. M., & Martins, P. A. F. (2017). Numerical and experimental analysis of coin minting. *Journal of Materials: Design and Applications,* https://doi.org/10.1177/1464420717709833.
18. Tekkaya, A. E., & Martins, P. A. F. (2009). Accuracy, reliability and validity of finite element analysis in metal forming: A user's perspective. *Engineering Computations, 26,* 1026–1055.
19. Barata Marques, M. J. M., & Martins, P. A. F. (1990). Three dimensional finite element contact algorithm for metal forming. *International Journal of Numerical Methods in Engineering, 30,* 1341–1354.
20. Nielsen, C. V., Zhang, W., Alves, L. M., Bay, N., & Martins, P. A. F. (2013). *Modelling of thermo-electro-mechanical manufacturing processes with applications in metal forming and resistance welding.* London, UK: Springer.
21. Alves, L. M., Nielsen, C. V., & Martins, P. A. F. (2011). Revisiting the fundamentals and capabilities of the stack compression test. *Experimental Mechanics, 51,* 1565–1572.

Chapter 4
Gradation, Dispersion, and Tribological Behaviors of Nanometric Diamond Particles in Lubricating Oils

Kai Wu, Bo Wu, Chuan Li and Xianguo Hu

Abstract To improve the dispersion characteristics of nanometric diamond in lubricating oil and optimize its tribological properties, the gradation, dispersion, and tribological behaviors of nanometric diamond particles in lubricating oils were studied in this chapter. The dispersion characteristics of modified nanometric diamond in lubricating oils were observed by centrifugal and static methods. The four-ball tribometer was performed to study the tribological properties of the modified nanometric diamond as additive in hydraulic fluid. The morphology, particle size, surface functional groups, and structure composition of diamond powders before and after modification were analyzed and compared by means of FE-SEM, Zeta, FTIR, and XRD. Scanning electron microscopy (SEM), energy-dispersive spectroscopy (EDS), and Raman spectroscopy were selected to characterize the worn scar surface after rubbing process. The results showed that the modified nanometric diamond particle was fined, the surface functional group number was increased, and the steric effect in lubricating oil was produced, so that it exhibited excellent dispersion stability. Modified nanometric diamond as lubricating additives has good anti-wear and friction-reducing performance in hydraulic oil is mainly attributed to a large amount of small nanometric diamond particles, some fined particles fill and repair the rough surfaces of the friction pairs, and part of them enters the interface between the friction pairs to form rolling effect, which improves the anti-wear and friction-reducing properties.

Keywords Nanometric diamond · Lubricating oil · Dispersion
Tribological behaviors

K. Wu · B. Wu · C. Li · X. Hu (✉)
Institute of Tribology, School of Mechanical Engineering,
Hefei University of Technology, Hefei 230009, China
e-mail: xghu@hfut.edu.cn

© Springer International Publishing AG, part of Springer Nature 2018
J. P. Davim (ed.), *Introduction to Mechanical Engineering*, Materials Forming,
Machining and Tribology, https://doi.org/10.1007/978-3-319-78488-5_4

1 Introduction

With the rapid development of modern industry, numerous mechanical equipment are developing toward smaller size, greater load, higher precision, and longer service life. Lubricating oil, as an indispensable material for mechanical operation, plays a very important role in maintaining the operation of the mechanism and improving the mechanical life. With the continuous progress of modern industry, the higher and higher requirement of the operating life of the mechanism is required, and the operating environment is becoming harsher [18]. All requires the lubricating oil to have more excellent comprehensive performance. The types of lubricants are various, and the hydraulic oil selected in this study is one of the most frequently used among a variety of lubricating oils. Hydraulic oil is mainly applied in hydraulic system. The basic components of the hydraulic system comprise a hydraulic motor, hydraulic pump, actuator, and hydraulic oil. Hydraulic oil is equivalent to the blood of the hydraulic system, which has the functions of energy transmission, lubrication, sealing, rust prevention, corrosion prevention, and cooling in the hydraulic system. Because hydraulic system has the characteristics of stable transmission and easy operation, hydraulic technology is applied in every industry field of productions. According to the advantages of broad application field, the hydraulic oil occupies a very important role in the international lubricating oil market. With the continuous development of hydraulic oil specifications, high-performance and environmental-friendly hydraulic oil are new requirements for the development of hydraulic oil market.

Nanoparticles, as lubricating additives, play an increasingly important role in reducing emissions and improving fuel economy in recent years. Compared with organic additives, nanoparticles are considered to be thermally stable at high temperatures and thus become favorable lubricant additives [4]. Some nanomaterials have emerged as potential anti-friction (AF) and anti-wear (AW) additives to a variety of base lubricants [23]. Nanodiamond is one of the most promising nanocolloidal additives [12, 13, 19, 27]. Nanodiamond has good energy-saving effect and environmental protection characteristics. As a lubricant additive, it can improve the friction, lubrication, and overall performance of the traditional lubricating oil and show a good application prospect in the field of lubrication [25, 30, 33, 36]. However, the dispersion ability of nanodiamond in the oil- or water-based products is poor and it is easy to agglomerate, which makes it greatly to be restricted in the process of application.

The modified method for the dispersion of nanodiamond powders in lubricating oils was discussed in order to accelerate the application of nanodiamond in lubricating oil in this chapter. Moreover, the tribological properties of the modified nanodiamond in hydraulic oil were tested, and the lubrication mechanism was analyzed and characterized.

2   Tribological Behavior of Nanometric Diamond

2.1   Fundamental of Nanometric Diamond Particle

Diamond is of a solid network structure (Fig. 1). The primary particle size of nanodiamond particles is around 7–10 nm. Nanodiamond has several advantages such as high hardness, attractive optical, high thermal conductivity, high wear resistance, excellent chemical stability, and favorable biocompatibility, which are not equipped to most other nanoparticles [3, 6, 15, 20, 32, 34]. These properties would facilitate its use in some potential applications including abrasive pastes and suspensions for high precision polishing [6], nanofluids for thermal conduction [5, 28], and nanodiamond–polymer composites for wear-resistant surface coatings, thermal conduction coatings, and transparent coatings [6, 8–10, 28]. However, the nanodiamond particles are of large specific surface and high surface area, as well as in thermodynamically unstable state, which makes it easy to agglomerate and forms larger particles, and present micrometer size. Nanodiamonds cannot effectively play its unique function in real application due to its agglomeration.

2.2   Survey of Tribological Behavior of Nanometric Diamond

The tribological behavior of nanometric diamond has been studied widely and deeply in recent years. Shen et al. [29] conducted a research on the tribological properties of ultrafine diamond powder additives by the principle of relative light intensity. The results showed that, in the film lubrication region, the thickening of the mixed oil film was caused by the increase of the adhesion of the spherical

Fig. 1 Crystal structure of nanodiamond

nanodiamond particles, and the microrolling effect caused the friction force to be decreased. The friction process is the process of continuous dispersion and aggregation of diamond particles with time effects. Elomaa et al. [7] compared the tribological properties of uniformly distributed and agglomerated nanocrystalline diamond on the contact surface of steel. It was found that the wear rate of agglomerated nanodiamond was lower than that of uniform distribution. The possible cause was that the diamond enters the friction film and reduced the friction and wear. Chou and Lee [1] studied the tribological behavior of nanodiamond dispersed in lubricating oils on the effect of carbon steel and aluminum alloy. The reason that reducing the wear of aluminum alloy is mainly due to the change in oil viscosity, the reason for reducing the wear of the carbon steel is not only due to the change of oil viscosity, but also the embedding of the diamond causes a change in the roughness of the contact surface. Kim et al. [14] studied the lubricating properties of nanodiamond dispersed in liquid paraffin. It is considered that the octagonal structure of nanodiamond is similar to spherical feature, and the rolling ball action reduces the friction coefficient, and the high thermal conductivity increases the wear life. Lee et al. [17] treated the nanodiamond with oleic acid, and the obtained nanodiamond was dispersed in oil with a stable state for more than 10 days. Moreover, the tribological experiments showed that nanodiamond with an optimal addition of 0.05 wt% could be provided with a significant reduction in the friction coefficient and low wear.

In general, it can be realized that the lubrication mechanism of nanometric diamond as a lubricant additive is various under different conditions. The dispersion state and stability in lubricating oil have important influence on their tribological behavior. Therefore, a stable dispersion state is an important prerequisite for the development of better anti-wear and anti-friction properties.

3   Gradation and Dispersion of Nanometric Diamond Particles in Lubricating Oil

In order to overcome the problem of agglomeration and dispersion of nanodiamond, the present section introduces the modification methods of nanodiamond and makes the diamond nanoparticles to be achieved as state of gradation and dispersion well. Finally, a well dispersion stability of the diamond nanoparticles was obtained, and the related characterization of nanodiamond was performed before and after modification process.

3.1 Introduction to Gradation and Dispersion

Nanodiamond is of high surface energy and specific surface area and can be easily agglomerated into micron size particles. Therefore, the precondition of the dispersion is to prevent aggregation and reduce the particle size. The most effective way to solve the agglomerate problem and improve the dispersibility of nanodiamond is to modify the nanodiamond surface. The surface modification of nanodiamond refers to the action of surface active agents which can cause chemical reaction or physical action on the nanodiamond surface, thus changing its surface state and improving dispersibility. The methods of surface modification of nanodiamond include surface coating, surface chemical, mechanochemical.

Surface coating modification is to make surface modifiers adsorb on the surface of particles and form a stable adsorption layer with a certain thickness, which result in a strong repulsion due to steric effect. On the one hand, the mutual repulsion between particles is increased by increasing the absolute value of the surface potential of particles. On the other hand, by enhancing the wettability of the medium, the surface solvation membrane force between the particles is increased and the repulsion force is increased, which make agglomeration between particles extremely difficult. Surface modification process is usually co-operated with ultrasonic method together. Although the process is simple and workable, there it maybe lead to some problems in the application of nanodiamond particles in lubricating oils. For instance, it is difficult to get a better dispersion effect. At the same time, the agglomerated particles size is also relatively large and easy to be deposited. In addition the type, dosage and usage of surface modifiers will have a great impact on the application of products.

Surface chemical modification is achieved by chemical reaction or adsorption of the particle surface and surface modifier. One of the most effective ways of surface chemical modification is that the surface of nanodiamond is grafted with different functional groups, so as to achieve the purpose of modification. The effect of this method will be more obvious compared with the modified method of surface coating. The aggregation size of nanodiamond decreased significantly, and the stability and dispersion time in the solution lasted for a long time. However, this method was only suitable for the diamond with relatively initial small average particle size.

Mechanochemical modification refers to the surface activation of powders by ultrafine grinding and other mechanical forces. The crystal structure, chemical adsorption, and reaction activity of the particle surface undergo a certain degree of change to cause the particles to take a reaction or adhesion each other for the purpose of surface modification. Compared with surface coating modification, mechanical-chemical modification may lead to better dispersion effect. Because it combines the mechanical action with chemical force acting on the agglomerated nanodiamond particles, it then modifies the nanodiamond particles surface by reducing the combination among agglomerated nanodiamond particles.

3.2   Experiment

In our experiment, the method of acid treatment combined with heat treatment at high temperature is employed to achieve the agglomeration of nanometric diamond particles to obtain smaller particle size. With further treatment of surfactant, a stable dispersion system of nanometric diamond and hydraulic oil was obtained. The specific steps to be taken in the experiment are as follows.

Nanodiamond was added into the solution of sulfuric acid and nitric acid (3:1 in volume), and then heated and stirred to ensure them oxidation sufficiently. The process was employed to exert a remarkable oxidation effect on nanodiamond due to the strong oxidizing property of the mixed acid of sulfuric acid and nitric acid. In order to remove the residual acid on the nanodiamond surface, the mixture was filtered and the obtained solid was washed with excess amounts of distilled water until the filtration pH reached about 7. The final product was obtained and then marked as ND-1 after drying at 50 °C.

Next, the obtained nanodiamond (ND-1) was treated at high temperature. The treatment procedures included as follows: ND-1 powder was heated at 600 °C under the flowing air atmosphere inside a high-temperature tube furnace for several hours, and the obtained final product was marked as ND-2. The role of high-temperature treatment was that the particle size of nanodiamond was decreased and the agglomeration tendency was achieved. On the other hand, the surface functional groups of nanodiamond were purified by high-temperature oxidation, and the rich oxygen functional groups were prepared for further modification and dispersion. Before ND-2 was dispersed in the lubricant, surfactant like oleylamine was selected to treat ND-2. After the completion of nanodiamond treated with surfactant, the lipophilic nanodiamond was obtained and marked as ND-3. Finally, ND-3 as additive was added in the hydraulic oil with a stable and homogeneous system.

3.3   Characterization

In order to know the modification effect and the dispersing mechanism of nanodiamond clearly, the raw ND particles and modified ND particles were characterized using the following methods.

3.3.1   TEM Characterization

In order to view the size and agglomeration of nanometric diamond, the micromorphology was observed by high-resolution field-emission electron microscope (SEM, Hitachi, SU8020).

Fig. 2 SEM image of raw nanometric diamond

Figures 2 and 3 show the SEM images of raw and treated nanometric diamond (ND-2) particles, respectively. The raw nanometric diamond and the treated nanometric diamond were observed and compared clearly as using the magnification of 30 thousand times. The raw nanometric diamond (Fig. 2) was exhibited with larger size and uneven distribution due to agglomeration. Compared with the raw nanometric diamond, the degree of agglomeration of the treated nanometric diamond (Fig. 3) is reduced, and the particle size of the nanometric diamond becomes smaller, and the particle size and morphology became more uniform.

3.3.2   Particle Size Distributions

Figure 4 shows the particle size distributions of the untreated NDs and treated NDs measured by Zeta potentiometer (Nano-ZS90, Malvern UK). The particle size of the raw NDs was distributed from 200 to 700 nm with an average size of 418.0 nm, which resulted from severe aggregation of the primary nanoparticles.

Fig. 3 SEM image after acid-heat-treated nanometric diamond (ND-2)

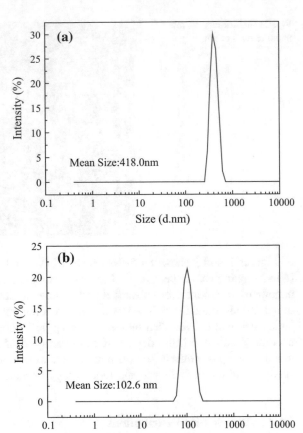

Fig. 4 Particle size distributions of raw nanodiamonds (ND) (**a**) and acid-heat-treated nanodiamonds (ND-2) (**b**)

It was clearly demonstrated from Fig. 4b that after acid and heat treatment, the ND aggregates were broken into smaller sized particles in the range from 60 to 200 nm and the average particle diameter of the treated NDs decreased to 102.6 nm. This indicates that the acid and heat treatment together are highly effective in breaking down the ND aggregates into smaller sized particles, which results in the size of diamond has been reduced obviously.

3.3.3 FTIR Analysis

Figure 5 shows the FTIR spectra of raw ND, ND-2, ND-3, respectively. FTIR spectrum of raw ND (Fig. 1a) displayed a relatively weak C–O stretching band at 1092 cm^{-1}, C=O stretching vibration band at 1773 cm^{-1}, and O–H stretching band at 3439 cm^{-1}, implying the existence of carbonyl and hydroxyl groups on ND surface. Another broad absorption band at 1630 cm^{-1} was attributed to the –OH deformation vibration of absorbed water. A weak absorption band at 2932 cm^{-1} belonged to C–H asymmetry stretching vibration.

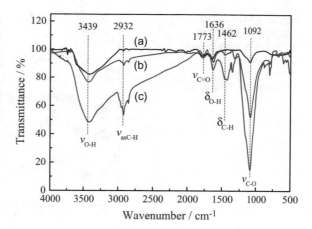

Fig. 5 FTIR spectra of raw ND (**a**), acid-heat-treated ND (ND-2) (**b**), and further modified ND (ND-3) (**c**)

After a strong acid oxidation and high-temperature treatment, the FTIR spectrum of nanodiamond (ND-2) was shown in Fig. 5b. The weak peaks of the intermediate position of 1700–2800 cm^{-1} disappear obviously, which indicated that some impurity groups were removed on the diamond surface after treatment. Moreover, the peaks of oxygen functional groups of C–O and O–H at 1092 and 3439 cm^{-1} were enhanced evidently. It reveals that the amount of surface oxygen groups increases, which indicates that the modification of the surface oxidation of nanodiamond has been achieved. The appearance of C–H stretching band at 1462 cm^{-1} and the stronger absorption peaks of asymmetric stretching vibration of hydrocarbon group at 2932 cm^{-1} implied the hydrogen bonds or oxygen bridge bonds totally to be destroyed under high-temperature condition. Apparently, the surface of nanodiamond particles was adsorbed with oxygen active groups, hydroxyl groups, carboxyl groups, ether groups, carbonyl groups, and ester groups and the existence of these surface groups provided the possibility for dispersing diamond nanoparticles in the medium [11, 22, 35].

Figure 5c shows the FTIR of ND-3 obtained by further treatment with dispersant to ND-2. The peak strengths of C–O stretching band at 1092 cm^{-1}, C–H bending in-plane vibration at 1462 cm^{-1}, C–H stretching band at 2932 cm^{-1}, and O–H stretching band at 3439 cm^{-1} were further strengthened sharply, which was observed that the surface of final modified nanodiamond with rich functional groups.

3.3.4 XRD Characterization

X-ray diffraction (XRD, D/MAX2500V Japan) was used to analyze the diamond before and after treatment. The result was shown in Fig. 6.

The powder XRD pattern of raw ND was shown in Fig. 6a. It was clear that the raw nanodiamond peaks of ND were more obvious. These patterns present crystalline planes of (111) and (220) with 2θ of 43.69° and 75.36°, respectively [31], and confirmed the presence the active structure. Besides, there was a small weak

Fig. 6 XRD patterns of raw
nanodiamond (ND) (**a**),
acid-treated nanodiamond
(ND-1) (**b**), and
acid-heat-treated
nanodiamond (ND-2) (**c**)

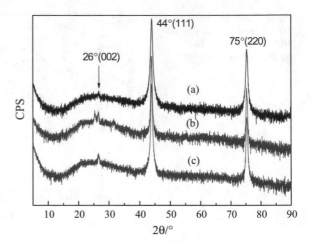

peak representing graphite diffraction peaks of (002) with 2θ of $26°$. After acid treatment, the XRD pattern of ND-2 was shown in Fig. 6b. The characteristic peaks of nanodiamond (ND-1) remained without any change, but graphite diffraction peaks of (002) became slightly stronger than before, and a little peak of impurity occurred at about $25°$. It illustrated that the degree of surface graphitization increased by treating diamond with acid. After further heat treatment of ND-1, the XRD pattern of ND-2 was shown in Fig. 6c. The peak of graphite carbon (002) was more obvious than ND-1. It indicates that the high-temperature treatment further alters the graphite carbon on the crystal surface of nanodiamond.

3.4 Dispersion Stability

Both experiments of the centrifugal stability and the static stability were selected to investigate in order to observe the dispersion stability of modified nanodiamond in $32^{\#}$ hydraulic oil. The suspension system of 0.1 wt% nanodiamond in $32^{\#}$ hydraulic oil was put into the 10 ml centrifugal tube. Then, the high-speed centrifuge was used to test centrifugal stability of the suspension system at 3000 rpm and 4000 rpm, respectively. After 10-min centrifugation, the bottom precipitation was shown in Fig. 7. It can be seen that there is no any precipitates at the bottom under 3000 rpm and 4000 rpm centrifugation for 10 min. The results show that the suspension system of hydraulic oil with nanodiamond has excellent centrifugal stability.

In addition, the suspension system of 0.1 wt% nanodiamond in hydraulic oil was stored in the test tube and observed at the bottom of test tube for a long period, as shown in Fig. 8. It was found that the modified nanodiamond was dispersed in hydraulic oil well and could exist stably for 120 days. Obviously, the results prove that the suspension system of nanodiamond and hydraulic oil has a high-degree

Fig. 7 Images of centrifugal stability of modified nanodiamond (0.1 wt%) dispersed in lubricating oil

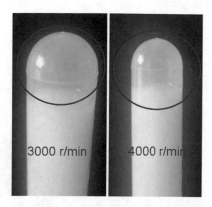

Fig. 8 Picture of static stability of modified nanodiamond (0.1 wt%) dispersed in lubricating oil

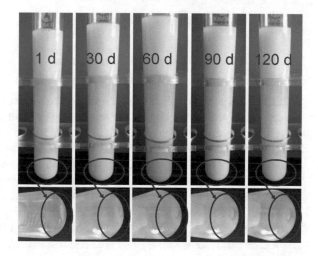

static stability. Due to the modification of nanodiamond, the particle size decreased and the oxygen functional groups increased, which was beneficial to disperse in the oily medium [21]. What's more, the surfactant can be adsorbed on the surface of nanodiamond particles to form steric hindrance and prevents it from aggregation. Generally, these favorable conditions make nanodiamond achieve a state of good dispersion in lubricating oil.

4 Tribological Behaviors of Nanodiamond as Additive

4.1 Tribometer

The tribological properties were evaluated on a four-ball tribometer. In this process, three balls remained stationary and one ball was rotated around the three balls.

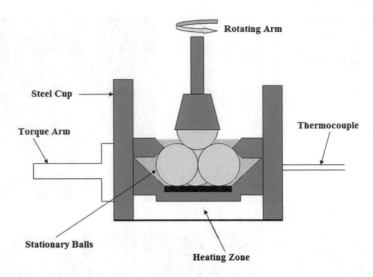

Fig. 9 Schematic diagram of the four-ball tribometer

A MQ-800-type four-ball tester was used for this experiment. All the balls were washed with petroleum ether and then blow dry with a blower. Three stationary balls were fixed in a steel cup, and 10 ml of test oil was poured on the cup. The three balls were covered with the test oil, with a depth of 3 mm covering the ball surface. Another ball, which was adjusted with a collet, was rotated around these stationary balls. Figure 9 shows the schematic diagram of the four-ball tribometer.

The effects of modified nanodiamond with different additions on the tribological properties of $32^{\#}$ anti-wear hydraulic oil provided by Anhui Runpu Nanometer Scientific Co. Scientific Company were carried out in this experiment. Before the friction test, ultrasonic cleaning of the steel ball and oil box used in the experiment was carried out in the petroleum ether. The experimental conditions are as follows: load 392 N, rotational speed 1450 r/min, duration 30 min, and test temperature 25 °C. The material of the four balls is AISI E52100 (φ12.7 mm, provided by Sinopec Research Institute of Petroleum Processing). After the four-ball test was completed, the wear scar diameter of the stationary balls was measured by an optical microscope. The friction coefficient is automatically generated.

4.2 Results

The friction and wear properties of modified nanodiamond lubricating oil added with a rate of 0, 0.02, 0.04, 0.06, 0.08, and 0.1 wt% were studied, respectively. The results are shown in Figs. 10 and 11.

As shown in Fig. 10, the average friction coefficient of pure hydraulic oil is the highest. When the diamond nanoparticles with different amounts are added,

Fig. 10 Variation of average friction coefficients with addition concentration of nanodiamond

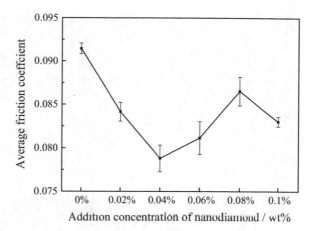

the average friction coefficient decreases. Moreover, the friction coefficient decreases firstly and then increases with the addition of nanodiamond. When the addition of nanodiamond was up to 0.04 wt%, the friction coefficient was shown with the lowest value and decreased by 13.8%. The results showed that the friction reduction of hydraulic oil was improved when the nanodiamond was used with 0.04 wt%. When the addition of nanodiamond continued to increase, the average friction coefficient began to increase. Therefore, the too high concentration of nanodiamond is disadvantage to the friction reduction effect. As shown in Fig. 11, when the hydraulic oil was lubricated, the diameter of wear scar was the largest. With the addition of different concentrations of nanodiamond, their wear scar diameters decreased. The wear scar diameter decreased to minimum values at the addition of 0.04 wt%, which had a decrease of 17.8%. As the addition amount of nanodiamond was over 0.04 wt%, it caused an increase of the wear scar diameter. Obviously,

Fig. 11 Variation of WSDs with addition concentration of nanodiamond

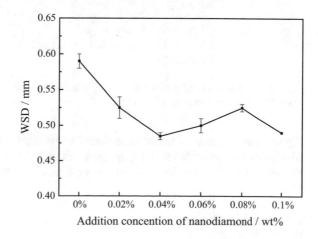

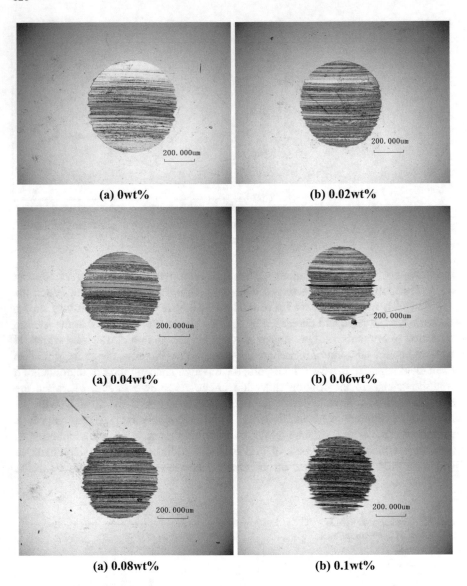

Fig. 12 Surface topography of the worn surface of the lower ball lubricated with $32^{\#}$ hydraulic oil containing different additions of nanodiamond

higher concentration of nanodiamond may cause reduce the anti-wear ability of oil. The corresponding detailed topography taken by 3D surface profilometer was shown in Fig. 12 in order to show the wear scar directly.

4.3   Friction and Wear Mechanisms

To further study the influence of tribological properties of modified nanodiamond in the lubricating oils, the surface morphology after friction was analyzed in detail, aiming to find the friction and wear mechanism of nanodiamond as additive. Therefore, the friction region was observed and analyzed by means of scanning electron microscopy, energy-dispersive spectroscopy, and Raman spectroscopy.

4.3.1   Characterization of Wear Scar

Figure 13 shows the SEM micrographs of the wear zones of the upper ball and the corresponding energy-dispersive spectrometry analysis. The worn surface area was shown in Fig. 13a for the pure hydraulic oil. It can be seen that the worn surface furrows are revealed obviously, which exposes the main form of wear is abrasive wear. The content of characteristic elements in the worn surface was analyzed by EDS (Fig. 13a). The content of C element on the worn zones is less, and a part of the C element comes mainly from the friction pairs, and partly from hydrocarbons in the lubricant.

As shown in Fig. 13b, when the 0.04 wt% nanodiamond was added to the hydraulic fluid, the worn surface was relatively smooth and taken on a shallower furrow. EDS result showed that the proportion of C elements on the worn surface was 3.19%, which was slightly higher than lubricant without nanodiamond. This may be because the particles are smaller or the smaller particles occupy a higher proportion when the nanodiamond added by 0.04 wt%. Since the surface roughness caused by friction increases, the nanodiamond with smaller particle size partially filled and repaired the concave and convex surfaces of the friction pairs, thus reducing the surface roughness and improving the hardness of the contact surface of the contact pair [2]. Besides, another part of nanodiamonds forms ball rolling between the friction surfaces, thereby reducing friction and wear between the friction pairs [14, 24, 29, 37].

It is shown in Fig. 13c that when the concentration of nanodiamond added to hydraulic oil is increased to 0.1 wt%, the wear zones is no longer smooth, and the deep furrow appeared. Furthermore, both abrasive and adhesive wear modes were observed at the wear track edges, and there were a few spalling pits on the local area. EDS results showed that the proportion of C elements on the surface of the wear zones increased to 7.92%. The reason may be that the concentration of nanodiamond is too high. Consequently, the occurrence of particle agglomeration results in the decrease of the rolling effect and the decreases of friction resistance of lubricant. On the other hand, a portion of the nanodiamond and the abrasive are agglomerated to form hard particles during the rubbing process, embedded in the friction pairs, and form a furrow effect.

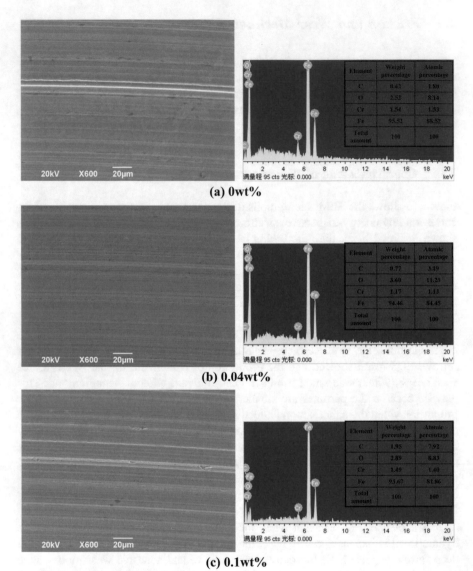

Fig. 13 SEM images of the wear zones of upper ball and the corresponding energy-dispersive spectrometry analysis

4.3.2 Raman Analysis

The Raman analysis is very predominant analysis for carbon material. Since the friction pairs, lubricating oil, and nanodiamond contain carbon elements, the Raman analysis is performed for wear zones of the upper ball lubricated with hydraulic oil containing 0, 0.04, and 0.1 wt% addition of nanodiamond, respectively.

Fig. 14 Raman spectra of wear zones of the upper ball lubricated with hydraulic oil containing different additions of nanodiamond **a** 0 wt%, **b** 0.04 wt%, **c** 0.1 wt%

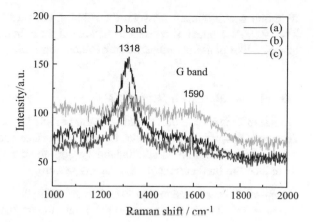

The results are shown in Fig. 14. The sharp peaks at 1318 and 1590 cm^{-1} refer to D band and G band of ND, which are attributed to the vibration of the disordered sp^3—hybridized carbon and the in-plane vibrations of the sp^2—bonded carbon, respectively [16, 26]. The I_D/I_G ratio of the two peak strengths reflects the degree of graphite ordering of the carbon material. The graphitization degree of the material becomes higher with the I_D/I_G ratio decreasing.

Also from Fig. 14, as the addition of nanodiamond increased from 0 wt% to 0.1 wt%, it caused a decrease of D band intensity. It can be known that the I_D/I_G ratios of 0.04 wt% and 0.1 wt% additions are 1.51 and 0.99, respectively, which are smaller than the ratio (I_D/I_G = 1.75) of 0 wt%. This indicates that the graphitization degree of the wear zones surface becomes high with increasing the concentration of nanodiamond. The Raman studies showed the structural ordering of wear zones surface in graphite-like phase as the nanodiamond addition increased.

5 Conclusion

The size of the modified nanometric diamond powder treated by strong acid and high-temperature oxidation was reduced obviously, and the surface functional group numbers of the powder were also increased, and the steric hindrance effect was easy to be formed in the lubricating oil, which provided favorable conditions to show excellent stability and dispersion. As the modified nanometric diamond was added in hydraulic oil with a rate of 0.04 wt%, the anti-wear and friction reduction of hydraulic oil were significantly improved, the average scar diameter decreased by 17.8%, and the average friction coefficient decreased by 13.8%. The lubricating mechanism was considered that smaller ND particles with an optimum concentration filled into the rubbing interfaces and repaired the rough surfaces of the friction pairs, and reduced the surface roughness and improved the hardness of the contact surface of tribo-pair. Meanwhile, the action of rolling balls of nanometric diamond was produced between the rubbing surfaces to reduce friction and wear. The Raman studies showed the structural ordering of wear zones surface in graphite-like phase as the addition of nanometric diamond increased.

List of Questions with Responses

Question 1: Is it right that the smaller the particle is, the easier it is to agglomerate?
Answer: Yes, it is. The particles have great surface energy due to their small size, which make them easy to spontaneously agglomerate and form secondary particles to reduce the total surface energy of the system.

Question 2: Is dispersion equivalent to solubility?
Answer: No, it isn't. From a chemical point of view, the dispersion is a substance in the form of particles dispersed in the liquid, but dissolution is in the form of molecules or ions dissolving with the solution as a whole. From the perspective of the material phase, the dispersion is formed by two phases, namely, dispersions and dispersants, usually form suspensions or emulsions. After dissolution, a phase is formed, which is a solution of the certain substance.

Question 3: Is it right that larger specific surface area results in better dispersion.
Answer: Yes, it is. Because the larger the specific surface area is, the smaller the particle size is and the smaller the particle size is, the better the dispersibility is.

Question 4: High-temperature treatment can disagglomerate diamond nanoparticles. So is it right that the higher the temperature is, the more sufficient the agglomeration become?
Answer: Yes, it is. But the temperature is too high, which will lead to the change of diamond structure and deep oxidation in the air.

Question 5: Is it right that the better dispersion stability of nanoparticles is in lubricant, the better the friction and wear properties is?
Answer: No, it isn't. But a good dispersion and stability has an important influence on friction and wear properties.

Question 6: Is it right that wear patterns usually exist in several forms simultaneously during actual wear phenomenon?
Answer: Yes, it is. Moreover, a kind of abrasion often induces other forms of wear.

Question 7: What are the forms of abrasive wear?
Answer: The forms of abrasive wear are mainly as follows: (a) The abrasion caused by the relative movement of abrasive particles along a solid surface is called two-body abrasive wear; (b) in a group of friction pairs, the rough peaks on the hard surface play a role in abrasive wear on the soft surface, which is also a kind of two-body wear. It is usually abrasive wear at low stress; (c) the external abrasive particles move between two friction surfaces, similar to the abrasive action, called three-body abrasive wear.

Question 8: What are the factors involved in abrasive wear?
Answer: In addition to the relative hardness of the friction pairs, abrasive wear resistance is also related to the hardness, strength, shape, sharpness, and particle size of abrasives.

Question 9: Is it right that the higher the concentration of nanodiamond is, the stronger the ball effect is in the process of friction?
Answer: No, it isn't. When the concentration of nanodiamond becomes higher, the particles become increasingly easy to reunite, which make the large particles difficult to achieve the ball effect.

Question 10: Does it necessarily result in graphitization of carbon materials in the process of friction?
Answer: No, it doesn't. Graphitization occurs only at high temperatures. When the temperature increases, the graphitization is intensified, but the temperature is too high, not only the graphitization does not occur, but the formed graphite is combined with ferrite to form cementite.

Acknowledgements This work was supported by the National Natural Science Foundation of China (Grant No. 51675153).

Glossary

Surface energy Surface energy is a measure of the destruction of intermolecular bonds when material surfaces are created.

Disaggregation The agglomeration structure of powder was broken by screening, mechanical grinding, preparation of high dispersion suspension and adding surfactant and so on.

Adsorption An exotic molecule or atom or ion from an environment adheres to the surface of a solid by physical or chemical action.

Steric hindrance The steric hindrance effect mainly refers to the spatial hindrance caused by the proximity of certain atoms or groups in the molecule to each other.

Dispersibility Flocculation or droplets of solid particles in water or other homogeneous liquid medium can be dispersed into small particles suspended in the dispersion medium without precipitation.

Rubbing pair A system consisting of two objects that move relatively in contact with the surface.

Wear scar Damage marks left on the friction surface after the friction and wear of the solid surface, which are the important basis for evaluating the type of wear.

Abrasive wear The wear caused by the act of extruding or moving along the surface of a solid surface by hard particles or rigid bodies. The wear caused by the hard particles or hard body on the solid surface extrusion and along the surface movement.

Ploughing The softer surface of the two surface of the relative motion results in groove damage due to plastic deformation.

Graphitization Graphitization refers to the orderly transformation of carbon atoms which are thermodynamically unstable from disordered structure to graphite crystal structure.

References

1. Chou, C. C., & Lee, S. H. (2010). Tribological behavior of nanodiamond-dispersed lubricants on carbon steels and aluminum alloy. *Wear, 269*(11–12), 757–762.
2. Chu, H. Y., Wen, C. H., & Lin, J. F. (2010). Scuffing mechanism during oil-lubricated block-on-ring test with diamond nanoparticles as oil additive. *Wear, 268*(11), 1423–1433.
3. Cui, X., Liu, X., Tatton, A. S., Brown, S. P., Ye, H., & Marsh, A. (2012). Nanodiamond promotes surfactant-mediated triglyceride removal from a hydrophobic surface at or below room temperature. *ACS Applied Materials & Interfaces, 4*(6), 3225–3232.
4. Dai, W., Kheireddin, B., Gao, H., & Liang, H. (2016). Roles of nanoparticles in oil lubrication. *Tribology International, 102,* 88–98.
5. Davidson, J. L., & Bradshaw, D. T. (2005). Compositions with nano-particle size diamond powder and methods of using same for transferring heat between a heat source and a heat sink, U.S. Patent 6858157B2.
6. Dolmatov, V. Y. (2001). Detonation synthesis ultradispersed diamonds: Properties and applications. *Russian Chemical reviews, 70*(7), 687–708.
7. Elomaa, O., Oksanen, J., Hakala, T. J., Shenderova, O., & Koskinen, J. (2014). A comparison of tribological properties of evenly distributed and agglomerated diamond nanoparticles in lubricated high-load steel–steel contact. *Tribology International, 71,* 62–68.
8. Grichko, V., Hens, S., & Walch, J. (2007). Detonation nanodiamonds as UV radiation filter Diam. *Diamond and Related Materials, 16*(12), 2003–2008.
9. Grichko, V., Tyler, T., Grishko, V. I., & Shenderova, O. (2008). Nanodiamond particles forming photonic structures. *Nanotechnology, 19*(22), 225201.
10. Gruen, D. M., Shenderova, O. A., & Vul, A. Y. (2005). *Synthesis properties and applications of ultrananocrystalline diamond*. Dordrecht, The Netherlands: Springer.
11. Huang, Z. J., Fei, Y. W., & Shang, Z. F. (2005). The application of nanoparticles as additive in lubricants and in lubricants and its development trend. *Lubricating Oil, 20*(2), 21–25.
12. Ivanov, G., Kharlamov, V. V., Buznik, V. M., Ivanov, D. M., Pavlyshko, S. G., & Tsvetkov, A. K. (2004). Tribological properties of the grease containing polytetrafluorethylene and ultrafine diamond. *Friction & Wear, 25*(1), 99–103.
13. Ivanov, M. G., Pavlyshko, S. V., Ivanov, D. M., Petrov, I., & Shenderova, O. (2010). Synergistic compositions of colloidal nanodiamond as lubricant-additive. *JVST B, 28*(4), 869–877.
14. Kim, H. S., Park, J. W., Park, S. M., Lee, J. S., & Lee, Y. Z. (2013). Tribological characteristics of paraffin liquid with nanodiamond based on the scuffing life and wear amount. *Wear, 301*(1–2), 763–767.
15. Krueger, A. (2008). Diamond nanoparticles: Jewels for chemistry and physics. *Advanced Materials, 20*(12), 2445–2449.

16. Krüger, A., Liang, Y., Jarry, G., & Stegk, J. (2006). Surface functionalisation of detonation diamond suitable for biological applications. *Journal of Materials Chemistry, 16*(24), 2322–2328.
17. Lee, G. J., Park, J. J., Lee, M. K., & Rhee, C. K. (2017). Stable dispersion of nanodiamonds in oil and their tribological properties as lubricant additives. *Applied Surface Science, 415*, 24–27.
18. Liu, W. M., Xu, J., Feng, D. P., & Wang, X. P. (2013). The research status and prospect synthetic lubricating oils. *Tribology, 33*(1), 91–104.
19. Michail, I. G., Pavlyshko, V. S., Ivanov, D. M., Petrov, I., McGuire, G., & Shenderova, O. (2009). Nanodiamonds particles as additives in lubricants. *MRS Proceedings, 1203*, 89–94.
20. Mochalin, V. N., & Gogotsi, Y. (2015). Nanodiamond–polymer composites. *Diamond and Related Materials, 58*, 161–171.
21. Motahari, H., & Spectroscopic, Malekfar R. (2017). Investigation for purity evaluation of detonation nanodiamonds: experimental approach in absorbance and scattering. *Journal of Cluster Science, 28*(4), 1923–1935.
22. Mou, G. J., & Zhao, B. (2004). Research advances of inorganic nano additives in lubricating oil. *Lubricating Oil, 19*(1), 59–61.
23. Neville, A., Morina, A., Haque, T., & Voong, M. (2007). Compatibility between tribological surfaces and lubricant additives—How friction and wear reduction can be controlled by surface/lube synergies. *Tribology International, 39*(10), 1680–1695.
24. Novak, C., Kingman, D., Stern, K., Zou, Q., & Gara, L. (2014). Tribological properties of paraffinic oil with nanodiamond particles. *Tribology Transactions, 57*(5), 831–837.
25. Nunn, N., Torelli, M., McGuire, G., & Shenderova, O. (2017). Nanodiamond: A high impact nanomaterial. *Current Opinion in Solid State and Materials Science, 21*(1), 1–9.
26. Osswald, S., Yushin, G., Mochalin, V., Kucheyev, S. O., & Gogotsi, Y. (2006). Control of sp2/sp3 carbon ratio and surface chemistry of nanodiamond powders by selective oxidation in air. *Journal of the American Chemical Society, 128*(35), 11635–11642.
27. Puzyr, A. P., Selyutin, G. E., Vorobyov, V. B., & Fedorova, E. N. (2006). Prospects for detonation nanodiamonds with colloidal stability in the technical field. *Nanotechnique, 4*, 96–105.
28. Red'Kin, V. E. (2004). Lubricants with Ultradisperse diamond–graphite powder. *Chemistry and Technology of Fuels and Oils, 40*(3), 164–170.
29. Shen, M. Z., Luo, J. B., & Wen, S. Z. (2001). Influence of diamond nano-particles on the tribological properties. *Chinese Journal of Mechanical Engineering, 37*(1), 14–16.
30. Shi, P. J., Xu, Y., & Xu, B. S. (2006). Application of nano friction-reducing lubricant on engine. *Lubrication Engineering-Huangpu, 7*, 73–76.
31. Sundar, L. S., & Singh, M. K. (2016). Experimental thermal conductivity and viscosity of nanodiamond-based propylene glycol and water mixtures. *Diamond and Related Materials, 69*, 49–60.
32. Taylor, A. C., Edgington, R., & Jackman, R. B. (2015). Patterning of nanodiamond tracks and nanocrystalline diamond films using a micropipette for additive direct-write processing. *ACS Applied Materials & Interfaces, 7*(12), 6490–6495.
33. Wang, F. F., Zhang, L. Q., Peng, J., & Zhou, W. J. (2015). Research progress in nano-diamond surface modification. *Superhard Material Engineering, 27*(3), 36–39.
34. Xie, H., Yu, W., Li, Y., & Chen, L. (2011). Discussion on the thermal conductivity enhancement of nanofluids. *Nanoscale Research Letters, 6*(1), 1–12.
35. Yang, C. J., Chen, G. X., & Zhao, L. T. (2009). Research progress of soft metal nanoparticles as self-repairing lubricating oil additive. *Lubrication Engineering, 34*(5), 115–117.
36. Zhang, D., Hu, X. G., Tong, Y., & Hung, F. (2006). The research development of nanodiamond as a lubricating additives. *Lubricating Oil, 21*(1), 50–54.
37. Zhang, J. X., Liu, K., & Hu, X. G. (2002). Effect of ultra-dispersed diamond nanoparticles as additive on the tribological properties of 15W30 engine oil. *Tribology, 22*(1), 44–48.

Part II
Thermal Engineering

Chapter 5
Basics and Applications of Thermal Engineering

T. K. Gogoi and U. S. Dixit

Abstract This chapter presents the basics of three major topics of thermal engineering, viz. fluid mechanics, heat transfer, and thermodynamics. The three main parts of fluid mechanics, viz. fluid statics, kinematics, and dynamics, are summarized. The three modes of heat transfer, conduction, convection, and radiation, are briefly explained with examples. Finally, the basics of thermodynamics are presented in detail. The important thermodynamic properties and laws of thermodynamics are introduced. Energy and exergy balances for thermodynamic systems are discussed. Some application areas of thermal engineering are highlighted.

Keywords Fluid mechanics · Fluid kinematics · Fluid dynamics
Irrotational flow · Laminar flow · Turbulent flow · Conduction
Convection · Radiation · Entropy · Exergy · Energy · Open system
Closed system · Temperature

1 Introduction

Thermal engineering is an important part of mechanical engineering comprising three important topics, viz. thermodynamics, fluid mechanics, and heat transfer. The knowledge of these topics is helpful for the design and analysis of thermal systems. Thermal systems deal with transfer, conversion, and storage of energy. They may consist of a number of components that function together as a whole unit. Some household examples of thermal system are pressure cooker, oven, refrigerator, air conditioner, water heater, electric bulbs, electric iron, computer, and television. The human heart that pumps blood to all body parts is also an example of a thermal system. In fact, the whole human body can be considered as a thermal system because

T. K. Gogoi
Department of Mechanical Engineering, Tezpur University, Tezpur, India

U. S. Dixit (✉)
Department of Mechanical Engineering, IIT Guwahati, Guwahati, India
e-mail: uday@iitg.ernet.in

© Springer International Publishing AG, part of Springer Nature 2018
J. P. Davim (ed.), *Introduction to Mechanical Engineering*, Materials Forming,
Machining and Tribology, https://doi.org/10.1007/978-3-319-78488-5_5

some form of energy conversion takes place inside the body cells. Thermal comfort in human body is maintained through metabolic heat rejection. To adjust heat transfer rate from our body with changing climatic condition, people wear different types of clothes in different seasons. Industrial thermal systems include automotive engines, jet engines, rockets, power plants, refrigeration and air-conditioning plants, refineries, chemical processing plants, food processing and preservation units, and manufacturing industries. Among these, the automotive engines, jet propulsion engines, and thermal power plants involve conversion of energy in fossil fuels for getting the useful work. Figure 1 shows some common examples of the thermal systems.

Window air conditioner   Heat exchanger

Solar water heater   Thermal power plant

Household Refrigerator   Diesel engine generator

Fig. 1 Some common examples of thermal systems

This chapter provides a brief introduction of three important subjects of thermal engineering, viz. fluid mechanics, heat transfer, and thermodynamics. Their importance in some practical engineering applications is emphasized. Historically, fluid mechanics is the oldest subject among these three subjects. It started around third century BC during the period of Archimedes. Archimedes (circa 287 BC–212 BC) has done extensive work in hydrostatics. His principle of buoyancy is still taught in a course on fluid mechanics. Later on several physicist and mathematicians worked on and contributed to fluid mechanics. Notable among them are Pascal (1623–1662), Daniel Bernoulli (1700–1782), Navier (1785–1836), and Stokes (1819–1903). The study of heat transfer started since the beginning of eighteenth century with the pioneering work of Sir Isaac Newton (1642–1727). Josef Stefan (1835–1893), Ludwig Boltzmann (1844–1906), Ludwig Prandtl (1875–1953), and Wilhelm Nusselt (1882–1957) contributed a lot to this subject. This subject has also attracted a lot of mathematicians. Thermodynamics is the youngest subject among the three; it started in early nineteenth century with the pioneering book of Sadi Carnot, entitled 'Reflections on the Motive Power of Fire,' published in 1824. Sadi Carnot (1796–1832) is considered as the father of the thermodynamics. Thermodynamics is the core subject of thermal engineering.

In this chapter, basic concepts in fluid mechanics and heat transfer are discussed, albeit avoiding mathematical details. Later on thermodynamics is discussed in relatively detailed manner. The objective of this chapter is to provide foundation for the more detailed study of thermal engineering and to clarify several intricate concepts of this field.

2   Fluid Mechanics

Fluid mechanics is the study of fluids and their properties in static as well as moving condition. Fluids include liquids and gases. Fluid mechanics consists of three parts —fluid statics, fluid kinematics, and fluid kinetics. Study of fluid behavior at rest is called fluid statics (also known as hydrostatics), while fluid kinematics and kinetics deal with the study of fluid behavior in motion.

By definition, a fluid remains in static equilibrium under the action of shear stresses. When the fluid is at rest, a typical fluid particle experiences gravitational force and normal stress (hydrostatic pressure) [1, 2]. Some examples, where fluid statics is used for analysis and design, are as follows:

(i)   Pressure measuring devices such as manometers;
(ii) Velocity and flow rate measuring devices, e.g., pitot tube, venturimeter, orifice meter, and rotameter;
(iii) Partially and fully submerged objects such as boats, ships, and submarines;
(iv) Dams, retaining walls, and other hydraulic systems.

Fluid dynamics comprises fluid kinematics and fluid kinetics. Fluid kinematics studies fluid motion without considering the nature of forces that causes the motion [2]. Knowledge of fluid kinematics is essential in understanding fluid kinetics. In fluid kinetics, the dynamic behavior of fluids is analyzed along with the forces governing the fluid motion. Fluid dynamics associated with the motion of air or gas is specifically known as aerodynamics, while the study of motion of a liquid is called hydrodynamics.

Governing equations of motion are the key to fluid dynamics. They are derived from physical laws by considering mass and force balance either on a differential fluid element or on a control volume (CV). Accordingly, the differential and integral forms of mass and momentum conservation equations are obtained. These governing equations are solved either analytically or numerically to determine the velocity and pressure field in various flow geometries. There are different classifications of a flow; some of them are described in the following subsections [3].

2.1 Viscous and Non-viscous (Inviscid) Flows

Viscosity plays a significant role in fluid flow. It is a fluid property by virtue of which it offers resistance to the flow of one fluid layer over another. Almost all the fluids in nature are viscous, and flow associated with viscous fluids is called viscous flow. However, in case of very low-viscosity fluids, the flow analysis can be done by neglecting the viscous effect of the fluids. Flows in which viscosity effects are neglected are called non-viscous or inviscid flows.

2.2 Steady and Unsteady Flows

A steady flow is one in which the velocity and fluid properties may vary at different points but remain constant with time at a fixed point in the flow field. On the other hand, the flow is unsteady if the velocity and fluid properties change with time at a given location in the flow.

2.3 Uniform and Non-uniform Flows

At a given instant of time, if the flow velocity remains same everywhere in the flow field, it is called uniform flow. If the fluid flow velocity is different at different points, then it is called a non-uniform flow.

2.4   Rotational and Irrotational Flows

The flow in which the infinitesimal fluid elements also rotate about their own axis while moving along a certain path is called rotational flow. A fluid element undergoes rotation due to angular momentum in the flow field. When the angular velocity vector is zero, the fluid particles do not undergo any rotation and the flow is said to be irrotational. The combined translational and rotational motion of an elemental fluid particle ABCD is shown in Fig. 2a, where the fluid element while moving in a straight path is also rotating about its own axis, and hence, it is rotational flow. In Fig. 2b, the fluid element ABCD although is moving in an elliptical path, but there is no rotation of the fluid element about its own axis; hence, it is an irrotational flow.

2.5   Compressible and Incompressible Flows

During fluid flow, if the fluid density changes significantly in the flow field, then it is termed as compressible flow. All kinds of gas flows are compressible except in case of gas flowing at low velocity with Mach number less than 0.3. Mach number is defined as the ratio of the velocity of gas to the velocity of the sound in the surrounding medium. Flow in which the density variation is negligible is called incompressible flow. Flow of liquid is often considered as incompressible because change in liquid density with pressure is almost negligible.

2.6   Laminar and Turbulent Flows

Laminar flow is a flow in which the fluid particles move in parallel layers in a very organized manner. Highly viscous fluids such as oils moving at low velocity usually exhibit laminar flow. On the other hand, the flow in which the movement of

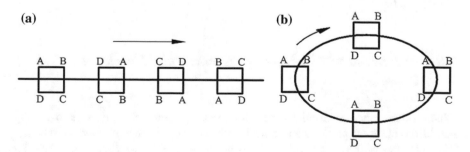

Fig. 2 Movement of a fluid element in **a** rotational flow and **b** irrotational flow

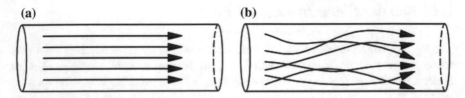

Fig. 3 A typical flow with arrows indicating the trajectory of the fluid particles: **a** laminar flow and **b** turbulent flow

fluid particles is highly irregular and chaotic in nature is called turbulent flow. Turbulent flowis associated with formation of eddies and random fluctuations of velocity and pressure with time and space. The low-viscosity fluids such as air moving at high velocity typically display turbulent behavior. The difference between laminar and turbulent flows is depicted in Fig. 3. In turbulent flow, considerable mixing occurs among the fluid particles. Sometimes, the turbulent flow is desired. For example, the combustion chamber in a diesel engine is designed to cause a turbulent flow in order to obtain a good mixing between the fuel and air. In fluidized bed boiler, turbulence is promoted by fluidization of fine solid particles that assists in uniform mixing between fuel and air. This is basically done for generating evenly distributed heat at lower temperature that minimizes harmful NO_x emission. It also allows the use of low-quality fuel in the boiler furnace.

2.7 Internal and External Flows

Analysis of fluid flow through circular and non-circular ducts (rectangular, square, triangular, and elliptical) is very common in fluid mechanics. These are internal flow where the fluid particles move inside the duct. In external flow, for example in flow past a cylinder, the fluid particles flow over the solid surface. Internal flow can be classified into open-channel and pipe flows. In an open-channel flow, liquids flow in a channel with a free surface open to atmosphere. Flow of water in rivers and irrigation canals are examples of open-channel flow. Pipe flow is a flow in a closed conduit.

2.8 Hydrodynamically Developing and Fully Developed Flows

When a fluid enters a circular pipe with certain uniform velocity, due to no-slip condition, the fluid particles in contact with the pipe surface attain zero velocity. The velocities of the adjacent fluid layers also decrease due to resistance offered by the slow-moving layers at the bottom. In this process, the fluid layer at the pipe center will attain the maximum velocity to satisfy the mass balance and the central velocity will also be greater than the original uniform velocity at the pipe entrance. This velocity variation gives rise to the development of velocity boundary layers along the tube. In the boundary layers, there is sufficient amount of velocity gradient, while outside the boundary layer, the velocity is more or less uniform. These two velocity boundary layers keep growing in thickness until they merge at the center at certain length from the pipe entrance. This length is called entry length, and the flow in this entrance region is called hydrodynamically developing flow. Beyond the entrance region, the velocity profile is fully developed; hence, it is called hydrodynamically fully developed flow. In the fully developed region, the velocity profile is the same at all locations along the flow direction.

2.9   One-, Two-, and Three-Dimensional Flows

During fluid flow, velocity and other flow properties actually vary in all the three directions, i.e., they are functions of (x, y, z) in Cartesian and (r, θ, z) in polar coordinate systems. Under this situation, fluid flow is considered as three-dimensional (3-D); 3-D flow analysis is usually complicated. However, under certain conditions, velocity and property variation can be neglected in certain directions. Accordingly, the fluid flow analysis can be done assuming it to be two-dimensional (2-D) or one-dimensional (1-D). 1-D and 2-D assumptions make fluid flow analysis simple. Exact solutions of the Navier–Stokes equations (momentum balance equations in the three directions) that are otherwise difficult to obtain for 3-D flow can be obtained without much difficulty for 1-D and 2-D flow in certain flow geometries. For example, fully developed laminar flow between two parallel plates (Fig. 4) can be approximated as one-dimensional if it is assumed that only the x velocity component (u) is the non-trivial one and it varies only in the y-direction.

Knowledge of fluid mechanics and fluid flow behavior under various forces and at different flow conditions helps mechanical, civil, and chemical engineers in solving many practical engineering flow problems. All kinds of power plants (hydroelectric, thermal, and nuclear), hydraulic machines, automobiles involve application of fluid mechanics theory and principles.

Fig. 4 1-D flow between two parallel plates where the velocity component u varies only in y-direction

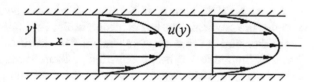

3 Fundamentals of Heat Transfer

Heat transfer deals with the rate of energy transfer (energy transfer per unit time) by virtue of temperature difference. However, thermodynamics is concerned only with the amount of heat transfer and not with the heat transfer rate. Thermodynamics gives no idea about the time that is required to complete the process. Heat transfer takes place by three modes—conduction, convection, and radiation. In most of the heat transfer problems, all the three modes are active, for example, the solar water heating systems that use different types of solar collectors.

3.1 Conduction

Conduction in a medium refers to heat transfer through molecular interactions from particles that are more energetic relative to the adjacent particles. All substances in solid, liquid, and gas phases support conduction. The rate of heat conduction is normally the highest in solid phase, followed by liquid and gaseous phases. Apart from temperature difference, the rate of heat conduction through a medium also depends on its geometry, thickness, and material of the medium. Conduction in liquids and gases occurs due to collision and diffusion mechanism of the molecules during their random motion, while in solids it occurs due to vibration of the molecules in the lattice and transport of energy by the free electrons [3].

Consider a plane wall of thickness Δx with surface area normal to the thickness direction as A. Assume that temperature varies along the thickness direction only. The conduction heat transfer rate is given by

$$\dot{Q}_{\text{cond}} = -kA\frac{\Delta T}{\Delta x},\tag{1}$$

where k is the thermal conductivity of the material that measures the ability to conduct heat. In the limiting case, when $\Delta x \to 0$, Eq. (1) is written in differential form as

$$\dot{Q}_{\text{cond}} = -kA\frac{\text{d}T}{\text{d}x}.\tag{2}$$

Equation (2) is known as Fourier law of heat conduction and $\text{d}T/\text{d}x$ as the temperature gradient. The negative sign in Eq. (2) implies that heat transfer occurs in the direction of decreasing temperature. The Fourier law represented by Eq. (2) is however limited to 1-D heat conduction problem. It can be generalized for 2-D and 3-D cases also. Heat transfer problems are often analyzed as steady or unsteady problems. In steady-state analysis, it is assumed that the temperature and the heat flux (heat transfer per unit area) at any location remain invariant with time although they may be different at different locations.

3.2   *Convection*

Convective heat transfer takes place between a solid surface and an adjacent fluid in motion. It involves the effects of diffusion, fluid bulk motion, or both. It can be in the form of forced convection, natural (or free) convection, and sometime in mixed mode of convection. In forced convection, the fluid is forced to flow over the solid surface by some external means such as a pump or a fan, while in natural convection the buoyancy force arising out of density difference caused by fluid temperature variation governs the fluid motion. In Fig. 5a, air at 18 °C is forced to flow over a hot plate at 300 °C by using an electric fan and heat transfer takes place from the hot plate to the air moving over the plate. This is forced convection. However, when the electric fan is switched off, the air ceases to flow over the plate but even then, the layer of air adjacent to the plate will be heated due to conduction of the plate to the air. Also, heat transfer by conduction will take place from this layer of air to the next outer layer and so on. Consequently, the temperature of the neighboring air particles increases and the density decreases. The lighter air particles rise up, and the empty space created by upward movement of the lighter air is filled up by the cold air that rushes from the top (Fig. 5b). This rising motion of lighter air and downward motion of cold air is called the natural or free convection current. Heat transfer by natural convection continues until the plate is cooled to the temperature of the surrounding air.

The heat transfer rate by convection is determined by using Newton's law of cooling:

$$\dot{Q}_{\text{convection}} = hA_{\text{s}}(T_{\text{s}} - T_{\infty}). \tag{3}$$

where A_{s} is the heat transfer surface area, T_{s} is the surface temperature, and T_{∞} is the temperature of the fluid amply far away from the surface. The heat transfer coefficient h is the most crucial parameter in convective heat transfer that can be determined either by experiment or theoretical analysis. The heat transfer coefficient h itself is not a fluid property; however, it depends upon the fluid properties and other flow variables such as flow regimes (laminar, transition, or turbulent), surface roughness, and geometry of the body. Value of h is typically high for liquid flow and forced convection compared to gas flow and free convection, respectively. It is also higher for turbulent than that for laminar flow.

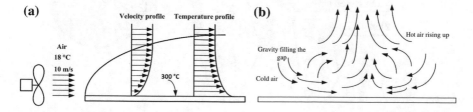

Fig. 5 Convection heat transfer from a hot plate to air **a** forced convection and **b** free convection

3.3   Radiation

In radiation, heat exchange between surfaces at different temperatures occurs in the form of electromagnetic waves or moving subatomic particles and unlike conduction and convection; it does not require any medium for heat transfer. Figure 6 shows the radiation heat transfers between the Sun at approximately 5500 °C and the Earth at atmospheric temperature, which occurs even when there are mediums (thermosphere, mesosphere, stratosphere, troposphere) in between, which are at lower temperatures than both of them.

Radiation exchange between bodies due to temperature difference is specifically known as thermal radiation. Radiation waves travel at the speed of light, and hence, it is the fastest among the three modes of heat transfer. All bodies above absolute zero temperature emit thermal radiation in all directions over a wide range of wavelengths. Radiation emitted from a surface at a specified wavelength depends on its surface temperature, material, and surface condition. Hence, it is obvious that the radiation emitted by two different bodies can never be the same even at the same temperature. Here comes the concept of blackbody, an idealized body that emits maximum possible radiation energy uniformly in all directions per unit area normal to direction of emission at a specified temperature and wavelength. A blackbody also absorbs all radiation, incident on its surface regardless of their wavelength and direction. Hence, a blackbody is both a perfect emitter and an absorber of radiation.

The rate of radiation energy emitted by a blackbody per unit surface area is known as blackbody emissive power. This is given by the Stefan–Boltzmann law as

$$E_b(T) = \sigma T^4. \tag{4}$$

where $\sigma = 5.67 \times 10^8$ W/m^{2}K^4 is the Stefan–Boltzmann constant and T is the absolute temperature of the surface in Kelvin. However, the radiation emitted by a real surface will always be less than the blackbody radiation. The ratio of the radiation emitted by a real surface at a given temperature to the radiation emitted by

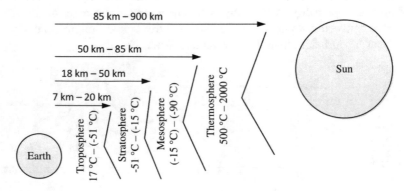

Fig. 6 Radiation heat transfer between the Sun and the Earth

a blackbody at the same temperature is defined as the emissivity of the surface. Emissive power of real surface is the product of emissivity ε and emissive power of the blackbody:

$$E(T) = \varepsilon E_b(T). \tag{5}$$

Understanding the mechanism of heat transfer is important because it finds its application in the design and development of many engineering systems such as power plants, automobiles, heat exchangers, cooling systems, electronic devices, and buildings. The real multimodal heat transfer problems are usually complicated and require the use of higher mathematics and computational techniques for solving the coupled momentum (fluid flow) and energy (heat transfer) equations. Combined fluid flow and heat transfer analysis is very much required for designing some thermal systems [4–7].

4 Thermodynamics

Thermodynamics is the study of science of energy transfer (or interaction). It mainly deals with energy and energy conversion principles along with the change in thermodynamic properties during state (property) change undergone by a system from one equilibrium state to another. Thermodynamic processes, cycles, and systems are mainly described in thermodynamics. In the following sections, some fundamental concepts and laws of thermodynamics, their importance, and application in analyzing some practical thermal systems are discussed.

4.1 Thermodynamic Systems and Their Classifications

A thermodynamic system is a quantity of matter or a volume in space described by its state variables (or properties) such as pressure, volume, temperature. Anything external to the system is called its surroundings, and the system and its surroundings are separated by the system boundary (Fig. 7).

Thermodynamic systems are mainly classified as closed, isolated, and open systems. A closed system is a system of fixed mass. Mass cannot enter or leave a closed system, although, there can be energy interaction between the system and surroundings. Closed systems are also referred to as non-flow system. An isolated system does not exchange or interact with its surroundings, neither through mass nor energy. In an open system, both mass and energy can cross the system boundary. Open systems are also known as flow systems and usually analyzed using the control volume concept. For example, a certain quantity of gas contained in a piston-cylinder assembly (Fig. 8) can be treated as a closed system if the inlet and exhaust valves at the piston top are closed. This is true particularly in internal

Fig. 7 Concept of system, surroundings, and system boundary

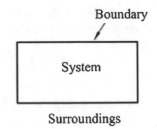

Fig. 8 Schematic of piston-cylinder assembly of an internal combustion engine

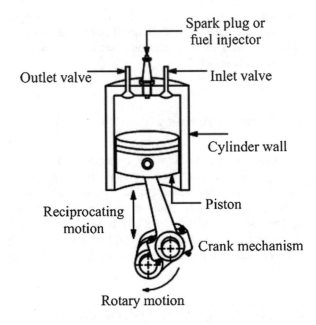

combustion engines where the piston moves upward and downward during the compression and expansion stroke, respectively, during which the inlet and exhaust valves remain closed. However, the inlet valve opens during the suction stroke and the outlet valve, during the exhaust stroke. Moreover, the system boundary on the piston side moves and there may be leakage of gas through the piston rings and crevices; therefore, the gas inside the engine does not constitute a pure closed system, rather it can be referred to as a semi-open system. Devices such as compressor, turbine, and nozzle involve both mass and energy transfer; hence, they are open systems (Fig. 9).

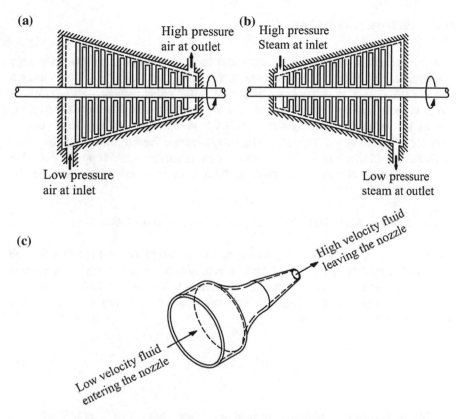

Fig. 9 Practical examples of open system: **a** air compressor, **b** steam turbine, and **c** nozzle

4.2 Thermodynamic State and Equilibrium

One needs to understand the meaning of equilibrium state in thermodynamics. The state of a system is described by its thermodynamic properties. At a given state, these properties will have fixed values; the property values change during a change of state in any operation. A system in equilibrium, if isolated from its surroundings, it does not experience any change of state. There are several types of equilibrium such as thermal, mechanical, chemical. A system in thermal equilibrium does not interact with the surroundings through heat transfer. Mechanical equilibrium implies absence of any unbalanced force within the system and also between the system and surroundings. Similarly, a system is in chemical equilibrium as long as its chemical composition is not altered by any chemical reaction occurring inside it. All these conditions of equilibrium need to be satisfied for a system to fulfill the condition of complete thermodynamic equilibrium.

4.3   Energy Interaction

Energy exchange between the system and its surroundings is known as energy interaction. Basically, it represents the energy gained or lost by a system during a process [8]. A system interacts with its surrounding either through heat or work. In open systems, energy interaction also occurs through mass transfer. Energy interaction by temperature difference is called heat transfer, and any other energy transfer is termed as work transfer. The basic modes of heat transfer were discussed in Sect. 3. Different forms of work transfer arise in thermodynamic applications due to interaction of system and surrounding. Some of them are briefly discussed below.

4.3.1   Moving Boundary Work or *pdV* or Displacement Work

Energy imparted to a system of gas by a moving piston in a closed cylinder (Fig. 8) is termed as *pdV* or displacement work. It is the most common form of mechanical work associated with the expansion and compression of gas inside a cylinder bounded by a piston. The displacement work transferred during a process 1–2 is mathematically calculated as

$$W_{1-2} = \int_{V_1}^{V_2} p\mathrm{d}V. \tag{6}$$

The general sign convention for heat and work interactions is as follows:

1. Heat flow into a system is positive, while heat outflow from a system is taken as negative.
2. Work transferred to a system is negative, while work done by the system is positive.

4.3.2   Shaft Work

The shaft of a pump or a compressor, if rotated by an electrical motor, is referred to as shaft work. Similarly, a turbine shaft rotated by water/steam/expanding gases moving through turbine blades is also shaft work.

4.3.3   Spring Work

If a spring, under the action of a force, is displaced from its undisturbed position, the total work done on the spring can be calculated as

$$W_{\text{spring}} = \frac{k}{2}\left(x_2^2 - x_1^2\right), \tag{7}$$

where k is the spring stiffness (force per unit displacement), x_1 and x_2 are the initial and the final displacements of the spring measured from the undisturbed position of the spring.

4.3.4   Other Types of Work Transfer

There are some non-mechanical types of work transfer. When work transfer occurs due to flow of electricity into and out of the system, then it is called electric work, for example electric work associated with electric motor and generator. It is the product of voltage and current. Electric polarization work is related to material polarization in the presence of an applied electric field. During electric polarization, the positive and negative charges within the material are displaced from to another and consequently, electric dipole moments (torque) are developed. The product of the electric field strength and total electric dipole moment of the molecules is defined as electric polarization work. Similarly, a material experiences magnetic dipole moment when it is placed in a magnetic field. Magnetic work is calculated by multiplying magnetic field strength with the total magnetic dipole moment.

Thermodynamic properties of a system are affected by electric or magnetic polarization. In some cases, the electric and magnetic properties are correlated with thermal properties to find out thermodynamic conditions for equilibrium of electric and magnetic dipole moments. Thermodynamics of electric and magnetic systems is a subject of practical interest such as thermoelectric generator, thermoelectric refrigerator, magnetic cooling systems.

4.4   Thermodynamic Processes and Cycles

When a system undergoes a change from an equilibrium state to another, the system undergoes a process. A system, while undergoing a process, passes through a sequence of intermediate states. The locus of all these states is called the path of the process. An ideal process that is almost static at every state is called quasi-static (or quasi-equilibrium) process. A quasi-static process is an infinitesimally slow process in which all the intermediate surroundings passed through by the system are equilibrium states. It is also called a reversible process. It is an ideal concept. An actual process can never be quasi-static process due to the presence of non-equilibrium effects, and hence, it is irreversible. However, some real processes can be modeled as quasi-static or reversible with negligible error. Reversible processes are easy to analyze, and they set standards to which actual processes can be compared [8].

Fig. 10 Thermodynamic processes represented in *P-v* diagram: **a** a constant pressure process and **b** a constant volume process

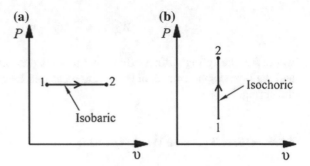

A process can be represented diagrammatically in thermodynamic coordinate systems using common properties such as pressure *P*, temperature *T*, and specific volume *v*. For example, a constant pressure (also called isobaric) and a constant volume process (also called isochoric) can be plotted in *P-v* coordinate system shown in Fig. 10a, b, respectively.

The process paths of the isobaric and isochoric processes shown in Fig. 10a, b can be indicated only when they are quasi-static or reversible. For non-quasi-equilibrium/irreversible processes, it is not possible to determine the equilibrium states at the intermediate points, and hence, process path for an irreversible process cannot be obtained. An irreversible process is represented by a dashed line between the initial and final states. There are other thermodynamic processes as well. An isothermal process is one in which the temperature *T* remains constant. A process with negligible heat transfer is called an adiabatic process.

A thermodynamic cycle is one in which the system after undergoing a number of processes returns back to the same initial state at the end. Thermodynamic cycles such as Otto cycle, Stirling cycle, Ericsson cycle, Rankine cycle, Brayton cycle, Atkinson cycle, Diesel cycle, and Dual cycle are very common in thermodynamics. These are all reversible cycles that are basically analyzed with the help of *P-v* and temperature–entropy (*T-s*) diagrams. The *P-v* diagrams of all these cycles are provided in [9] along with the years of their development.

4.5 Thermodynamics and Energy

Energy is the most fundamental concept of thermodynamics and an important aspect of thermodynamic analysis. The fundamental law of energy conservation is one of the most important building blocks of thermodynamics. The science of thermodynamics is well developed today with a much larger scope of dealing with a wide variety of energy systems, not only conventional but also other non-conventional energy systems such as solar, wind, biomass, tidal, geothermal.

4.5.1 First Law of Thermodynamics for a Closed System

The first law of thermodynamics is an expression of energy balance applied to a thermodynamic system. It provides the relationship among various forms of energy and its interactions. The general energy balance equation for a closed system undergoing a thermodynamic process is expressed as [8]

$$E_{in} - E_{out} = \Delta E_{system}, \tag{8}$$

where E_{in} is total energy input, E_{out} is the total energy output from the system, and ΔE_{system} is the change in total energy. The energy balance equation in rate form can be written as

$$\dot{E}_{in} - \dot{E}_{out} = \frac{dE_{system}}{dt}. \tag{9}$$

E_{system} in the above equations contain internal (U), kinetic, and potential energies and therefore,

$$E_{system} = U + \frac{1}{2}mV^2 + mgZ. \tag{10}$$

On unit mass basis, the specific total energy will be

$$e_{system} = u + \frac{1}{2}V^2 + gZ. \tag{11}$$

The first law for a closed system undergoing a process can also be written as

$$Q - W = \Delta E_{system}. \tag{12}$$

In this case, energy interactions involve heat transfer Q into the system and work transfer W from the system. If a closed system undergoes a thermodynamic cycle, the first law is expressed as

$$\left(\sum Q\right)_{cycle} = \left(\sum W\right)_{cycle}, \tag{13}$$

or

$$\oint \delta Q = \oint \delta W. \tag{14}$$

Symbol 'δ' is used instead of 'd' to indicate that Q and W are path functions; they depend on the path followed between two states. The energy is a thermodynamic property. It has a fixed value at a given state, and hence, change in energy during a process is not affected by the process path followed by the system. In fact, the first law does not provide any information about energy value at a given state; it only deals with the change in energy [8].

4.5.2 Flow Work and Total Energy of a Flow Stream

Energy imparted to a flow stream by means of a pump, compressor, or blower represents flow work. Flow work is associated with open systems. Since open systems involve mass flow, work needs to be supplied to maintain the continuous flow of mass into or out of the CV. Specific flow work (per unit mass) is defined as the product of pressure P and specific volume v:

$$\dot{w}_{\text{flow}} = Pv. \tag{15}$$

The total energy E that was previously expressed as the sum of internal, kinetic, and potential energies, now for a moving fluid, will additionally involve the flow energy term. Hence, the total specific energy of a flow stream will be

$$e_{\text{flow}} = Pv + u + \frac{1}{2}V^2 + gZ. \tag{16}$$

The summation of specific internal energy u and specific flow energy Pv together constitutes another property called specific enthalpy. It is denoted by h and is expressed as

$$h = u + Pv. \tag{17}$$

Specific flow energy thus becomes

$$e_{\text{flow}} = h + \frac{1}{2}V^2 + gZ. \tag{18}$$

Further, internal energy is a property associated with a closed system, while the enthalpy is a property linked with an open system.

4.5.3 First Law of Thermodynamics for an Open System

The formulation of first law of thermodynamics is different for an open system from that of a closed system. Since an open system involves mass flow across their boundaries, energy transfer by mass also needs to be considered. Thermodynamic processes involving open systems can be analyzed from two aspects, either as steady-flow or as unsteady-flow processes. During a steady-flow process, the fluid flows steadily through the control volume and the fluid properties including mass and energy remain constant with time. Hence, heat and work interactions between a steady-flow system and its surroundings do not change with time. The properties may however vary from point to point within the control volume. Many engineering devices such as pumps, turbines, compressors, nozzles, and diffusers operate under steady conditions; hence, they are often classified as steady-flow devices.

The energy balance equation for an open system or a control volume is expressed as

$$E_{in} - E_{out} = \Delta E_{CV}. \tag{19}$$

where E_{in} is the total energy transferred into the control volume by heat, work, and mass. E_{out} is the total energy transferred out of the CV. ΔE_{CV} is the change in energy of the CV.

In rate form, the energy balance equation can be written as

$$\dot{E}_{in} - \dot{E}_{out} = \frac{dE_{CV}}{dt}. \tag{20}$$

Thermodynamic analysis based on energy conservation principle is usually called first law or energy analysis.

For a steady-flow process, the total energy content of a control volume does not change with time implying $dE_{CV}/dt = 0$. Therefore, for a steady-flow process, the rate form of the energy balance equation reduces to

$$\dot{E}_{in} = \dot{E}_{out}. \tag{21}$$

The general steady-flow energy equation (SFEE) for an open system with multiple inlets and outlets can be written as [8]

$$\dot{Q}_{in} + \dot{W}_{in} + \sum_{in} \dot{m}\left(h + \frac{V^2}{2} + gZ\right) = \dot{Q}_{out} + \dot{W}_{out} + \sum_{out} \dot{m}\left(h + \frac{V^2}{2} + gZ\right). \tag{22}$$

For an open system with heat input $\dot{Q}$ into the system and work output $\dot{W}$ from the system, the SFEE becomes

$$\dot{Q} - \dot{W} = \sum_{out} \dot{m}\left(h + \frac{V^2}{2} + gZ\right) - \sum_{in} \dot{m}\left(h + \frac{V^2}{2} + gZ\right). \tag{23}$$

The SFEE for a single-stream (only one inlet and one outlet) further simplifies to the following form:

$$\dot{Q} - \dot{W} = \dot{m}\left[(h_2 - h_1) + \frac{1}{2}\left(V_2^2 - V_1^2\right) + g(Z_2 - Z_1)\right]. \tag{24}$$

The changes in kinetic and potential energies are neglected in most of the analyses; the SFEE under this assumption becomes

$$\dot{Q} - \dot{W} = \dot{m}(h_2 - h_1). \tag{25}$$

4.6   The Second Law of Thermodynamics

The first law of thermodynamics states about the energy conservation during a process. However, it does not indicate the direction in which the process is occurring. Also, it does not tell the direction of a spontaneous heat transfer process which occurs always from a hotter to a cooler body. A spontaneous process occurs naturally without outside intervention, just as water falls downward. It is also true that the spontaneous natural processes which occur in one direction are not spontaneous in the reverse direction. Therefore, this direction-related issue on energy transformation put a limit on what is stated by the first law regarding energy conservation. In other words, satisfying the first law alone is not sufficient to ensure the spontaneity of a process. This is where, the second law comes in and actually for a process to occur, it must also satisfy the second law of thermodynamics in addition to the first law of thermodynamics [8, 10].

4.6.1   Thermal Energy Reservoirs

Thermal energy reservoirs come into reference in the context of the second law of thermodynamics. These are bodies of infinite heat capacity (mass × specific heat) which can supply or absorb finite quantity of heat without suffering any significant change in temperature. A reservoir which supplies heat to a system is called a source, and a reservoir to which heat is rejected is called a sink. A furnace where large amount fuel is burnt continuously is typical example of a source. For example, an ocean can be treated as a sink.

4.6.2   Heat Engines

Enough discussion was provided earlier on energy interactions in the form of work and heat transfer. Work is a form of high-grade energy, while heat is low-grade energy. Work can be converted into heat completely, while direct conversion of heat into work is not possible [8, 10]. The device which can be used to convert heat into work is called a heat engine. A heat engine operating on a complete thermodynamic cycle is termed as a cyclic heat engine (CHE). For example, a simple steam power plant (Fig. 11) in which heat is produced in the furnace (source) is supplied to produce high-pressure steam in the boiler that drives the turbine to deliver mechanical work. The low-pressure steam at the turbine exit is condensed in the condenser as it rejects heat to the sink. The pump is used in the CHE for maintaining continuous circulation of water through the system. There are other heat engines such as open-cycle gas turbine, petrol engine, and diesel engine; however, the working fluid in these heat engines does not undergo a complete cycle. The combustion gases in these heat engines are not recycled; instead, they are exhausted into the atmosphere. In every cycle, a fresh air–fuel mixture is inducted into the engine, and hence, they are called non-cyclic heat engines. In Fig. 11, $\dot{Q}_{in}$ is the rate of heat supplied from the source, $\dot{Q}_{out}$ is the rate of heat rejected to sink, $\dot{W}_{out}$ is the amount of work delivered by steam turbine, and $\dot{W}_{in}$ is the amount of work supplied to drive the pump.

The performance of a heat engine is usually measured in terms of thermal efficiency, η_{th} that is defined as the ratio of net work output from the engine to the total heat input to the engine:

$$\text{Thermal efficiency} = \frac{\text{Net work output}}{\text{Total heat input}} \tag{26}$$

or

$$\eta_{th} = \frac{\dot{W}_{net,out}}{\dot{Q}_{in}}, \tag{27}$$

where

$$\dot{W}_{net,out} = \dot{W}_{out} - \dot{W}_{in} = \dot{Q}_{in} - \dot{Q}_{out}. \tag{28}$$

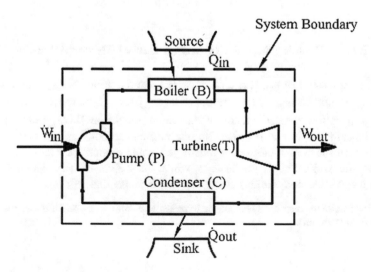

Fig. 11 Schematic of a simple steam power plant

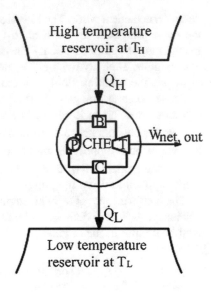

Fig. 12 Schematic of CHE

Figure 12 shows schematic of a CHE which operates between a source at high temperature T_H and a sink at low temperature T_L. If $\dot{Q}_H$ is the rate of heat supplied from the source and $\dot{Q}_L$ is the amount of heat rejected to the sink, then the net work output from the CHE and its thermal efficiency are expressed as

$$\dot{W}_{net,out} = \dot{Q}_H - \dot{Q}_L, \tag{29}$$

$$\eta_{th} = 1 - \frac{\dot{Q}_L}{\dot{Q}_H}. \tag{30}$$

4.6.3 Kelvin–Planck Statement of Second Law of Thermodynamics

From the expression of η_{th}, it is clear that for a fixed amount of $\dot{Q}_H$ supplied to the engine, η_{th} would be high if the amount of heat rejected $\dot{Q}_L$ is reduced. In the limit, when $\dot{Q}_L$ becomes zero, efficiency of the engine would be 100%, which however cannot be achieved in real practice. No heat engine can convert heat completely to work even in the best condition. It must reject some amount of heat to the sink to complete the cycle. This is the basis on which the Kelvin–Planck statement of the second law of thermodynamics is defined and it states that [8]:

> It is impossible to construct a heat engine to produce net work in a cycle if it receives heat only from the source.

4.6.4   Refrigerators and Heat Pumps

Refrigerators and heat pumps are also cyclic devices that exchange heat between a low- and a high-temperature reservoir. However, unlike a heat engine, which is a work-producing device, refrigerators and heat pumps are work-consuming devices. Moreover, the direction of heat transfer is opposite; i.e., heat is transferred from the low-temperature medium to the high-temperature one. This is made possible by supplying external work to the system. Both refrigerators and heat pumps work on the same principle but with a different purpose. Refrigerators are used to maintain low temperature in the refrigerated space by removing heat from it, while the function of a heat pump is to maintain high temperature in a space by supplying heat into it.

A cyclic refrigerator (also a cyclic heat pump) uses a refrigerant as working fluid and operates in a vapor compression system shown in Fig. 13. A vapor compression system comprises a compressor, a condenser, an expansion/throttle valve, and an evaporator. The compressor increases the pressure of the refrigerant vapor from the evaporator to the condenser pressure. During compression, the temperature of the refrigerant also increases above the temperature of the surroundings. The refrigerant enters the condenser where it is condensed through heat rejection to a surrounding medium. The refrigerant is then throttled to the evaporator pressure by passing it through an expansion valve or a capillary tube. The low-temperature refrigerant then enters the evaporator, where it evaporates by absorbing heat from the source. The refrigerant vapor then enters the compressor and completes the cycle. Thus, in a vapor compression system, the refrigerant absorbs heat in the evaporator and it rejects heat in the condenser. When used as a refrigerator, the system is operated in a way such that low temperature is maintained in the refrigerated space during heat absorption from the source. Heat rejection in the condenser is necessary to operate the system, but it is not the purpose of a refrigerator. The sole purpose of a heat pump is however to discharge heat to a medium, and the heat rejected in the condenser is used to heat a room or a house during winter season. Figures 14 and 15 depict the schematics of refrigerator and heat pump, respectively.

Fig. 13 Schematic of a vapor compression system

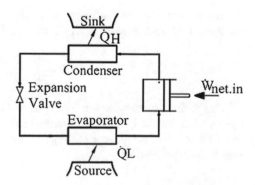

Fig. 14 Refrigerator

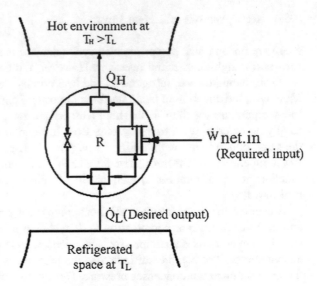

Fig. 15 Heat pump

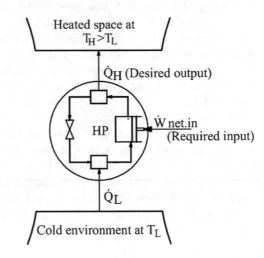

The performance of a refrigerator or a heat pump is usually expressed in terms of coefficient of performance (COP) that is defined as

$$COP = \frac{\text{Desired output}}{\text{Required input}} \tag{31}$$

The desired output for a refrigerator is the heat removed from the refrigerated space, while for the heat pump, it is the heat supplied to the heated space. Therefore, the COP of the refrigerator (COP_R) and the heat pump (COP_{HP}) can be mathematically expressed as

$$\text{COP}_R = \frac{\dot{Q}_L}{\dot{W}_{net,in}} = \frac{\dot{Q}_L}{\dot{Q}_H - \dot{Q}_L},\tag{32}$$

$$\text{COP}_{HP} = \frac{\dot{Q}_H}{\dot{W}_{net,in}} = \frac{\dot{Q}_H}{\dot{Q}_H - \dot{Q}_L}.\tag{33}$$

It is easy to see that if a refrigerator is used as a heat pump, the COP of the heat pump will be 1 plus the COP of the refrigerator.

4.6.5 Clausius Statement of Second Law of Thermodynamics

The Clausius statement of second law of thermodynamics arises from the working principle of refrigerators and heat pumps, and it is stated as follows:

> It is impossible to construct a device, which operates in a cycle and produces no effect other than transfer of heat from a body at low temperature to a body at high temperature.

Thus, heat cannot flow itself from a cold to a hot medium unless it is made possible by some external means. This is the reason that in a refrigerator (also in a heat pump), an electric motor is used to drive the compressor of the vapor compression system. Energy in the form of work must be supplied from an external power source to the system; otherwise, it would not be possible to transfer heat from a cooler to a hotter body.

4.6.6 Additional Remarks on Reversible and Irreversible Processes

It was discussed earlier that processes that occur spontaneously in a given direction are not spontaneous in the reverse direction, and once it occurs in certain direction, restoring the system to its initial state is not possible. Therefore, all real processes are irreversible. A reversible process on the other hand can be reversed back to its initial state, and while doing so, it would not leave any trace to show the occurrence of the process [10]. However, an irreversible process, if attempted to bring it back to its initial state, would produce some change in the system and its surroundings. An irreversible process can approach the corresponding reversible one as a limiting case. Additional work can be obtained from a work-producing device (such as turbine) if it is made to operate in conditions close to reversible ones. Similarly, the work consumed by a work-consuming device (such as pump, compressor) would also be less if it is operated in near reversible conditions.

4.6.7 What Makes a Process Irreversible?

The factors that contribute toward making a process irreversible are the lack of thermodynamic equilibrium and the presence of dissipative effects. Heat transfer between two mediums through a finite temperature difference, fluid flow due to pressure difference, free expansion of gases from one compartment to a neighboring evacuated compartment, mixing between two or more fluid streams, fuel combustion in combustion chambers of engines are examples of some irreversible processes where irreversibility is caused mainly by non-equilibrium effects. Dissipative effects arise from friction, viscosity, turbulence, inelasticity, electric resistance, and magnetic hysteresis. When these factors are present, the work done on a system is dissipated into heat without an equivalent increase in the energy of the system.

The extent of deviation of an irreversible process from its corresponding reversible process is measured by its irreversibility. Irreversibility arising out of dissipative effects internally within the system is called internal irreversibility. Similarly, the irreversibility caused by non-equilibrium effects at the system boundary is known as external irreversibility.

4.6.8 Carnot Cycle

In discussing heat engine with reference to a steam power plant (Fig. 11), the expressions for net work output and thermal efficiency were obtained. Efficiency of the heat engine increases with increase in net work output. Net work output can be maximized by maximizing the turbine work and minimizing the pump work. These can be achieved through use of reversible processes. Similarly, if the other two processes, i.e., production of high-pressure steam in the boiler and condensation of steam in the condenser, are also replaced with corresponding reversible processes, then all the processes of the cyclic heat engine will be reversible. Such a heat engine can be referred to as reversible heat engine. The efficiency of a reversible heat engine is always higher than an actual heat engine with irreversible processes. The reversible cycle referred here is known as the Carnot cycle and was proposed in 1824 by French engineer Sadi Carnot, who is called the father of thermodynamics [9]. The P-v diagram of the Carnot cycle is shown in Fig. 16. It comprises two reversible isothermal processes (1–2 and 3–4) and two reversible adiabatic processes (2–3 and 4–1).

In a Carnot cycle, if executed in a simple steam power plant, which is a steady-flow system, the thermodynamic processes should be carried out ideally in the following manner. Heat from the source (furnace) to water in the boiler should be supplied reversibly and isothermally at a constant temperature T_H. The expansion of steam in the steam turbine should be reversible adiabatic. Heat rejection by steam in the condenser to the sink should take place reversibly and isothermally at constant temperature T_L. Work needs to be supplied to the pump reversibly and adiabatically to run the pump with minimum power consumption.

Fig. 16 *P-v* diagram of the
Carnot cycle

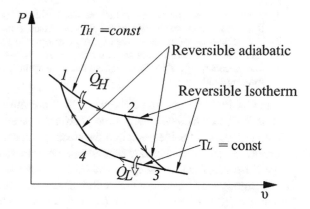

4.6.9   The Reversed Carnot Cycle

In a reversible cycle, it is possible to reverse the direction of the processes including the direction of heat and work interactions. If the direction of Carnot cycle is reversed, then we obtain the reversed Carnot cycle, the reversible cycle for refrigerator and heat pump. The *P-v* diagram of the reversed Carnot cycle is shown in Fig. 17. In this cycle, $\dot{Q}_L$ is the amount of heat absorbed from the low-temperature reservoir and $\dot{Q}_H$ is the amount of heat rejected to a high-temperature reservoir.

4.6.10   The Carnot Principles and Thermodynamic Temperature Scale

From the limitations imposed by Kelvin–Planck statement on heat engines, Carnot made the following conclusions that are known as Carnot's principles:

Fig. 17 *P-v* diagram of the
reversed Carnot cycle

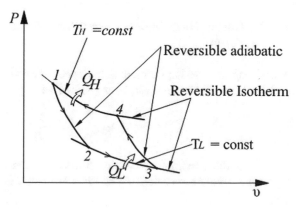

1. A reversible heat engine has the highest possible efficiency of any heat engine operating between a source and a sink, both held at constant temperatures.
2. The efficiencies of all reversible heat engines operating between a pair of source and sink held at prescribed constant temperatures are always the same.
3. The efficiency of a reversible heat engine is independent of the nature and amount of working fluid.

The Celsius (°C) and the Fahrenheit (°F) are the common temperature scales used in SI and British systems for temperature measurement, respectively. In these temperature scales, the ice point and the steam point are used as references to assign temperature values. A thermodynamic temperature scale, on the other hand, is a temperature scale that is independent of the properties of the working substance. The thermodynamic temperature scales in SI and British systems are known as Kelvin scale (K) and Rankine scale (R), respectively. The temperatures on this scale are called absolute temperatures. On this scale, the lower limit is zero and the upper limit is infinity.

The Kelvin and Celsius scale is related by

$$T(\text{K}) = T(°\text{C}) + 273.15. \tag{34}$$

The Rankine and Fahrenheit scale is related by

$$T(\text{R}) = T(°\text{F}) + 459.67. \tag{35}$$

Thermodynamic calculations are made easy by a thermodynamic temperature scale. Through the application of Carnot principles to reversible heat engines, a thermodynamic temperature scale is developed by the following relation:

$$\left(\frac{\dot{Q}_\text{H}}{\dot{Q}_\text{L}}\right)_\text{rev} = \frac{T_\text{H}}{T_\text{L}}. \tag{36}$$

Thus, in a reversible cycle, the heat transfer ratio $\dot{Q}_\text{H}/\dot{Q}_\text{L}$ is replaced with T_H/T_L and it is independent of the physical properties of any substance.

4.6.11   Carnot Heat Engine, Carnot Refrigerator, and Carnot Heat Pump

A heat engine working on reversible Carnot cycle is known as Carnot heat engine, and its thermal efficiency can be expressed as

$$\eta_\text{th,rev} = 1 - \frac{T_\text{L}}{T_\text{H}}. \tag{37}$$

This is the maximum possible efficiency attained when operating between a high-temperature reservoir at T_H and low-temperature reservoir at T_L. Refrigerators and heat pumps working on reversed Carnot cycle are called Carnot refrigerators and Carnot heat pumps. The highest possible COPs with respect to Carnot refrigerators and Carnot heat pumps operating between the temperature limits T_H and T_L can be expressed as

$$COP_{R,rev} = \frac{1}{\frac{T_H}{T_L} - 1},$$ (38)

$$COP_{HP,rev} = \frac{1}{1 - \frac{T_L}{T_H}}.$$ (39)

4.7 Clausius Inequality, Entropy, and Entropy Principle

R. J. E. Clausius (1822–1888) proved that for all thermodynamic cycles,

$$\oint \frac{dQ}{T} \leq 0.$$ (40)

The equality sign in the above equation applies to reversible cycles and the inequality to irreversible cycles. Thus,

$$\oint \left(\frac{dQ}{T} \right)_{rev} = 0.$$ (41)

However, a quantity whose cyclic integral is zero is a property and Clausius defined this property as entropy and gave the following definition:

$$ds = \left(\frac{dQ}{T} \right)_{rev}.$$ (42)

For an irreversible process,

$$ds > \left(\frac{dQ}{T} \right)_{irrev}.$$ (43)

For a system undergoing a process i-f, the change in entropy can be determined as follows:

$$\int_i^f ds = s_f - s_i = \int_i^f \left(\frac{dQ}{T} \right)_{rev}.$$ (44)

An important point here is that the above integration can be done only for a reversible process; hence, the irreversible process i-f has to be replaced with a corresponding reversible one for performing the integration [10].

The entropy principle is obtained from the definition of entropy and Clausius inequality. It states that the entropy of the universe (system and surrounding together) is ever increasing. That is

$$dS_{sys} + dS_{surr} \geq 0. \tag{45}$$

s in Eq. (44) is specific entropy (kJ/kgK), while S in Eq. (45) is the total entropy (kJ/K). Change in entropy occurs due to both internal irreversibility and external irreversibility. During heat supply to a system, its entropy increases and vice versa.

4.8   Quality of Energy, Exergy (Availability), Dead State, and Second-Law Efficiency

The second law of thermodynamics emphasizes that energy has both quantity and quality. The measure of usefulness of energy is termed as exergy. It is a Greek word first used in 1956 and is derived from *ex* (external) and *ergos* (work). Whenever there is no exchange of energy and other parameters, i.e., the equilibrium condition is attained by the system with the surrounding, it is said to have attained a dead state. Rudolf Clausius in 1865 proclaimed that if an increase in entropy is associated with decreased ability to perform work, the universe reaches a dead state. A useful parameter to characterize the performance along with first law is defined by the second-law efficiency.

4.8.1   Energy Quality

The second law of thermodynamics not only indicates direction of processes; it also defines the quality concept of energy. A system of gas at higher temperature possesses superior quality to a system at lower temperature because more work can be extracted from the gas at higher temperature. Energy quality can be accessed from the concept of exergy or availability.

4.8.2   Exergy, Dead State, and Exergy Balance

Exergy (also called availability) of a system is a combined property linked with the surroundings, and it indicates the extent of departure of a system from the equilibrium state. It refers to the maximum useful work that can be obtained from a system when it reaches complete thermodynamic equilibrium with the reference environment. A system in complete thermodynamic equilibrium with its surroundings is said to be at dead state where it has zero exergy with negligible kinetic and potential energy relative to the reference environment. The properties of a system at the dead state are symbolized by subscript zero, for example P_0, T_0, h_0, u_0, and s_0. The pressure and temperature of the reference environment are usually taken as 25 °C and 1 atmosphere (101.325 kPa) unless otherwise stated.

The specific exergy of a closed and an open system at a given state can be calculated using the following equations [8]:

$$\text{Specific exergy of a closed system: } \phi = (u - u_0) + P_0(V - V_0)$$
$$- T_0(s - s_0) + \frac{V^2}{2} + gZ. \quad (46)$$

$$\text{Specific exergy of an open system: } \psi = (h - h_0) - T_0(s - s_0) + \frac{V^2}{2} + gZ. \quad (47)$$

Like energy, exergy is also transferred due to heat and work interaction and also due to mass flow. The exergy transfer rate associated with heat, work, and mass are given as follows:

$$\text{Exergy transfer rate by heat: } \dot{X}_{\text{heat}} = \left(1 - \frac{T_0}{T}\right)\dot{Q}, \quad (48)$$

$$\text{Exergy transfer rate by work: } \dot{X}_{\text{work}} = \dot{W}, \quad (49)$$

$$\text{Exergy transfer rate by mass: } \dot{X}_{\text{mass}} = \dot{m}\psi. \quad (50)$$

In the way in which the energy balance equation is applied to find out energy loss in a device, in a similar manner, the exergy balance equation is applied to find out exergy destroyed in a process.

The general exergy balance for any system undergoing a process can be expressed as

$$X_{\text{in}} - X_{\text{out}} - X_{\text{destroyed}} = \Delta X. \quad (51)$$

The exergy balance equation in rate form can be written as

$$\dot{X}_{\text{in}} - \dot{X}_{\text{out}} - \dot{X}_{\text{destroyed}} = \frac{dX_{\text{system}}}{dt}. \quad (52)$$

In case of a steady-flow device,

$$\dot{X}_{\text{in}} - \dot{X}_{\text{out}} - \dot{X}_{\text{destroyed}} = 0. \quad (53)$$

Exergy destroyed represents the irreversible losses present during a process. It is also called irreversibility I and expressed in rate form as

$$\dot{I} = \dot{X}_{\text{destroyed}} = T_0 S_{\text{gen}}, \tag{54}$$

where S_{gen} is the total entropy generated during the process.

4.8.3 Second-Law Efficiency

Thermal efficiencies of heat engines and COP of refrigerators or heat pumps are actually first-law-based performance parameters and hence referred to as first-law efficiencies. The first-law efficiency in its definition ignores the best possible performance of a system, and hence, it is not an accurate measure of actual system performance. The second-law efficiency is defined to correlate the first-law-based system performance with the performance under reversible conditions.

For work-producing devices (heat engines), it is defined as

$$\eta_{\text{II}} = \frac{\eta_{\text{th}}}{\eta_{\text{th,rev}}}. \tag{55}$$

For refrigerators and heat pumps, it is defined as

$$\eta_{\text{II}} = \frac{\text{COP}}{\text{COP}_{\text{rev}}}. \tag{56}$$

For other devices, the following general definition is provided [8]:

$$\eta_{\text{II}} = \frac{\text{Exergy recovered}}{\text{Exergy supplied}} = 1 - \frac{\text{Exergy destroyed}}{\text{Exergy supplied}}. \tag{57}$$

The basic concepts and understanding on energy, energy interactions, energy balance, exergy, exergy balance, and second-law efficiency are important because these are often used in developing thermodynamic models for performance analysis of thermal systems and energy conversion devices.

5   Exergy Analysis: Its Importance in Performance Analysis of Thermal Systems

Thermodynamic analysis in which exergy balance is applied to determine exergy destruction (or irreversibility) in various components of a thermal system is called exergy analysis. Since exergy analysis is based on second law of thermodynamics, it is also called second law analysis. First-law-based energy analysis provides a quantitative measurement of energy balance in a device or a system. Through energy analysis, it is not possible to identify processes that cause unrecoverable energy degradation of the working substance. It is the exergy analysis which assigns quality to energy of a system at a given state and thus provides the theoretical base for evaluating the irreversible losses and the second-law efficiency of thermal systems. Say for example, in a steam power plant, the information regarding heat supplied in the boiler, net power developed, the energy loss in the condenser, and the plant's thermal efficiency (first-law efficiency) can be obtained from an energy analysis. Energy analysis would indicate that the energy loss in the condenser is mainly responsible for low efficiency of the plant because a major portion of energy is lost in the condenser in transferring heat to the cooling water. This information obtained from energy analysis in no way gives any idea about the real utility of the low-temperature cooling water and the exergy loss in the condenser. It is possible, only through exergy analysis, to quantify that there is hardly any exergy destruction in the condenser compared to the boiler irreversibility due to fuel combustion and heat transfer through huge temperature difference between combustion gas and water.

Exergy analysis is more appropriate for performance assessment and efficiency of processes, energy conversion devices, and thermal systems. It is mainly applied to design, evaluate, and optimize thermal energy systems to improve their performance. Often energy and exergy analyses are performed together for analyzing thermal systems. Combination of energy and exergy analysis is a better approach of performance assessment as it gives a complete depiction of system characteristics.

6   Analysis of Thermal Energy Systems

The performance of a thermal system is quantified on the basis of its thermal efficiency, specific fuel consumption, and other thermodynamic performance parameters. Performance analysis involves the application of energy and exergy balance in determining the performance parameters in addition to energy and exergy losses. Exergy analysis necessities cost to account for energy quality. By applying the concept of cost, thermodynamic analysis and economy are combined, known as thermoeconomic analysis. Additionally, numerous optimization techniques are applied to get better efficiency of a thermodynamic system. This section also describes about the identification of unknown parameters appearing in an analysis of systems, known as inverse analysis.

6.1   Exergy and Exergoeconomic Analysis

So far as application of laws of thermodynamics is concerned, a lot of studies have been carried out in the past to evaluate energetic and exergetic performance of thermal systems. Some application areas where energy and exergy analyses are very common are given below:

1. Thermal power plants such as fossil-fuel-fired steam power plants [11, 12], gas turbine, renewable energy plant, and combined cycle (CC) power plants [13–15], integrated gasification CC plants [16, 17];
2. Cogeneration [18, 19], trigeneration [20, 21], and multi-generation [22, 23] systems;
3. Heating [24, 25], cooling [26, 27], and air-conditioning [28] systems;
4. Internal combustion engines [29, 30];
5. Fuel cells [31, 32];
6. Solar [21, 33, 34] and geothermal energy [35] systems;
7. Organic Rankine cycle-based energy systems [36, 37].

Energy and exergy analyses are performed on thermal systems mainly to

(i)   Analyze the effect of various operating parameters on energy and exergy losses (irreversibility);
(ii) Determine magnitudes, location, and causes of irreversibility;
(iii) Identify those aspects of the processes that are significant to overall system performance.

In energy-system-related researches, often new innovative thermal systems are conceptualized and energy and exergy studies are performed to evaluate their system performance under various operating conditions. The current research trend is to develop and perform thermodynamic analysis on renewable-energy-based hybrid multi-generational systems.

Exergoeconomics is another exergy-based cost reduction approach that combines thermodynamics (exergy) and economics (cost) to obtain better system design and cost-effective operation of thermal systems. In an exergoeconomic analysis, the principle of exergy costing and cost accounting methods are applied to allocate costs to the energy streams and to govern exergy destruction in each system component [37, 38]. The fuel–product–loss (F–P–L) is a very commonly used method for developing exergetic cost balance equations for the system components. In this method, each system component is characterized by its fuels, products, and losses. With respect to each system component, the fuels represent the exergy inputs and the products are the exergy outputs. The losses refer to the exergy loss in the system component. The cost balance equations are solved to determine cost rate of exergy destruction of the individual components and the overall system. Cost assessment of energy streams and processes in a complex energy system provides better understanding of the system economics in terms of the input resources and the final products.

6.2 Thermoeconomic Optimization Studies

In some analyses, the investment cost of each individual system component is determined along with the operating and maintenance cost to calculate the plant's total annual operating cost rate. The calculation of the cost terms involves system's design and thermodynamic operating parameters; hence, it combines the thermodynamic performance data in calculating the total annual cost rate of the system. These are called thermoeconomic analyses. Combined thermoeconomic analysis and optimization is another area where the concept of thermoeconomics is applied in optimizing the performance of thermal systems [39–41]. In such studies, different optimization techniques such as nonlinear programming, simplex search method, conjugate-gradient method, genetic algorithm, differential evolution, particle swarm optimization are used to identify the most interesting system design configuration through determination of optimal parameters. System analysis combined with thermodynamic and economic optimization is an effective tool as it indicates the economic feasibility, improving operational performance and efficiency of the best design configuration identified through optimization.

6.3 Inverse Analysis of Thermal Systems

Inverse analysis is another method that is used to estimate unknown parameters against some known values of the system performance parameters. Inverse techniques have been successfully applied in estimating parameters of many thermal systems [42, 43]. Inverse method also uses optimization methods in estimating parameters; however, it is different from normal single-/multi-objective optimization studies. In a general multi-objective optimization problem, the objective functions are not known a priori and they are optimized (some are maximized and some are minimized) simultaneously to determine the optimal decision variables (parameters). In an inverse analysis, however, it is always a minimization of the objective functions which are assumed to be known and accordingly the unknown parameters are estimated. The inverse method is quite useful in finding operating parameters against some set objectives of designed performance criteria. Say for example, when a power plant is designed, it is always designed to fulfill certain criterion of net power output. In this case, the inverse method can be used for estimating the plant's operating parameters such as boiler pressure, condenser pressure, fuel flow rate, number of heaters, and their operating pressures against a set objective (net power). Any value of the net power can be set, and accordingly, the unknown operating parameters can be estimated using an appropriate inverse method. A major advantage of inverse analysis is that multiple sets of parameters can be obtained [42, 43]. To perform inverse analysis, first the basic thermodynamic model (forward model) is developed and then it is coupled with the inverse model. The inverse and the forward model together offers lot of flexibility to the designer in selecting the most suitable parameters satisfying a given set of objective functions [42, 43]. In the modeling of manufacturing processes, the inverse analysis can be especially useful when it is difficult to measure certain input parameters [44].

7   Conclusions

Thermal engineering is a vast subject. A brief overview of the fundamental concepts of fluid mechanics, heat transfer, and thermodynamics, the three major topics of thermal engineering, is provided. For each, important preliminary concepts and applications are discussed. Special attention is given on the basics of thermodynamics and application of laws of thermodynamics, particularly the first and the second laws. Energy and exergy balance equations are shown for their application in thermodynamic system analysis. Importance of energy and exergy is highlighted in the context of thermodynamic performance analysis. Some application areas are discussed to give the readers an idea about the kind of studies that are carried out in the field of thermodynamics. Some current trends of research are discussed at the end, particularly the combined use of thermodynamic, thermoeconomic, optimization, and inverse methods. It is expected that this chapter would give the readers some basic ideas about fluid mechanics, heat transfer, and thermodynamics and importance of these topics in some practical engineering applications.

Ten (10) Review Questions (T & F Type)

   (i) A flow in which the fluid flow rate is constant is called steady flow.
        (T)
  (ii) The volume of a liquid increases as the pressure is increased at a constant temperature.
        (F)
 (iii) Fluid flow through a circular pipe is laminar when Reynolds number (Re) is less than 2000.
        (T)
  (iv) Conduction of heat occurs only in solids.
        (F)
   (v) Hot air rises and cold air always moves downward in a forced convection.
        (F)
  (vi) Thermal radiation includes the entire visible light and infrared radiation and a part of the ultraviolet radiation.
        (T)
 (vii) It is possible to convert entire available heat into work.
        (F)
(viii) It is impossible to transfer heat from a cold reservoir to a hot reservoir.
        (F)
  (ix) Entropy of system plus surrounding always increases.
        (T)
   (x) The second law of thermodynamics provides a qualitatively useful *inequality* ("Clausius"), but it does not yield a quantitatively useful *balance* (conservation) equation.
        (T)

Glossary

Aerodynamics Study of flow of gases.

Blackbody A hypothetical body that emits maximum radiation uniformly in all directions at a given temperature and wavelength. A blackbody also absorbs radiation of all wavelengths falling on its surface from all directions.

Closed (non-flow) system A thermodynamic system that allows only energy interaction through the system boundary but no mass can enter or leave the system.

Cogeneration Combined generation of power and heating (or cooling) from the same thermal system.

Compressible flow Flow in which fluid density changes in the flow field.

Conduction Heat transfer by molecular motion.

Convection Heat transfer due to bulk motion of fluid over a solid surface.

Cyclic heat engine A work-producing device that delivers net work operating in a thermodynamic cycle exchanging heat with a source and a sink.

Cyclic heat pump A work-consuming device that produces heating effect operating in a thermodynamic cycle exchanging heat with a source and a sink.

Cyclic Refrigerator A work-consuming device that produces cooling–heating effect operating in a thermodynamic cycle exchanging heat with a source and a sink.

Emissivity The ratio of radiation energy emitted by a real surface at a given temperature to the radiation emitted by a blackbody at the same temperature.

Energy interaction Energy exchange between system and its surroundings in the form of heat and work.

Energy analysis Analysis based on first law of thermodynamic that applies energy balance equation in determining energy losses of system components.

Enthalpy A thermodynamic property associated with an open (or flow) system defined as $h = u + Pv$ where h is the specific enthalpy, u is the specific internal energy, P is the pressure, and v is the specific volume.

Entropy A thermodynamic property that measures the degree of energy degradation or irreversibility in a process. It also indicates disorderliness and randomness of the system.

Equilibrium state The state of a system with fixed property values not undergoing any change due to system–surrounding interaction.

Exergy Maximum useful work obtained from a system when it comes in equilibrium with the surroundings.

Exergy analysis Analysis based on second law of thermodynamic that applies exergy balance equation in determining exergy destruction (irreversibility) of system components.

Exergoeconomic analysis Thermodynamic analysis based on exergy costing theory.

External flow Fluid flow over an unbounded surface such as a plate or a pipe.

First law of thermodynamics The law of energy conservation applied to any system undergoing a change of state or a cycle.

Flow field A region where velocities are specified at each and every location at all times.

Fluid It is the state of matter which does not have a fixed shape and cannot remain static under the application of a shear force.

Fluid dynamics The branch of fluid mechanics that studies fluid motion along with the governing forces (inertia force, pressure force, viscous force, etc.).

Fluid Kinematics The branch of fluid mechanics that studies fluid motion without considering the nature of forces.

Fluid statics The branch of fluid mechanics that studies fluids at rest.

Governing equations of motion Conservation equations of mass and momentum. The mass conservation equation is called continuity equation, and the momentum conservation equation is called momentum equation.

Hydrodynamics Study of flow of liquids.

Incompressible flow Flow of constant density fluids (all liquid flows).

Internal energy Total energy (sum of molecular internal energy, potential energy and kinetic energy) of a system.

Internal flow Fluid flow bounded inside a solid surface such as a pipe or a duct.

Irrotational flow Flow in which rotational components in the flow field are zero.

Isolated system A thermodynamic system in total isolation from its surroundings with no mass and energy transfer.

Laminar flow Fluid flow in which fluid particles move in parallel layers smoothly over one another.

Navier–Stokes Equations Momentum equations for incompressible, isotropic, Newtonian fluid of constant viscosity.

Non-uniform flow Flow in which velocity and other hydrodynamic parameters are different at different locations at a given instant of time.

Open (flow) system A thermodynamic system that allows both mass and energy transfer across the system boundary.

Radiation Heat transfer due to propagation of electromagnetic waves through vacuum.

Reversible process An ideal process in which all the states passed through by a system are equilibrium states.

Reynolds Number Ratio of inertia to viscous force defined mathematically as $\text{Re} = \frac{VL_c}{\nu}$, where V is the average flow velocity, L_c is the characteristic length (e.g., diameter of a circular pipe, hydraulic diameter of non-circular ducts), and ν is the kinematic viscosity of the fluid. Reynolds number is a non-dimensional parameter that is used to identify laminar and turbulent flow.

Rotational flow Fluid flow with rotation of fluid particles about their own axis.

Second law of thermodynamics The law that indicates the direction of energy transfer. If the first law of thermodynamics is called the law of energy conservation, then second law can be termed as the law of energy degradation.

Steady flow Flow in which the velocity and the fluid properties do not vary with time in the flow field.

Stefan–Boltzmann law The law that governs radiation heat transfer rate emitted by a blackbody at a given temperature.

Thermal energy reservoir A body of infinite heat capacity whose temperature does not change upon receiving and rejecting heat.

Thermodynamic cycle A series of processes undergone by a system in which the initial and final states are identical.

Thermodynamic process A change of state undergone by a system in which the properties of the system change.

Thermoeconomic analysis Combined thermodynamic performance and cost (capital, operating and maintenance) analysis.

Trigeneration Generation of power heating and cooling from the same thermal system.

Turbulent flow Irregular and chaotic flow with random fluctuations of velocity and fluid properties with time and space coordinates.

Uniform flow Flow in which velocity and other hydrodynamic parameters assume constant values at all locations at a given time.

Unsteady flow Flow in which the velocity and the fluid properties change with time in the flow field.

Work transfer All modes of energy transfer except heat transfer governed by temperature difference.

References

1. White, F. M. (2009). *Fluid mechanics* (6th ed.). New Delhi: Tata McGraw Hill.
2. Som, S. K., & Biswas, G. (2008). *Introduction to fluid mechanics and fluid machines* (2nd ed.). New Delhi: Tata McGraw Hill.
3. Cengel, Y. A., & Ghazar, A. J. (2014). *Heat and mass transfer: Fundamentals and applications* (5th ed.). Boston, MA: Tata McGraw Hill.
4. Jang, J. Y., & Chen, L. K. (2004). Numerical analysis of heat transfer and fluid flow in a three-dimensional wavy-fin and tube heat exchanger. *International Journal of Heat and Mass Transfer, 47,* 4327–4338.
5. Lee, P. S., & Garimella, S. V. (2006). Thermally developing flow and heat transfer in rectangular microchannels of different aspect ratios. *International Journal of Heat and Mass Transfer, 49,* 3060–3067.
6. Karmare, S. V., & Tikekar, A. N. (2010). Analysis of fluid flow and heat transfer in a rib grit roughened surface solar air heater using CFD. *Solar Energy, 84,* 409–417.
7. Rashidi, M. M., Rastegari, M. T., Sadi, M., & Bég, O. A. (2012). A study of non-newtonian flow and heat transfer over a non-isothermal wedge using the homotopy analysis method. *Chemical Engineering Communications, 199*(2), 231–256.
8. Cengel, Y. A., & Boles, M. A. (2006). *Thermodynamics: An engineering approach* (5th ed.). Boston, MA: Tata McGraw Hill.
9. Dixit, U. S., Hazarika, M., & Davim, J. P. (2017). *A brief history of mechanical engineering.* Switzerland: Springer.
10. Nag, P. K. (2010). *Basic and applied thermodynamics* (2nd ed.). New Delhi: Tata McGraw Hill.
11. Sengupta, S., Datta, A., & Duttagupta, S. (2007). Exergy analysis of a coal-based 210 MW thermal power plant. *International Journal of Energy Research, 3,* 14–28.
12. Ganapathy, T., Alagumurthi, N., Gakkhar, R. P., & Murugesan, K. (2009). Exergy analysis of operating lignite fired thermal power plant. *Journal of Engineering Science and Technology Review, 2,* 123–130.
13. Woudstra, N., Woudstra, T., Pirone, A., & van der Stelt, T. (2010). Thermodynamic evaluation of combined cycle plants. *Energy Conversion and Management, 51,* 1099–1110.
14. Shi, X., Agnew, B., & Che, D. (2011). Analysis of a combined cycle power plant integrated with a liquid natural gas gasification and power generation system. *Proceedings of the Institution of Mechanical Engineers, Part A: Journal of Power and Energy, 225,* 1–11.
15. Sarmah, P., & Gogoi, T. K. (2017). Performance comparison of SOFC integrated combined power systems with three different bottoming steam turbine cycles. *Energy Conversion and Management, 132,* 91–101.
16. Kanniche, M., & Bouallou, C. (2007). CO_2 capture study in advanced integrated gasification combined cycle. *Applied Thermal Engineering, 27,* 2693–2702.
17. Romano, M. C., Campanari, S., Spallina, V., & Lozza, G. (2011). Thermodynamic analysis and optimization of IT-SOFC-based integrated coal gasification fuel cell power plants. *Journal of Fuel Cell Science and Technology, 8*(041002), 1–11.

18. Hycienth, I., Onovwiona, V., Ugursal, I., & Fung, A. S. (2007). Modeling of internal combustion engine based cogeneration systems for residential applications. *Applied Thermal Engineering, 27,* 848–861.
19. Talukdar, K., & Gogoi, T. K. (2016). Exergy analysis of a combined vapor power cycle and boiler flue gas driven double effect water-LiBr absorption refrigeration system. *Energy Conversion and Management, 108,* 468–477.
20. Khaliq, A. (2009). Exergy analysis of gas turbine trigeneration system for combined production of power heat and refrigeration. *International Journal of Refrigeration, 32,* 534–545.
21. Al-Sulaiman, F. A., Dincer, I., & Hamdullahpur, F. (2011). Exergy modeling of a new solar driven trigeneration system. *Solar Energy, 85,* 2228–2243.
22. Khalid, F., Dincer, I., & Rosen, M. A. (2015). Energy and exergy analyses of a solar-biomass integrated cycle for multigeneration. *Solar Energy, 112,* 290–299.
23. Al-Ali, M., & Dincer, I. (2014). Energetic and exergetic studies of a multigenerational solar-geothermal system. *Applied Thermal Engineering, 71,* 16–23.
24. Hawlader, M. N. A., Chou, S. K., & Ullah, M. Z. (2001). The performance of a solar assisted heat pump water heating system. *Applied Thermal Engineering, 21,* 1049–1065.
25. Ayompe, L. M., & Duffy, A. (2013). Thermal performance analysis of a solar water heating system with heat pipe evacuated tube collector using data from a field trial. *Solar Energy, 90,* 17–28.
26. Jain, V., Kachhwaha, S. S., & Sachdeva, G. (2013). Thermodynamic performance analysis of a vapor compression–absorption cascaded refrigeration system. *Energy Conversion and Management, 75,* 685–700.
27. Llopis, R., Sánchez, D., Sanz-Kock, C. S., Cabello, R., & Torrella, E. (2015). Energy and environmental comparison of two-stage solutions for commercial refrigeration at low temperature: Fluids and systems. *Applied Energy, 138,* 133–142.
28. Davirana, S., Kasaeiana, A., Golzari, S., Mahian, O., Nasirivatan, S., & Wongwises, S. (2017). A comparative study on the performance of HFO-1234yf and HFC-134a as an alternative in automotive air conditioning systems. *Applied Thermal Engineering, 110,* 1091–1100.
29. Rakopoulos, C. D., & Giakoumis, E. G. (2006). Second-law analyses applied to internal combustion engines operation. *Progress in Energy and Combustion Science, 32,* 2–47.
30. Gogoi, T. K. (2013). Exergy analysis of a diesel engine operated with koroch seed oil methyl ester and its diesel fuel blends. *International Journal of Exergy, 12,* 183–204.
31. Park, J., Li, P., & Bae, J. (2012). Analysis of chemical, electrochemical reactions and thermo-fluid flow in methane-feed internal reforming SOFCs: Part I—modeling and effect of gas concentrations. *International Journal of Hydrogen Energy, 37,* 8512–8531.
32. Silveira, J. L., Leal, E. M., & Ragonha, L. F., Jr. (2001). Analysis of a molten carbonate fuel cell: Cogeneration to produce electricity and cold water. *Energy, 26,* 891–904.
33. Yagoub, W., Doherty, P., & Riffat, S. B. (2006). Solar energy-gas driven micro-CHP system for an office building. *Applied Thermal Engineering, 26,* 1604–1610.
34. Siqueira, A. M. O., Gomes, P. E. N., Torrezani, L., Lucas, E. O., & Pereira, G. M. C. (2014). Heat transfer analysis and modeling of a parabolic trough solar collector: An analysis. *Energy Procedia, 57,* 401–410.
35. Coskun, C., Oktay, Z., & Dincer, I. (2011). Performance evaluations of a geothermal power plant. *Applied Thermal Engineering, 31*(2011), 4074–4082.
36. Tuo, H. (2013). Energy and exergy-based working fluid selection for organic Rankine cycle recovering waste heat from high temperature solid oxide fuel cell and gas turbine hybrid systems. *International Journal of Energy Research, 37,* 1831–1841.
37. Sahoo, P. K. (2008). Exergoeconomic analysis and optimization of a cogeneration system using evolutionary programming. *Applied Thermal Engineering, 28,* 1580–1588.
38. Khaljani, M., Saray, R. K., & Bahlouli, K. (2015). Comprehensive analysis of energy, exergy and exergo-economic of cogeneration of heat and power in a combined gas turbine and organic Rankine cycle. *Energy Conversion and Management, 97,* 154–165.

39. Valdes, M., Duran, M. D., & Rovira, A. (2003). Thermoeconomic optimization of combined cycle gas turbine power plants using genetic algorithms. *Applied Thermal Engineering, 23,* 2169–2182.
40. Quoilin, S., Declaye, S., Tchanche, B. F., & Lemort, V. (2011). Thermo-economic optimization of waste heat recovery Organic Rankine Cycles. *Applied Thermal Engineering, 31,* 2885–2893.
41. Modi, A., Kærn, M. R., Andreasen, J. G., & Haglind, F. (2016). Thermoeconomic optimization of a Kalina cycle for a central receiver concentrating solar power plant. *Energy Conversion and Management, 115,* 276–287.
42. Gogoi, T. K. (2016). Estimation of operating parameters of a water–LiBr vapor absorption refrigeration system through inverse analysis. *ASME Journal of Energy Resources Technology, 138,* 022002.
43. Sarmah, P., Gogoi, T. K., & Das, R. (2017). Estimation of operating parameters of a SOFC integrated combined power cycle using differential evolution based inverse method. *Applied Thermal Engineering, 119,* 98–107.
44. Yadav, V., Singh, A. K., & Dixit, U. S. (2015). Inverse estimation of thermal parameters and friction coefficient during warm flat rolling process. *International Journal of Mechanical Sciences, 96–97,* 182–198.

Chapter 6
Alternate Fuels for IC Engine

Shailendra Kumar Shukla

Abstract This chapter describes the production and use of biofuels/alternate fuels to run the internal combustion engine (IC Engine). Two methods, using biodiesel blends with diesel and using syngas obtained from gasification system, for running the IC engine in dual fuel mode have been analyzed. Describing the first method, the objectives enumerate to select the vegetable oil plants for biodiesel production. To accomplish this, the energy input and output analysis of biodiesel plants for 20-year plantation have been done. The biodiesel plants, namely jatropha, mahua, neem, palm, coconut, karanja, jojoba, and tung, have been identified for this purpose. This analysis includes energy input during oil extraction, cultivation, and biodiesel production. The energy inputs are based on manpower, fossil fuel, electricity, fertilizers, plants protection, and water for irrigation, expeller used for oil extraction, agricultural machinery, methanol, catalyst (H_2SO_4 and NaOH/KOH) and a transesterification unit for biodiesel production. Net energy gain (NEG) and net energy ratio (NER) are calculated for different biodiesel plants for 20-year plantation. Palm and coconut consumed highest energy (117,122.32 and 122,832.84 MJ/ha) during cultivation. Maximum energy output is obtained for palm (123,206.4 MJ/ha) and minimum for tung after its maturity. Maximum net energy ratio is found for mahua (1.7164) and minimum for coconut (1.3034) oil plants. The highest net energy gain for palm (41,689.14 MJ/a) and lowest for jojoba (8494.92 MJ/ha) were found after maturity of plants. There are significant increments in energy output, net energy ratio, and net energy gain with the addition of coproduct (glycerin). This analysis with methanol (used for biodiesel production) recovery has also been done and found reduction in energy input during biodiesel production and also the improvement in net energy gain and net energy ratio. In the second method, the virgin biomass obtained from wood and cow dung is used to generate producer gas as the feedstock for gasifier and in turn the syngas. The producer gas combination and gasifier-engine system are operated in dual fuel mode operation and diesel, respectively. The emission characteristics and performance of the CI engine are

S. K. Shukla (✉)
Mechanical Engineering Department, Indian Institute of Technology
(Banaras Hindu University), Varanasi 221005, India
e-mail: shuskla@gmail.com

© Springer International Publishing AG, part of Springer Nature 2018
J. P. Davim (ed.), *Introduction to Mechanical Engineering*, Materials Forming, Machining and Tribology, https://doi.org/10.1007/978-3-319-78488-5_6

analyzed by running the engine in dual fuel mode operation and liquid fuel mode operation with respect to maximum diesel savings at different load conditions in the dual fuel mode operation. It is found that in the dual fuel mode of operation, the specific energy consumption is found to be the higher side at various load conditions. As comparison to dual fuel mode operation, the brake thermal efficiency using diesel is higher. In the dual fuel, the NO_x emission was found to be very low which is highly advantageous over diesel fuel alone, but HC and CO emissions are found to be higher than the diesel.

Keywords Alternate fuels · Dual fuel engine · Gasification · Biodisel Transesterification

1 Introduction on Biodiesel as Alternate Fuel

The declination of crude oil deposits and its adverse effects on environment have prompted to discover renewable source and environmental friendly of fuel. There is some substitute for petroleum-based fuel-like vegetable oil, biogas, biodiesel, and alcohol. The vegetable oils are one of the energy sources, and also, it has unique compositions (fatty acids and glycerol) which provide a good base for making biodiesel. In India, various types of edible (coconut, soybean, mustard, peanuts, palm, etc.) and non-edible (neem, mahua, karanja, jojoba, jatropha, tung, etc.) vegetable oil are available for producing the biodiesel. The purpose of this study is to choose the suitable vegetable oil plants for the production of biodiesel. The energy input and output analysis of biodiesel plants have been done in this study for 20-year plantation. For this purpose, considered biodiesel plants, namely jatropha, mahua, neem, palm, coconut, karanja, jojoba and tung, are being used. This analysis includes energy input during oil extraction, cultivation, and biodiesel production. The energy inputs are based on manpower, fossil fuel, electricity, fertilizers, plants protection, and water for irrigation, expeller used for oil extraction, agricultural machinery, methanol, catalyst (H_2SO_4 and NaOH/KOH) and a transesterification unit for biodiesel production. Net energy gain (NEG) and net energy ratio (NER) are calculated for different biodiesel plants for 20-year plantation. Palm and coconut consumed highest energy (117,122.32 and 122,832.84 MJ/ha) during cultivation. Maximum energy output is obtained for palm (123,206.4 MJ/ha) and minimum for tung after its maturity. Maximum net energy ratio is found for mahua (1.7164) and minimum for coconut (1.3034) oil plants. The highest net energy gain for palm (41,689.14 MJ/a) and lowest for jojoba (8494.92 MJ/ha) were found after maturity of plants. There are significant increments in energy output, net energy ratio, and net energy gain with the addition of coproduct (glycerin). Above analysis has also done with methanol (used for biodiesel production) recovery and found reduction in energy input during biodiesel production and also the improvement in net energy ratio and net energy gain.

It has been evaluated that approximately 20% of energy should be achieved from biodiesel by 2017. For this, it is expected to require 12–13 million ha biodiesel feedstock plantation. Use of this oil to produce biodiesel is not feasible in India because of high gap between demand and supply of edible oil. As per Indian government policy, only non-edible oils are used for making biodiesel. The non-edible oilseeds, fruit, or nut contain 30% or more oil. Around 300 or more species of oil-bearing seeds trees are found in India [1, 2]. It has been identified that approximately 75 plant species contain 30% or more oil in their seeds/kernel. Generally, the poor people collect and sell the seeds. The biodiesel becomes more attractive because it is produced from renewable resources and has low effects on environments [3]. Among the various non-edible seed-bearing plants/trees such as neem, jatropha curcas, and karanja (*Pongamia pinnata*) are rich in oil. And also, it can be grown on all types of sands all over India. Non-edible oil can be only used as biodiesel under Indian condition, which is produced in appreciable quantity and can be grown in large scale on wastelands and non-cropped marginal lands. India has around 63 million ha of wasteland, of which only three categories, namely (1) grazing land and degraded pastures, (2) degraded land under plantation crop categories, and (3) underutilized degraded notified forest land. Essentially the alternative fuels are produced from renewable feedstocks with a lower negative environmental impact. Day-by-day improvement and innovation in CI engines have steered to excessive utilization of conventional fuel and acceleration of oil extraction from the resources. Excess use of fossil fuels leads to rapid increase in environmental pollution, greenhouse gasses and depletion of ozone layer. In the last few decades swiftly increasing prices and uncertainties in the availability of petroleum, therefore above problems have to lead to great interest on biodiesel produced from vegetable oil as a substitute for conventional diesel [1]. In the transportation sectors, petroleum fuels are hugely used. Due to inadequate use of these fossil fuels, the government is trying to find out substitute of these fuels [2]. To improve the energy security and sustainability, the wide range of investigations has been done about energy management in the fuel production system. The production of alternative fuel from vegetable oil should be investigated to achieve the above objectives. Worldwide, many countries are a huge consumer of petroleum fuel in the transportation sector [3]. The alternative fuels have become the subjects of great interest because of lower environmental pollution, availability of feedstock, and more consumption of fossil fuel. Among the various alternative sources such as biodiesel, biomass, hydrogen, and biogas, biodiesel shows the similar properties as diesel fuel [4]. The certain properties of biofuels like renewability, biodegradability, and low global warming potential make more attractive to replace petroleum fuels. These are obtained from biomass conversion. Biomass, which is extracted from organic or living biological material, can be used as fuel for industrial production in the energy production industry [5]. The main biofuels are biodiesel and bioethanol. Bioethanol, known as alcohol, is produced from carbohydrates (found in sugar or starch crops like sugarcane or corn) prepared by fermentation and distillation. There is a significant reduction in greenhouse gas emissions when bioethanol is used as a fuel [6, 7]. The biodiesel is an alternative fuel for diesel, produced from various feedstocks such as vegetable oil and animal fats. It has a lot of advantages such as low environmental pollutions and biodegradability, and use of these fuels reduced the dependency on foreign country for import of petroleum [8]. De Souza et al. [9] have studied the life-cycle emissions and energy balance of palm biodiesel. They have conducted wide range of investigation on life-cycle assessment and energy cost. The essential factors are considered such as energy balance, reduction in global warming potential, and biodiesel yields for life-cycle assessments. Sheehan et al. [8] have conducted the life-cycle inventory on energy consumption for biodiesel production from soybean oil in the USA. During the life-cycle inventory, they considered all inputs of biodiesel production from soybean oil. There are six stages for LCA: cultivation of soybean, transportation of soybean and oil, oil extraction, biodiesel production, distribution of biodiesel, and its uses in engines.

2 Materials and Methods

According to ISO standard [10], the study was performed to evaluate the energy life-cycle assessment of biodiesel plants for 20 years. For this purpose, we considered three stages of biodiesel production process: cultivation of crops, oil extraction, and biodiesel production. In the cultivation phase, various energy inputs include human labor, agricultural machinery, fossil fuel, fertilizer, chemical, seed, electricity, and water for irrigation. Oil extraction phase includes electricity, decorticator, oil expeller, and filter. Biodiesel production phase includes alcohol (methanol), catalyst (acid H_2SO_4 and base NaOH/KOH), and electricity. The data have been collected from various agricultural sites and research papers for the above study.

2.1 Methodology

This energy analysis is presented in terms of net energy gain (NEG) and net energy ratio (NER), and NER and NEG can be defined as the ratio of total energy output to total energy input and the differences between total energy output and input for one year. The energy equivalent coefficient for different inputs and outputs is given in Table 1. There are following parameters considered during the calculation of energy input and output of different biodiesel plants.

1. Consider only those plants which are cultivated for long time.
2. Jatropha, mahua, neem, coconut, palm, karanja, jojoba, and tung oil plants.
3. Cultivation time 20 year.
4. Plant survival 100% taken for all calculations.
5. Plantation area is 1 ha.
6. All the input and output are converted into its corresponding energy.

Table 1 Energy equivalent coefficient of inputs and outputs

Items (Unit)	Energy equivalent (MJ/unit)	Items (Unit)	Energy equivalent (MJ/unit)
A. Cultivation		B. Oil Extraction	
1. Human labor (MD)	10 [43]	1. Human labor (MD)	10 [43]
2. Machinery (h)	[44, 45]	2. Electricity (KWh)	11.93 [46, 47]
(a) Tractor	93.6	3. Diesel (L)	47.8 [48]
(b) Self propelled	87.63	C. Biodiesel production	
(c) Other machinery	62.7	1. Human labor (MD)	10 [43]
3. Chemical (kg)	[49]	2. Machinery (h)	8 [48]
(a) Herbicides	238	3. Electricity	11.93 [46, 47]
(b) Insecticides	101.2	4. Methanol (kg)	33.67 [50, 51]
4. Diesel (L)	47.8 [48]	5. Catalyst (kg)	
5. Fertilizers (kg)		(a) KOH	19.87 [8, 50]
(a) Nitrogen	66.14 [46]	(b) NaOH	23.30 [43]
(b) Phosphate (P_2O_5)	12.14 [46]	6. Hydrochloric Acid (kg)	21 [50]
(c) Potassium (K_2O)	11.15 [46]	7. H_2SO_4 (L)	3 [43]
(d) Sulfur (S)	1.12 [49]		
(e) Farmyard manure	0.3 [49]		
6. Poly bag usage (Number)	0.69 [43]		
7. Water for irrigation (m^3)	1.02 [44, 52]		
8. Electricity (KWh)	11.93 [46]		

7. The inputs are fertilizers, manpower, machinery, fossil fuel, methanol, catalyst, electricity, etc.
8. Biodiesel is final product.
9. Glycerin is the only coproduct.

2.2   Data Collection

Cultivation phase: This is the first stage of the energy life-cycle assessment for biodiesel production. Cultivation of crops comprises various activities such as seeding (by seeds, stem cutting, sowing in nursery beds, etc.), pit preparation, field preparation (tilling and weed removal), plantation practices (direct planting by seeds and transplanting), use of poly bags, fertilizers application (urea during nursery and then nitrogen, phosphorous, potassium, di-ammonium phosphate (DAP), cow dung (gobar), etc., are used), intercropping, irrigation, weeding, pruning, and harvesting. The agricultural input data for jatropha [11, 12], mahua

Table 2 Parameters for different seeds of biodiesel production

Vegetable plants	Plants/ ha	Maturity time (Years)	Oil contents (%)	Biodiesel conversion (%)	CV of biodiesel (MJ/kg)
Jatropha [11, 12]	2500	6	38 [25, 53]	97 [54]	37.27 [55]
Mahua [13, 14]	200	14	30–40 [13, 56]	98 [57]	37 [57]
Neem [15, 16]	400	13	20–30 [15, 58]	88 [59]	37 [43]
Coconut [21, 22]	177	10	63–65 [60]	98.4 [21]	38.1 [21]
Palm [23, 24]	143	11	30–60 [61, 62]	92 [63]	37.2 [64]
Karanja [17]	500	11	27–39 [17]	97 [65]	37.98 [66]
Jojoba [19]	2500	13	50 [19]	97 [67]	39.862 [67]
Tung [20]	500	13	30–40 [20]	88 [68]	39 [69]

[13, 14], neem [15, 16], karanja [17, 18], jojoba [19], tung [20], coconut [21, 22], and palm [23, 24] oil are adopted from National Oilseeds and Vegetable Oils Development Board (Ministry of Agriculture, Govt. of India). The production of seeds from plants depends on its gestation period and maturity time. The gestation period of jatropha, mahua, neem, palm, coconut, karanja, jojoba, and tung is 1–2, 8, 5, 3, 6, 4, 4, and 4 years, respectively [11–24]. The maximum seed yield after maturity of plants is 4000, 4500, 4876, 18,000 FFB (fresh fruit bunch), 10,620 nuts/ ha, 9000, 1125, and 3625 kg/ha/year for jatropha, mahua, neem, coconut, palm, karanja, jojoba, and tung, respectively.

Oil Extraction: Crushing of seeds for oil extraction is the second stage of energy life-cycle assessment for biodiesel production. Decortications and seed pressing are the process carried out during oil extraction phase. The harvested fruits were decorticated, i.e., extract the seeds from fruit shell. The decorticator is available in IIT (BHU) Varanasi of 1.5 kW with 150 kg/h capacity. Screw press expeller of 5.5 kW with capacity 50 kg/h is used for extraction of oil from seed. Filtering of oil after oil extraction is carried out with filter press of 1.5 kW with capacity of 600 kg/ h [25]. Oil and oil cake are obtained after crushing the seeds. The calculation has been done with 100% efficient screw press for all the seeds. The oil content in various seeds is given in Table 2.

Biodiesel production: Conversion of vegetable oil into biodiesel is the third stage of energy life-cycle assessment. Direct use or blends, microemulsion, pyrolysis, and transesterification are the biodiesel conversion technique to produce biodiesel from vegetable oils or animal fats [26]. Among the above process, transesterification reaction of vegetable oil or animal fats is a universal technique to produce biodiesel (known as methyl or ethyl ester). There are two steps for biodiesel conversion, esterification and transesterification. Esterification is the pretreatment of oil with alcohol (30% by volume) and the acid catalyst (1% H_2SO_4 by volume) to reduce the fatty acid, and this leads to avoid the soap formation and also increase the yield. Transesterification reaction has been carried out at 65 °C temperature for 2 h with alcohol (methanol or ethanol 30%, generally by volume) in the presence of catalyst (NaOH/KOH, 0.5–1% wt/volume) [26–29]. The catalysts are used to improve the reaction rate and also biodiesel yield. Table 2 shows the biodiesel conversion efficiency and calorific value of biodiesel.

3   Results and Discussion

3.1   Energy Input and Output

Figure 1 shows the energy input of different biodiesel plants (jatropha, mahua, neem, palm, coconut, karanja, jojoba, and tung) for 20 years. The figure depicts initially (from seedling to gestation period) lower input, and then after, its value increases and reaches maximum at maturity of plants. The energy input is increasing (for coconut and palm from) and decreasing (for jatropha, mahua, neem, and tung) order from seedling to gestation period of plants. This is because the material (fertilizers) requirement increases for coconut and palm up to the gestation period and the reverse trend for remaining plants. After completion of the gestation period, plant starts fruiting, and after that, material inputs (fertilizers) are totally stopped or reduced that is why there are lower energy inputs in the cultivation of crops. The yields increase from gestation period to its maturity and on the same time oil extraction and biodiesel also started, and hence, total energy input increases up to the plant stabilization periods. The increasing of energy input due to the more chemical (methanol, NaOH and H_2SO_4) is needed during biodiesel productions when plants produce high yield. Increasing order of energy input during the first year is of the plantation are mahua, neem, karanja, tung, coconut, jojoba, jatropha

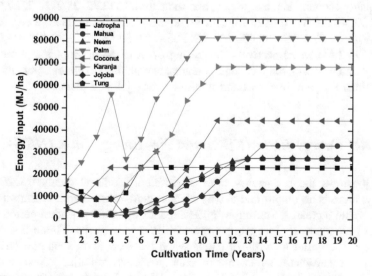

Fig. 1 Energy input for 20 year

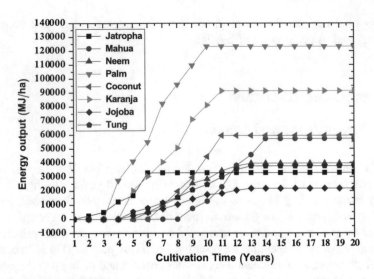

Fig. 2 Energy output for 20 year

and palm, respectively. Initially, higher energy inputs for palm, jatropha, coconut, and jojoba are due to the higher requirement of fertilizers and chemical.

Fig. 2 shows the energy output for different biodiesel plants for 20 years. It is clear from the figure that the initially (from seedling to gestation period of plants) there is no output because of only cultivation of crops. After completion of the gestation period, plants start yielding, and its value increases up to its maturity. Therefore, maximum output achieved at maturity of plants. Figure 3 shows the energy input and output after maturity of plants. Energy inputs for jatropha, mahua, neem, palm, coconut, karanja, jojoba, and tung are 23,373.75, 33,271.93, 27,623.93, 81,517.26, 44,502.43, 68,597.72, 13,254.78, 27,485.55 MJ/ha, and energy outputs are 32,970.53, 57,109.5, 39,690.64, 123,206.4, 59,309.81, 91,180.485, 21,749.70, and 37,696.2 MJ/ha, respectively. Increasing order of energy inputs and outputs is jojoba, jatropha, neem, tung, coconut, karanja, and palm. Maximum input and output are obtained for palm biodiesel and minimum for jojoba oil plants.

3.2   Net Energy Ratio (NER) and Net Energy Gain (NEG)

Figure 4 shows the net energy ratio of different biodiesel plants for 20 years. Initially, there is no output that is why NER is zero; when the plant starts fruiting, then its value increases. Starting of NER depends upon the gestation period of that plant. All the plants achieve its maximum NER at its maturity. Figure 5 shows the net energy ratio after maturity of plants when biodiesel is only output. NER after maturity for jatropha, mahua, neem, palm, coconut, karanja, jojoba, and tung biodiesel plants is 1.4106, 1.7164, 1.4368, 1.5114, 1.3034, 1.3292, 1.6409, and

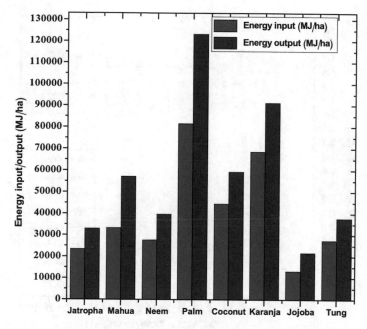

Fig. 3 Energy input/output after maturity

1.3715, respectively. Maximum NER is obtained for mahua and minimum for coconut biodiesel plants. Increasing order of NER is coconut, karanja, tung, jatropha, neem, palm, jojoba, and mahua. Figure 6 shows the NER after maturity of plants when the product (biodiesel) and coproduct (glycerin) both are considered as output. NER for above biodiesel plants with considering coproducts are 1.5904, 1.8641, 1.5993, 1.5991, 1.4084, 1.4627, 1.7822, and 1.5032 respectively. With the addition of coproduct (glycerin) there is significant increase in NER has been found. Net energy ratio increased by 12.75, 9.19, 11.31, 5.80, 8.05, 10.04, 8.61, and 9.60% for jatropha, mahua, neem, palm, coconut, karanja, jojoba, and tung, respectively, with the addition of coproduct (glycerin).

Figure 7 shows the variation of net energy gain of different biodiesel plants for 20 years. The NEG decreases from seedling to certain period and constant up to the gestation period and after that increases up to the maturity of the plants. Initially, there is no output that is why there is negative net energy gain for all biodiesel plants. The net energy gain is in decreasing order from seeding to gestation period when fruiting started, the NEG is in increasing order up to the maturity value. Requirement of fertilizers increases from seedling to growing period due to this very deep curve for palm and coconut but reverse for remaining biodiesel plants.

Figure 8 shows the net energy gain (NEG) after maturity of different biodiesel plants. Net energy gain for jatropha, mahua, neem, palm, coconut, karanja, jojoba, and tung is 9596.78, 23,837.57, 12,066.71, 41,689.14, 13,807.38, 22,582.77, 8494.92, and 10,210.68 MJ/ha after maturity of plants. Increasing order of NEG is

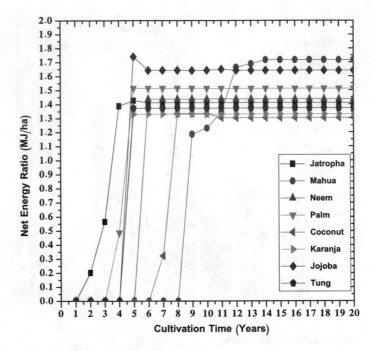

Fig. 4 NER for 20-year plantation

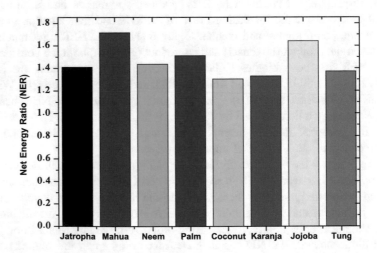

Fig. 5 NER after maturity of plants

jojoba, jatropha, tung, neem, coconut, karanja, mahua, and palm, respectively. Net energy gain is maximum for palm and minimum for jojoba biodiesel plants.

Figure 9 shows the net energy gain when the product (biodiesel) and coproduct (glycerin) both are considered as output. With the addition of glycerin, net energy gain

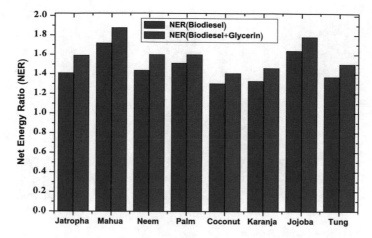

Fig. 6 NER after maturity of plants with coproduct

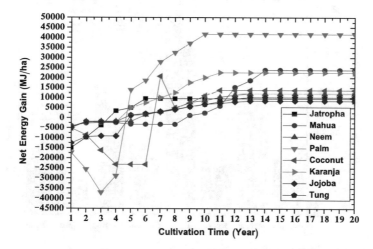

Fig. 7 NEG for 20 year plantation

increased by 43.79, 22, 28.03, 17.14, 24.38, 40.55, 22.05, and 35.47%, respectively, for jatropha, mahua, neem, palm, coconut, karanja, jojoba, and tung after maturity.

3.3   Energy Input, Output, NER, and NEG Analysis with Methanol Recovery

During the life assessment of biodiesel plants, methanol shares very high energy input. The esterification and transesterification have been done for biodiesel

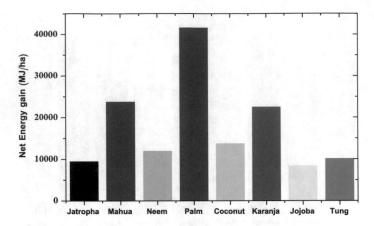

Fig. 8 NEG after maturity of plants

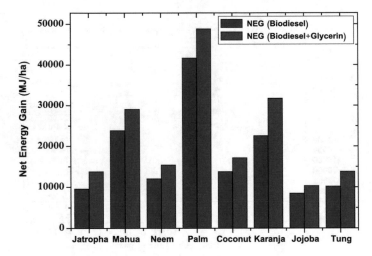

Fig. 9 NEG after maturity of plants with coproduct

production. The recovery of methanol during biodiesel production for castor, mahua, neem, coconut, and karanja is 50, 52.70, 55.81, 52.27, and 52.67%, respectively. For the energy calculations, we have taken average value of 52.69% methanol recovery for above results to the remaining oil (jatropha, palm, jojoba, and tung). The recovered methanol after esterification and transesterification is reused.

Figure 10 shows the energy input (with or without methanol recovery) and output of different biodiesel plants after maturity for one year. There is a significant reduction in energy input when methanol is reused for biodiesel productions. The reduction in energy input for jatropha, mahua, neem, palm, coconut, karanja,

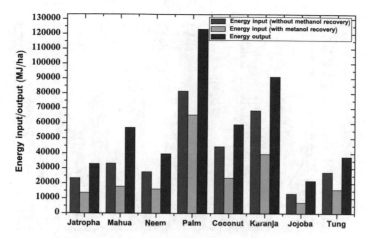

Fig. 10 Variation in ratio of energy input and output

jojoba, and tung is 41.53, 46.19, 41.45, 19.46, 46.92, 42.24, 45.17, and 42.11%, respectively. Increasing order of energy input reduction after methanol recovery is palm, neem, jatropha, tung, karanja, jojoba, mahua, and coconut biodiesel plants.

Figure 11a shows the net energy ratio for different biodiesel plants after maturity of plants with and without methanol recovery. Net energy ratios are 2.413, 3.19, 2.454, 1.877, 2.511, 2.301, 2.993, and 2.369, respectively, for jatropha, mahua, neem, palm, coconut, karanja, jojoba, and tung with methanol recovery. Net energy ratio increased by 71.06, 85.85, 70.79, 24.19, 92.65, 73.11, 82.38, and 72.76%, respectively, for jatropha, mahua, neem, palm, coconut, karanja, jojoba, and tung with methanol recovery.

Figure 11b shows the net energy gain for different biodiesel plants after maturity of plants with and without methanol recovery. Net energy gains are 19,306.6, 39,205.51, 23,517.91, 57,553.26, 35,688.98, 51,562.25, 14,482.41, and 21,786.5 MJ/ha, respectively, for jatropha, mahua, neem, palm, coconut, karanja, jojoba, and tung with methanol recovery. Net energy gain increased by 101.17, 64.47, 94.89, 38.05, 158.47, 128.32, 70.48, and 113.37% for jatropha, mahua, neem, palm, coconut, karanja, jojoba, and tung with methanol recovery.

3.4   Pugh Matrix Analysis of Different Biodiesel Plants

To compare the vegetable oil plants for production of biodiesel, the Pugh matrix is used with the use of some selection criteria. The jatropha oil plant is selected as "base line," and it scored S against all of the selected criteria. The mahua, neem, palm, coconut, karanja, jojoba, and tung vegetable oil plants are compared for each of the criteria. If the selection criteria are:

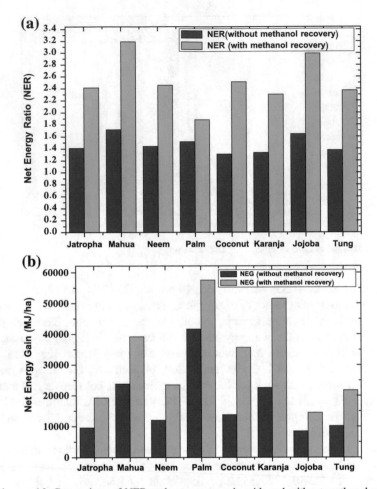

Fig. 11 a and **b** Comparison of NER and net energy gain with and without methanol recovery

- Better than the baseline, a "+" is entered in the appropriate cell;
- Worse than the baseline, a "−" is entered in the appropriate cell;
- The same than the baseline, a "S" is entered in the appropriate cell.

Table 3 indicates that the total score ("9") is maximum for mahua and neem as compared to other vegetable oil plants, which concluded that both are highly preferred for making the biodiesel. Palm, coconut, and tung show lower total score and lowest for jojoba oil plants.

Table 3 Pugh matrix

Pugh concept selection matrix

Selection Criteria	Biodiesel plants							
	Jatropha	Mahua	Neem	Palm	Coconut	Karanja	Jojoba	Tung
Plants type (Edible/non-edible)	S	S	S	–	–	S	S	S
Life of spam of plants (year)	S	+	+	–	+	+	+	+
Seed yield (kg/ha)	S	+	+	+	+	+	–	–
Oil extraction (kg/ha)	S	+	+	+	+	+	–	+
Biodiesel production (kg/ha)	S	+	+	+	+	+	–	+
Energy input (MJ/ha)—during cultivation of crops	S	+	+	–	–	+	–	+
Energy output (MJ/ha)	S	+	+	+	+	+	–	+
Net energy ratio	S	+	+	+	–	–	+	–
Net energy gain	S	+	+	+	+	+	–	+
Availability in India	S	+	+	–	S	S	–	–
Total +	0	9	9	6	6	7	2	6
Total –	0	0	0	4	3	1	7	3
Total Score	0	9	9	2	3	6	–5	3

4 Conclusion

The following conclusions are drawn based on the present study:

1. The total energy inputs calculated during cultivation of jatropha, mahua, neem, palm, coconut, karanja, jojoba, and tung are 32,376.02, 24,877.32, 11,951.48, 117,122.32, 122,832.84, 11,455.89, 40,393.23, and 10,710.62 MJ/ha, respectively. Palm and coconut oil plants have higher energy input as compare to other plants and lowest for neem and tung. Palm oil has highest average energy input (for 20 years) than other biodiesel plants.
2. The total energy outputs calculated after plant maturations are 32,970.53, 57,109.5, 39,690.64, 12,3206.4, 59,309.81, 91,180.485, 21,749.70, and 37,696.2 MJ/ha, respectively. Maximum energy output is obtained for palm biodiesel and minimum for jojoba oil plants after its maturity.
3. Mahua and jojoba have higher net energy ratio than those of remaining biodiesel plants. With addition of glycerin significant increment in net energy ratio has been found for all plants.
4. Total net energy gains for one year after maturity of plants are 9596.78, 23,837.57, 12,066.71, 41,689.14, 13,807.38, 22,582.77, 8494.92, and 10,210.68 MJ/ha for jatropha, mahua, neem, palm, coconut, karanja, jojoba, and tung, respectively. Net energy gain is highest for palm and lowest for jojoba. With addition of coproduct (glycerin), NEG increased by 43.79, 22, 28.03, 17.14, 24.38, 40.55, 22.05, and 35.47%, respectively. Average percentage of net energy gain is maximum for mahua and neem plants.
5. There is significant reduction of energy input with recover and reuse of methanol during esterification and transesterification process. The maximum reduction has been reported for coconut and mahua oil.
6. Net energy ratio increased by 71.06, 85.85, 70.79, 24.19, 92.65, 73.11, 82.38, and 72.76%, respectively, for jatropha, mahua, neem, palm, coconut, karanja, jojoba, and tung with methanol recovery.
7. Net energy gain increased by 101.17, 64.47, 94.89, 38.05, 158.47, 128.32, 70.48, and 113.37% for jatropha, mahua, neem, palm, coconut, karanja, jojoba, and tung with methanol recovery.

5 Introduction of Syngas as Alternate Fuel

Due to fuel crisis, alternative fuels have been interesting option for everyone in country. To fulfill energy requirements, there has been growing interest in alternative fuels like methyl alcohol, biogas, biodiesels, hydrogen, and producer gas ethyl alcohol to provide a suitable diesel alternative for internal combustion engines. For a very efficient method, gasification ion is used to extract energy from different types of organic materials and also has applications as a clean waste disposal technique. A mixture of combustible gases is produced through a solid biomass with the gasification process. Hence, by substituting considerable amount of diesel fuels, producer gas can act as a promising alternative fuel, especially for diesel engines. In India, there are so many alternative fuels, for example, biodiesel, gas fuel, and others. Producer gas is the one of them which is produced from biomass such as agricultural waste products. For the electricity generation system mostly, the producer gas is used as the fuel. A mixture of dual fuel in which the diesel is main fuel while producer gas is the secondary fuel is used for this system. In this experiment, with the help of the dual fuel system, to produce the producer gas, downdraft gasifier is used and wood chips has been used as biomass feedstock for downdraft gasifier [30]. Through the process of gasification, the producer gas generated from biomass such as wastes from, wood chips, agricultural products, groundnut shell, coconut shells etc. can be used as alternative fuel for IC engines. The carbonaceous materials, such as biomass, converts into carbon monoxide and hydrogen by reacting the raw material with a controlled amount of oxygen and at high temperatures, through gasification process. A producer gas is produced as a resulting gas mixture and is itself a fuel. For extracting energy from many different types of organic materials, gasification is a very suitable method and also has many applications such as a clean waste disposal technique [31]. To meet the energy requirements, there has been growing interest in alternative fuels like methyl alcohol, biodiesels, biogas, ethyl alcohol, producer gas, and hydrogen to give a suitable diesel oil substitute for internal combustion engines [32]. A biological material generated by means of photosynthesis containing energy stored in organic compounds is considered as biomass [33]. It includes both aquatic matter and terrestrials such as aquatic plants, night soil, industrial refuse sewage solid wood, disposable garbage herbaceous plants, algae, and residues, corncobs, like straw, husks saw-dust, cow dung, wood shaving etc. [33]. The energy density of biomass is too small compared with that of fossil fuel, to use these as an energy source [34]. To utilize energy from it in a correct way, gasification of such biomass is a good option. Compression ignition (CI) engine could be operated with the following fuels in either the form of mixture or alone. A huge amount of biomass feedstock is available in a large variety. For decentralized power generation in many parts of the country, biomass gasifier-based power generation may be an appropriate option as these are available locally [35]. Power generation from few kilowatts up to several hundred kilowatts has been successfully developed indigenously with biomass gasifier-based system. For the conservation of diesel, the utilization of producer gas in the diesel engine in dual fuel operation is an established technology. Without any modification in the engine, producer gas could be used in CI engine. But diesel cannot be completely replaced by this. In the dual fuel mode, diesel replacements up to 70–90% have been achieved. Some minimum amount of diesel is required to start the ignition because of delay ignition and poor ignition characteristics. On the other hand, for CI engine the use of plant oil as fuel is not new. Except volatility and viscosity, the all properties of plant oils were near to the diesel. To resolve these problems, various methods were adopted. It included esterification of plant oils, blending of oils with diesel, before injecting into the combustion chamber of engine

Table 4 Composition of producer gas

Biomass	Producer gas composition (%)					
	H_2	CO	CH_4	CO_2	O_2	N_2
UK liptus	14.45	16.38	1.07	10.4	8.6	49.1
Wood	15.54	18.4	1.34	8.9	7.9	47.92
Cow dunk	13.36	15.45	1.03	17.6	7.6	46.96

the heating of plant oils [36]. To extract energy from many different types of organic materials, gasification method is used. Gasification is the decentralized energy conversion system which operates economically even for small scale [37]. Partial combustion of biomass in the gasifier generates producer gas that can be used for heating purposes in internal combustion engines. Producer gas could be used in CI engine, without any modification in the engine. In present study, an attempt is made to utilize the biomass residue as feedstock for biomass gasification. The diesel replacement with the use of producer gas is also analyzed. The emission and performance characteristics of the engine were studied for various gas flow rates at different loads. When operated on dual fuel mode, a reduction in the consumption of diesel fuel was observed though there was a decrement in the brake thermal efficiency. In the dual fuel, NO_x emission was found to be very low which is a main benefit of dual fuel mode over diesel fuel alone, but the emission of HC and CO for dual fuel mode was found to be greater than diesel fuel. A typical composition of producer gas generated by biomass gasification on volumetric basis is given in Table 4.

6 Methodology

A double-cylinder, constant speed (1500 rpm), four-stroke, direct injection with rated output of 5-kW water-cooled, CI engine was used for the study. For generation of producer gas, a downdraft gasifier is selected in this study, as this configuration reduces particles and the amount of high molecular weight hydrocarbons as compared to the updraft configuration [38]. The schematic diagram of experimental setup and gasifier-engine assembly is shown in Fig. 12 (Tables 5, 6, and 7).

The experimental setup consists of gas filter, loading device, diesel engine, biomass gasifier, gas cooler. For the production of producer gas, a downdraft gasifier is used. The feedstock for gasification, small pieces of SHF, and 25 cm³ were used. At a regular interval of time, the gasifier receives the feedstock. Producer gas (PG) is generated due to partial combustion of biomass in the gasifier that enters the gas cooler. Gas filter is used to remove dust, and the flow rate of producer gas is controlled by valves. The flow of air is measured by manometers. At different load conditions, performance and emission tests are carried out in dual fuel and liquid mode. The operation of dual fuel mode is carried out by supplying

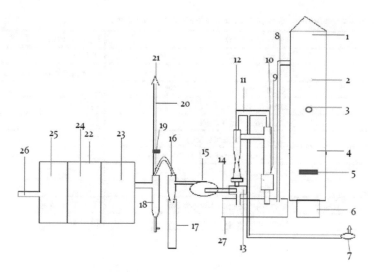

Fig. 12 Schematic layout of the experimental test with gasifier unit

Table 5 List of various parts of layout of experimental gasifier unit shown in Fig. 1

(1) Hopper	(14) Blower
(2) Reactor first	(15) Cyclone separator
(3) Air nozzle	(16) Cyclone drain pipe
(4) Reactor second	(17) Catch pot
(5) Grid	(18) Fire valve
(6) Water seal	(19) Chimney
(7) Water pump	(20) Gas checking point
(8) Thermocouple	(21) Three way gas filter
(9) Down draft	(22) Dry filter using rice husk
(10) Up draft	(23) Dry filter using wood dust
(11) Steam jet control valve	(24) Cotton filter
(12) Venture	(25) Producer gas outlet
(13) Butter fly valve or gas supply	(26) Tar collecting pot

the PG–air mixture to the combustion chamber of the engine through inlet manifold. On the volumetric basis, the liquid fuel flow rate was measured. For the measurement of emission concentration, AVL 444 digas analyzer was used. To measure the flow rate of producer gas, a gas flow rate meter was used. Producer gas control valve (PG-valve) controls the producer gas supply rate. There was no control on aspirated air. PG-valve is slowly opened till the engine started faltering, to know the maximum intake of PG by engine. Later PG-valve is turned back slightly, and the intake of PG at that point was taken as the maximum replacement in the engine for each operating condition. At five different loading conditions 1, 2,

Table 6 Specifications of gas producer engine system under test

S. No.	Engine (Item)	Description
1	Model	PV-14
2	Type	4-stroke diesel engine
3	Number of cylinders	2
4	Stroke/Bore (mm)	110/87.5
5	Compression Ratio	18.8:1
7	Cylinder capacity (cc)	1323
8	Crank shaft speed (rated), rpm	1500
9	Rated capacity(kW)	11
10	S.F.C. (gm/kWh)	251
11	Method of cooling system	Water cooled
15	Fuel tank capacity	11.0 L

Table 7 Specifications of gas producer gasifier unit system under test

S. No.	Gasifier Unit (Item)	Description
1	Type of gasifier	Downdraft
2	Material of construction	Generally M.S except hearth and air nozzle
3	Rated capacity	60 kg
4	Air nozzle	2 Nos.
5	Hopper capacity	60 kg
6	Fuel used	Waste woody biomass
7	Size of fuel (wood chips)	25 cm^3
8	Gas Temperature	Ambient {after cooling}
9	Tar and soot in Gas	Negligible
10	Test Time Duration	20 min
11	Gas cooling medium	Water cooled
14	Filter media	Pebble, cotton yarn waste, etc.
15	Scrubber	Direct contact, co-current water jet

3, 4, 5 kW using diesel, experiments were carried out on the engine, diesel + PG, respectively, as a fuel to the engine. The calorific value of producer gas was calculated from the gas composition, and the composition of producer gas was determined using AVL 444 digas analyzer. The different parameters such as specific energy consumption (SEC), thermal efficiency (BTE), speed of engine are calculated and compared. The data for emissions from the engine were recorded during these trials and compared with the baseline data using diesel fuels.

7   Results and Discussions

7.1   Brake Thermal Efficiency (BTE)

The performance of the engine is mainly projected by calculating the brake thermal efficiency of the engine. It shows the behavior of combustion of the engine. Figure 13 shows the brake thermal efficiency (BTE) and variation of brake power at different load conditions. The BTE of the diesel fuel is higher than that operated under dual fuel mode. It has been observed that there is a relative reduction in brake thermal efficiency of the engine with the engine operated under dual fuel mode at all loads. In dual fuel mode, the efficiency achieved 18.68% is maximum so far and 17.96% for D+PG (CD) and D+PG (W), whereas the maximum efficiency achieved by diesel fuel is 20.93%. The main reason for the decrement in the BTE is the lower calorific value of producer gas, which holds more combusted mixture that enters into the engine. Producer gas at higher temperature is evolved from the engine so that the density of producer gas is decreased, which leads to reduction in the mass flow rate of air and producer gas required for combustion. For this type of combustion, low oxygen level is needed. The cause of incomplete combustion in the combustion chamber is the insufficient oxygen [39]. It was found that when the engine was operated with a large fraction of the fuel energy from producer gas particularly the detonation or knocking occurred at high load. It was also noticed that the sound-like knocking from engine reducers as intake of producer gas reduces.

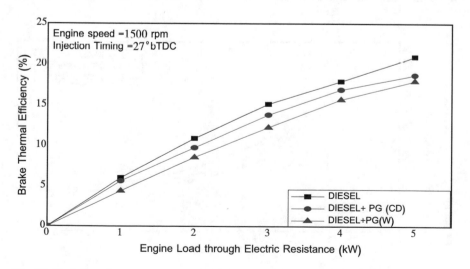

Fig. 13 Variation of engine load on brake thermal efficiency in dual fuel mode

7.2   Brake Specific Energy Consumption (BSEC)

The brake specific energy consumption for the engine running on mixture of diesel
and producer gas has been used to analyze the performance of CI engine at different
loading conditions and diesel. The variation of specific energy consumption, brake
power, brake specific fuel consumption, to compare the two fuels having different
calorific values and density, may be weak parameter; however, it has been depicted
in Fig. 14. The brake specific energy consumption to compare the performance of
CI engine in dual fuel mode is calculated from the fuel consumption and calorific
value of diesel and producer gas. It has been revealed that specific energy consumption in dual fuel is higher than that of diesel mode at all load conditions. The
relation between specific energy consumption and brake thermal efficiency seems to
be inversely proportional, and hence, the value of brake thermal efficiency reduces
by using the producer gas with air. The same may be attributed to the more energy
content with increased gas flow rate of wood (W) and cow dung (CD) producer gas
in the combustion chamber.

7.3   Carbon Monoxide (CO) Emission

The formation of CO emmision is mainly attributed due to the mainly attributed to
insufficient time in the cycle for completion of combustion and the incomplete
combustion due to oxygen supplied is not sufficient to complete the combustion in
combustion chamber. Figure 15 shows the variation of CO emission of the engine
with dual fuel and DF mode. It has been observed that CO emission increases with
increase in the load. In the dual fuel, the values of CO emissions recorded are much

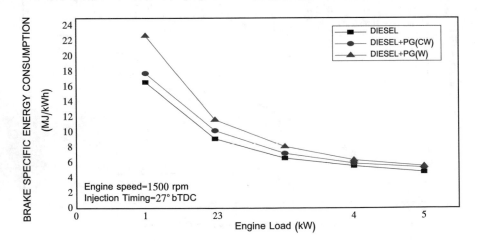

Fig. 14 Variation of brake power on BSEC in dual fuel mode

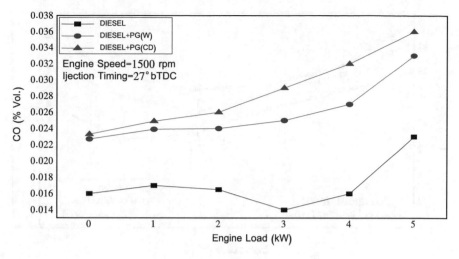

Fig. 15 Effect of BP on CO emission at varying load in dual fuel mode

higher as compared to diesel fuel mode. Reduction in the oxygen amount required for complete combustion reduces due to high-temperature mixture of producer gas and air, flowing to the engine. This phenomenon increases the CO emission due to incomplete combustion [40].

7.4 Hydrocarbon (HC) Emission

With the increment in loads, there is possibility that an increase in hydrocarbon (HC) emission occurs always as depicted in Fig. 16. The direct result of incomplete combustion is unburnt hydrocarbon emissions. The volatile organic compounds (VOCs) are unburned hydrocarbons and their derivatives that readily vaporize. The VOCs and oxides of nitrogen react to form oxidants and photochemical smog in the presence of sunlight [41]. When some portion of fuel induced into the engine causes escapes combustion, the emission arises. The ignition delay period is responsible to contribute to higher HC emissions when a fuel–air mixture becomes too rich to ignite and combust.

7.5 Nitric Oxide (NO$_x$) Emission

The formation of NO$_x$ attributed to mainly two reasons, higher temperature and availability of oxygen. Nitrogen is inert at low temperature; however, in higher temperatures such as 1000 °C, nitrogen reacts with oxygen and forms oxides of nitrogen [42]. Thus, the emission of NO$_x$ depends upon temperature in combustion

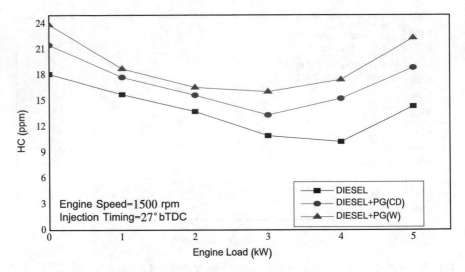

Fig. 16 Variation of brake power on HC emission at varying load in dual fuel mode

chamber and also varies with the applied load. From Fig. 17, it is observed that NO_x emission increases with increase in load for all the fuels, i.e., D+PG fuel (wood, cow dung) and diesel. The nitrogen with oxygen causes formation of NO_x at high temperature in combustion chamber at high load. Insignificant difference in NO emission was recorded at low brake power, while operating the engine on dual fuel mode and diesel fuel mode, but the emission of NO_x increases as brake power increases for combinations of diesel and producer gas and diesel fuel. The excellent advantage of dual fuel mode is that there is no possibility of lower emission in dual fuel mode than diesel mode.

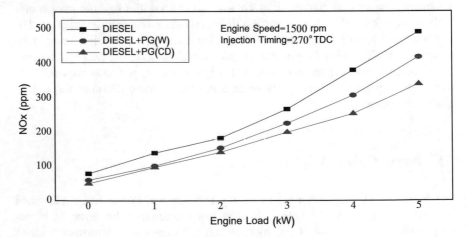

Fig. 17 Effect of brake power on NO_x emission at varying load in dual fuel mode

8 Conclusion

Some important findings on the engine emission characteristics and performance were investigated experimentally in double-cylinder 4-stroke water-cooled CI diesel engine operating cow dung as biomass in the downdraft gasifier and with downdraft gasifier on dual fuel mode operation with wood. The main outcomes are listed below.

> The savings of up to 60% of diesel oil were achieved at a load of 5 kW with the use of gasification gas from wood and cow dung as the main fuel.
> With increased emissions at all load conditions, the engine performance decreases in dual fuel mode operation.
> It is found that the increase in load on the engine lowered SEC and increases the BTE for the existing engine set up.
> With little or no modification in the gasifier, the cow dung as biomass materials is suitable for the engine.
> In dual fuel mode, CO emission is found to be higher which leads to insufficient oxygen in combustion chamber.
> In dual fuel mode, HC emission is found to be greater which shows an indication of insufficient oxygen in combustion chamber.
> The great advantage of dual fuel mode over diesel fuel alone is lower emission of NO_x.

Objective Question and Answers

(1) Only edible oil can be used as biodiesel under Indian condition. (No)
(2) Without any modification in the engine producer gas could be used in CI engine. (Yes)
(3) There are two major cause of formation of CO emission: The first one is the incomplete combustion due to insufficient amount of oxygen supplied in combustion chamber, and the second one is the insufficient time in the cycle for completion of combustion. (Yes)
(4) Palm and coconut consumed highest energy (117,122.32 and 122,832.84 MJ/ha) during cultivation. (Yes)
(5) In the dual fuel mode of operation, "the specific energy consumption is found to be the lower at various load conditions". (No)
(6) In the dual fuel, the NO_x emission was found to be very low which is highly advantageous over diesel fuel alone, but HC and CO emission found to higher than the diesel. (Yes)
(7) In India, there are various types of non-edible (coconut, soybean, mustard, peanuts, palm, etc.) and edible (neem, mahua, karanja, jojoba, jatropha, tung, etc.). (No)
(8) Maximum NER for mahua and minimum for coconut biodiesel plants. (Yes)

(9) Among the above process, transesterification reaction of vegetable oil or animal fats is a universal technique to produce biodiesel (known as methyl or ethyl ester). (Yes)

(10) The NO_x formation takes place at elevated temperature of combustion chamber in the presence of oxygen for longer duration. (Yes)

Glossary

Biomass Renewable organic matter such as agricultural crops, crop-waste residues, wood, animal and municipal wastes, aquatic plants, fungal growth used for the production of energy.

Biodiesel Diesel fuel produced from biomass. Biodiesel cannot be used in standard engines without modification as it corrodes rubber seals and gaskets. It also has a lower gelling point than petrodiesel, making it unsuitable for use in colder climates. Biodiesel is often blended with petrodiesel.

Bioenergy Any renewable energy made from biological sources. Fossil fuels are not counted because, even though they were once biological, they are long dead and have undergone extensive modification.

Biofuel Any fuel derived from biological carbon fixation, including solid fuels, bioethanol, and other bioalcohols, biodiesel, etc.

Carbon dioxide A molecule made of one carbon atom double bonded to two oxygen atoms (one of each side of the carbon). It is a colorless, odorless gas at standard temperature and pressure and is widely implicated as one of the major causal agents in greenhouse warming.

Carbon monoxide A molecule made of one carbon atom bonded to a single oxygen atom via a triple bond. It is a product of inefficient combustion of hydrocarbon compounds. It is a pollutant and is toxic to humans at concentrations above 50 ppm during long-term exposure or 667 ppm during short-term exposure.

Cogeneration The production of electric energy along with a second for energy (often heat).

Combustion Commonly referred to as burning. This is the process by which a fuel and an oxidant react to product heat (energy) and other compounds (CO_2 and H_2O in ideal hydrocarbon combustion).

Combustion Gases The gases released during a combustion process, that is, similar to emissions.

Compression ignition engine An internal combustion engine in which the fuel is ignited by heat generated from compressing the gas to high pressures rather than from a spark. Most diesel engines work this way.

Diesel fuel Any liquid fuel used in a diesel engine.

Emissions The gaseous or particulate components expelled during a combustion reaction. The term commonly refers to the mix of gases and particulate that exits the exhaust of an internal combustion engine.

Energy balance In regard to biofuel, this term refers to the amount of energy required to produce the fuel versus the amount of energy derived from the fuel.

Energy content Also referred to as heating value, energy content is a measure of the number of British Thermal Units obtained by burning a set volume of fuel. Because it relies on volume, energy content can change with temperature and pressure.

Energy crop A low-cost, low-maintenance plant gown exclusively for use as fuel.

Energy density Generally, the amount of energy stored in a given region of space per unit volume. Specifically, the amount of energy obtained from a specified mass of biofuel. Useful for comparing various types of biofuels in a standardized manner.

Energy efficiency ratio A comparison of the energy stored in a fuel and the energy needed to produce, transport, and distribute the fuel.

Ethanol An alcohol composed of two carbons. The formula is C_2H_4O.

FAME biodiesel Fatty acid methyl ester biodiesel.

Feedstock The raw material from which a biofuel is produced. Feedstock is generally a plant itself, but in the case of algae, the feedstock is any source of carbon (often carbon dioxide).

First-generation biofuel Any biofuel derived from sugar, starch, or vegetable oil. In general, these fuels are considered a threat to food supply chains.

Flash point The lowest temperature at which a flammable liquid produces enough vapor to ignite. For most flammable liquids like gasoline, it is the vapor and not the liquid itself that is combustible.

Flexible Fuel Vehicle A vehicle with an internal combustion engine that can run on more than one fuel. Usually, the vehicle is designed to run on pure gasoline or a defined blend of gasoline/ethanol or gasoline/methanol. In some cases, the vehicle can run on pure ethanol.

Fossil fuel A fuel formed by the anaerobic decomposition, over thousands or millions of years, of dead biomass. Petroleum, coal, and natural gas are major categories of fossil fuels.

Gasification Any chemical or heat process used to convert a feedstock to a gaseous fuel.

Transesterification A process in which organically derived oils or fats are combined with alcohol (ethanol or methanol) in the presence of a catalyst to form esters (ethyl or methyl ester).

References

1. Balat, M., & Balat, H. (2008). A critical review of bio-diesel as a vehicular fuel. *Energy Conversion and Management, 49,* 2727–2741.
2. Basha, S. A., Gopal, K. R., & Jebaraj, S. (2009). A review on biodiesel production, combustion, emissions and performance. *Renewable and Sustainable Energy Reviews, 13,* 1628–1634.
3. IEA, International Energy Agency. (2012). http://www.iea.org/stats/.
4. Najafi, G., Ghobadian, B., Tavakoli, T., Buttsworth, D. R., Yusaf, T., & Faizollahnejad, M. (2009). Performance and exhaust emissions of a gasoline engine with ethanol blended gasoline fuels using artificial neural network. *Applied Energy, 86,* 630–639.
5. Ghobadian, B., Najafi, G., Rahimi, H., & Yusaf, T. (2009). Future of renewable energies in Iran. *Renewable and Sustainable Energy Reviews, 13,* 689–695.
6. Bounds, A. (2007). OECD warns against biofuels subsidies. *Financial Times.* Available at: www.ft.com/cms/s/0/e780d216-5fd5-11dc-b0fe-0000779fd2ac.html.
7. Woods, J., Brown, G., & Estrin, A. (2005). *Bioethanol greenhouse gas calculator.* London, UK: Home-Grown Cereals Authority. Biomass Energy Group, Centre for Environmental Policy Imperial College. www.publications.hgca.com/publications/documents/Bioethanol_Greenhouse_Gas_Calculator_Appendix_4.pdf.
8. Sheehan, J., Camobreco, V., Duffield, J., Graboski, M., & Shapouri, H. (1998). *Life cycle inventory of biodiesel and petroleum diesel for use in an urban bus. Final report.* Golden, CO (US): National Renewable Energy Lab. www.nrel.gov/docs/legosti/fy98/24089.pdf.
9. De Souza, S. P., Pacca, S., De Avila, M. T., & Borges, J. L. B. (2010). Greenhouse gas emissions and energy balance of palm oil biofuel. *Renewable Energy, 35,* 2552–2561.
10. ISO14044. (2006). *Environmental management-life cycle assessment and requirements and guidelines.* Geneva: International Organisation for Standardisation (ISO).
11. Kureel, R. S., Singh, C. B., Gupta, A. K., & Pandey, A. Jatropha: An alternate source for biodiesel. National Oilseeds & Vegetable Oils Development Board (Ministry of Agriculture, Govt. of India), website: www.novodboard.org.
12. Lele, S. (2007). The cultivation of *Jatropha curcas.* Available at: http://www.svlele.com/jatropha_plant.htm.
13. Kureel, R. S., Kishore, R., & Dutt, D. Mahua: A potential tree borne oilseed. National Oilseeds & Vegetable Oils Development Board (Ministry of Agriculture, Govt. of India), website: www.novodboard.org.
14. Lele, S. (2007). The cultivation of Mahua (*Madhuca indica*). Available at: http://www.svlele.com/mahua.htm/.
15. Kureel, R. S., Kishore, R., & Dutt, D. Neem: A tree borne oilseed. National Oilseeds & Vegetable Oils Development Board (Ministry of Agriculture, Govt. of India), website: www.novodboard.org.
16. Lele, S. (2007). The cultivation of Mahua (*Madhuca indica*). Available at: http://www.svlele.com/neem.htm.

17. Kureel, R. S., Singh, C. B., Gupta, A. K., & Pandey, A. Karanja a potential source of bio-diesel: An alternate source for biodiesel. National Oilseeds & Vegetable Oils Development Board (Ministry of Agriculture, Govt. of India), website: www.novodboard. org.

18. Agarwal, A. K., & Rajamanoharan, K. (2009). Experimental investigations of performance and emissions of Karanja oil and its blends in a single cylinder agricultural diesel engine. *Applied Energy, 86,* 106–112.

19. Kureel, R. S., Kishore, R., & Dutt, D. Jojoba: A potential tree borne oilseed. National Oilseeds & Vegetable Oils Development Board (Ministry of Agriculture, Govt. of India), website: www.novodboard.org.

20. Kureel, R. S., Singh, C. B., Gupta, A. K., & Pandey, A. Tung: A potential tree borne oilseed. National Oilseeds & Vegetable Oils Development Board (Ministry of Agriculture, Govt. of India), website: www.novodboard.org.

21. How, H. G., Masjuki, H. H., Kalam, M. A., & Teoh, Y. H. (2014). An investigation of the engine performance, emissions and combustion characteristics of coconut biodiesel in a high pressure common-rail diesel engine. *Energy, 69,* 749–759.

22. https://www.nabard.org/english/plant_coconut.aspx.

23. Oil Palm Cultivation Practices: Directorate of Oil Palm Research (Indian Council of Agricultural Research) Pedavegi-534450, West Godavari District, Andhra Pradesh. Website: http://dopr.gov.in.

24. https://www.nabard.org/english/plant_oilpalm.aspx.

25. Pandey, K. K., Pragya, N., & Sahoo, P. K. (2011). Life cycle assessment of small-scale high-input Jatropha biodiesel production in India. *Applied Energy, 88,* 4831–4839.

26. Girard, P., & Fallot, A. (2006). Review of existing and emerging technologies for the production of biofuels in developing countries. *Energy for Sustainable Development, 10,* 92–108.

27. Maeda, H., Hagiwara, S., Nabetani, H., Sagara, Y., Soerawidjaya, T. H., Tambunan, A. H., et al. (2008). Biodiesel fuels from palm oil via the non-catalytic transesterification in a bubble column reactor at atmospheric pressure: A kinetic study. *Renewable Energy, 33,* 1629–1636.

28. Alamu, O., Waheed, M., & Jekayinfa, S. (2007). Biodiesel production from Nigerian palm kernel oil: Effect of KOH concentration on yield. *Energy for Sustainable Development, 11,* 77–82.

29. Noiroj, K., Intarapong, P., Luengnaruemitchai, A., & Jai-In, S. (2009). A comparative study of KOH/Al_2O_3 and KOH/NaY catalysts for biodiesel production via transesterification from palm oil. *Renewable Energy, 34,* 1145–1150.

30. Banapurmath, N. R., et al. (2009). Combust ion characteristics of a 4-stroke CI engine operated on Honge oil, Neem and Rice Bran oils when directly injected and dual fuelled with producer gas induction. *Renewable Energy, 34,* 1877–1884.

31. Ramadhas, A. S., et al. (2008). Dual fuel mode operation in diesel engines using renewable fuels: Rubber seed oil and coir-pith producer gas. *Renewable Energy, 33,* 2077–2083.

32. Banapurmath, N. R., & Tewari, P. G. (2009). Comparative performance studies of a 4-stroke CI engine operated on dual fuel mode with producer gas and Honge oil and its methyl ester (HOME) with and without carburetor. *Renewable Energy, 34,* 1009–1015.

33. Young, S., Yoon, S. J., Kim, Y. K., & Lee, J. (2011, July 5). Gasification and power generation characteristics of woody biomass utilizing a downdraft gasifier. Climate Change Technology Research Division, Korea Institute of Energy Research, 71-2 Jang-dong, Yuseong-gu, Daejeon 305-343, Republic of Korea.

34. *Book of wood gas as engine fuel, food and agriculture organization of the United Nation, M-38.* ISBN 92-5-102436.

35. Banapurmath, N. R., et al. (2008). Experimental investigations of a four-stroke single cylinder direct inject ion diesel engine operated on dual fuel mode with producer gas as inducted fuel and Honge oil and its methyl ester (HOME) as injected fuels. *Renewable Energy, 33,* 2007–2018.

36. Singh, R. N., Singh, S. P., & Pathak, B. S. (2007). Investigations on operation of CI engine using producer gas and rice bran oil in mixed fuel mode. *Renewable Energy, 32,* 1565–1580.
37. Kothadia, H. B. (2011, May). Effect of size of lignite on gas generation rate, calorific value of gas and gasifier efficiency of downdraft gasifier. M. Tech. dissertation, Department of Mechanical Engineering, Institute of Technology, Nirma University, Ahmedabad-382481.
38. McKendry, P. (2002). Energy production from biomass (part 2): Conversion technologies. *Bioresource Technology, 83,* 47–54.
39. Sharma, K. A. (2009). Experimental study on 75 kWth downdraft (biomass) gasifier system. *Renewable Energy, 34,* 1726–1733.
40. Murugan, S. (2010). Design and development of downdraft gasifier for operating CI engine on dual fuel mode, 53.
41. Hernandez, J. J., Aranda-Almsnsa, G., & Bula, A. (2010). Gasification of biomass wastes in an entrained flow gasifier: Effect of the particle size and the residence time. *Fuel Processing Technology, 91,* 681–692.
42. Pengmei, Lv, Zhenhong, Yuan, Ma Longlong, Wu, Chuangzhi, Chen Yong, & Jingxu, Zhu. (2007). Hydrogen-rich gas production from biomass air and oxygen/steam gasification in a downdraft gasifier. *Renewable Energy, 32,* 2173–2185.
43. Lokesh, A. C., Mahesh, N. S., Gowda, B., Kumar, R., & White, P. (2015). Neem biodiesel—A sustainability study. *Journal of Biomass to Biofuel, 1.* ISSN 2368-596.
44. Mousavi-Avval, S. H., Rafiee, S., Jafari, A., & Mohammadi, A. (2011). Optimization of energy consumption for soybean production using data envelopment analysis (DEA) approach. *Applied Energy, 88,* 3765–3772.
45. Canakci, M., Topakci, M., Akinci, I., & Ozmerzi, A. (2005). Energy use pattern of some field crops and vegetable production: Case study for Antalya Region, Turkey. *Energy Conversion and Management, 46,* 655–666.
46. Mohammadi, A., & Omid, M. (2010). Economical analysis and relation between energy inputs and yield of greenhouse cucumber production in Iran. *Applied Energy, 87*(1), 191–196.
47. Heidari, M. D., Omid, M., & Mohammadi, A. (2012). Measuring productive efficiency of horticultural greenhouses in Iran: A data envelopment analysis approach. *Expert Systems with Applications, 39,* 1040–1045.
48. Kitani, O. (1999). *CIGR handbook of agricultural engineering. Energy and biomass engineering* (Vol. 5). St. Joseph, MI: ASAE Publications.
49. Rafiee, S., Mousavi Avval, S. H., & Mohammadi, A. (2010). Modeling and sensitivity analysis of energy inputs for apple production in Iran. *Energy, 35,* 3301–3306.
50. Rajaeifar, M. A., Ghobadian, B., Safa, M., & Heidari, M. D. (2014). Energy life-cycle assessment and CO_2 emissions analysis of soybean based biodiesel: A case study. *Journal of Cleaner Production, 66,* 233–241.
51. Huo, H., Wang, M., Bloyd, C., & Putsche, V. (2008). Life-cycle assessment of energy use and greenhouse gas emissions of soybean-derived biodiesel and renewable fuels. *Environmental Science and Technology, 43,* 750–756.
52. Beheshti Tabar, I., Keyhani, A., & Rafiee, S. (2010). Energy balance in Iran's agronomy (1990–2006). *Renewable and Sustainable Energy Reviews, 14*(2), 849–855.
53. Kumar, M., & Dwivedi, K. N. (2009). *Jatroph: Ek parichay*. Kanpur: Chandrashekhar Azad University of Agriculture and Technology Press.
54. Achten, W. M. J., et al. (2010). Life cycle assessment of Jatropha biodiesel as transportation fuel in rural India. *Applied Energy, 87*(12), 3652–3660.
55. Kumar, S., Singh, J., Nanoti, S. M., & Garg, M. O. (2012). A comprehensive life cycle assessment (LCA) of Jatropha biodiesel production in India. *Bioresource Technology, 110,* 723–729.
56. Puhan, S., Vedaraman, N., Ram, B. V. B., Sankarnarayanan, G., & Jeychandran, K. (2005). Mahua oil (*Madhuca indica* seed oil) methyl ester as biodiesel-preparation and emission characterstics. *Biomass and Bioenergy, 28,* 87–93.
57. Raheman, H., & Ghadge, S. V. (2007). Performance of compression ignition engine with mahua (*Madhuca indica*) biodiesel. *Fuel, 86,* 2568–2573.

58. Ragit, S. S., Mohapatra, S. K., Kundu, K., & Gill, P. (2011). Optimization of neem methyl ester from transesterification process and fuel characterization as a diesel substitute. *Biomass and Bioenergy, 35*(3), 1138–1144.
59. Karmakar, A., Karmakar, S., & Mukherjee, S. (2012). Biodiesel production from neem towards feedstock diversification: Indian perspective. *Renewable and Sustainable Energy Reviews, 16,* 1050–1060.
60. Haas, M. J. (2005). Improving the economics of biodiesel production through the use of low value lipids as feedstocks: Vegetable oil soapstock. *Fuel Processing Technology, 86,* 1087–1096.
61. Yusup, S., & Khan, M. (2010). Basic properties of crude rubber seed oil and crude palm oil blend as a potential feedstock for biodiesel production with enhanced cold flow characteristics. *Biomass and Bioenergy, 34*(10), 1523–1526.
62. Queiroz, A. G., Franc, L., & Ponte, M. X. (2012). The life cycle assessment of biodiesel from palm oil ("dendê") in the Amazon. *Biomass and Bioenergy, 36,* 50–59.
63. Siregara, K., Tambunanb, A. H., Irwantoc, A. K., Wirawand, S. S., & Arakie, T. (2015). A comparison of life cycle assessment on Oil Palm (*Elaeis guineensis* Jacq.) and Physic nut (*Jatropha curcas* Linn.) as feedstock for Biodiesel production in Indonesia. Conference and Exhibition Indonesia—New, Renewable Energy and Energy Conservation (The 3rd Indo-EBTKE ConEx 2014). *Energy Procedia, 65,* 170–179.
64. Cho, H. J., Kim, J. K., Ahmed, F., & Yeo, Y. K. (2013). Life-cycle greenhouse gas emissions and energy balances of a biodiesel production from palm fatty acid distillate (PFAD). *Applied Energy, 111,* 479–488.
65. Naik, M., Meher, L. C., Naik, S. N., & Das, L. M. (2008). Production of biodiesel from high free fatty acid Karanja (*Pongamia pinnata*) oil. *Biomass and Bioenergy, 32,* 354–357.
66. Dhar, A., & Agarwal, A. K. (2014). Performance, emissions and combustion characteristics of Karanja biodiesel in a transportation engine. *Fuel, 119,* 70–80.
67. Shehata, M. S., & Razek, S. M. A. (2011). Experimental investigation of diesel engine performance and emission characteristics using jojoba/diesel blend and sunflower oil. *Fuel, 90,* 886–897.
68. Shang, Q., Jiang, W., Lu, H., & Liang, B. (2010). Properties of Tung oil biodiesel and its blends with 0$^{\#}$ diesel. *Bioresource Technology, 101,* 826–828.
69. Gadhave, S. L., & Ragit, S. S. (2016, July). Optimization of Tung oil methyl ester from transesterification process and fuel characterization as diesel substitute. *International Journal of Latest Trends in Engineering and Technology (IJLTET), 7*(2), 116–120. ISSN 2278-621X.

Part III
Robotics and Automation

Chapter 7
Robotics: History, Trends, and Future Directions

Shyamanta M. Hazarika and Uday Shanker Dixit

Abstract In this chapter, the history of robotics is traced. The emphasis is on highlighting significant moments in robotics history that had far-reaching consequences on the field. A brief presentation of the technical intricacies is followed by highlight of the recent trends. The chapter finally dwells on the direction robotics is surging and poised to change the world in more things than one could possibly imagine.

Keywords Robot · Robotics · Hmanoid · Sensors · Mechatronics
Kinematics · Dynamics · Entertainment robotics · Medical robotics
Actuators · Artificail Intelligence · History

1 Introduction

Talk of a *robot* and the image that immediately conjures in the mind is that of a *humanoid* catering to the daily chores! Possibly the basis of this image is the popular conceptualization of robots largely through science fiction. Over the ages, many science fiction writers imagined human-like mechanical beings with tremendous physical and intellectual capabilities. Starting with Czech playwright Karel Capek, these machines were mostly referred by the science fiction writers and moviemakers as robots. This is in spite of that fact that even the most sophisticated robots of today are nowhere near to what their imaginations created!

Robots and robotics are driven by the urge to create or synthesize machines that could replicate human function and capabilities. This involves use of mechanisms, sensors, and actuators and advanced levels of computer programming. A multitude of fundamental concepts ranging over a spectrum of different areas of engineering is involved. This chapter presents a basic understanding of a robot in the current perspective. Further, the current trends and future directions are presented.

S. M. Hazarika · U. S. Dixit (✉)
Mechanical Engineering, Indian Institute of Technology Guwahati, Guwahati 781039, India
e-mail: uday@iitg.ernet.in

© Springer International Publishing AG, part of Springer Nature 2018
J. P. Davim (ed.), *Introduction to Mechanical Engineering*, Materials Forming,
Machining and Tribology, https://doi.org/10.1007/978-3-319-78488-5_7

2 History of Robotics

Progress in science, engineering, and technology has influenced the growth of robotics. Growth and development in robotics are interwoven with the breakthroughs and significant advancements in science and engineering. A complete account of such evolvement is neither intended nor possible. The focus of this section is to present a brief history of robotics with particular focus on dreams and dreamers who had ushered in one of the greatest accomplishments in the history of mankind.

2.1 Early Automatons and Automatic Mechanisms

People have long imagined machines endowed with human abilities. Dreamt of automatons that could move on its own and mechanical devices that could perform rational reasoning. Human-like machines are described in many stories and are pictured in sculptures, paintings, and drawings. Not only automatons have been a fascination in the myths and legends across cultures, as early as fourth century BC many mechanized implements have been conceived. The Greek mathematician Archytas of Tarentum conceived a mechanical bird! Aristotle imagined abolition of slavery through a surge in automatons and wrote in *The Politics*:

> For suppose that every tool we had could perform its task, either at our bidding or itself perceiving the need.........shuttles in a loom could fly to and fro and a plucker play a lyre of their own accord, then master craftsmen would have no need of servants nor masters of slaves.

Aristotle

The Politics [1]

Around 1495, Leonardo da Vinci (1452–1519) sketched designs for a humanoid robot in the form of a medieval knight. It was supposed to be able to not only move its arms and head but also open its jaw! It is not clear whether Leonardo da Vinci or his contemporaries tried to build the humanoid. However, the seeds of an arduous journey looking for an artificial intelligent-mechanized human were sown. Thus, began the story of dreams and dreamers that drive the urge for looking beyond the mundane to design automatons which no one had imagined before.

In 1651, Thomas Hobbes (1588–1679) the patriarch of artificial intelligence published *Leviathan*, his book about social contract and the ideal state. Hobbes propounded that it might be possible to build artificial animal and stated:

> For seeing life is but a motion of limbs, the beginning whereof is in some principal part within, why may we not say that all automata (engines that move themselves by springs and wheels as doth a watch) have an artificial life? For what is the heart, but a spring; and the nerves, but so many strings; and the joints, but so many wheels, giving motion to the whole body...

Thomas Hobbes

Leviathan [2]

This led to a number of automatons and automatic mechanisms being built in and around 1700. These automatons were capable of drawing, flying, and few could act and play music. French inventor and engineer, Jacques de Vaucanson (1709–1782), created some of the most sophisticated automatons that moved in startlingly lifelike ways. His masterpiece *The Digesting Duck* was built in 1738. The mechanical duck could quack, flap its wings, paddle, drink water, and eat and digest grain!

2.2   Robots in Science Fiction

Some 17 years before, Joseph Capek wrote the short story *Opilec* describing automats, Frank Baum in 1900 invented one of the literary world's most beloved robots in *The Wonderful Wizard of Oz*. It is Tin Woodsman, a mechanical man in search of a heart! In 1921, Joseph's brother Karel Capek introduced the word robot in his play *R.U.R.* (*Rossum's Universal Robots*). The play centers around a mad scientist who tries to usurp the powers of God by proving that man has acquired the technology and intelligence to create life. The artificial human is a magnificent worker and a tireless laborer who does not complain! The term robot is derived from the Czech word "robota" that means servitude. Prior to *Rossum's Universal Robots*, the term *automaton* was used.

Science fiction writer Issac Asimov made robots popular through a collection of short stories published between 1938 and 1942. Asimov introduced the word "robotics." Asimov in his story *Runaround* describes robotics as the study of robots and enumerated three "laws of robotics"—the fundamental principles based on which he defines his robots and their interaction within the worlds he creates. Thereafter, robots have been a part of literary and cinematic fiction.

More often than not, robots are shown to take over the world and be bad guys rather than as man's friend! Most significant of these are the images of The Terminator (1984) and its numerous sequels. There are instances of robots being used as more than a simple killer. These include The Day the Earth Stood Still (1951 and 2008); Robbie in Forbidden Planet (1956); 2001: A Space Odyssey (1968); or A.I. Artificial Intelligence (2001).

2.3   Significant Moments in Robot History

Robots in real life are far away from what have been created by the science fiction writers in their imagined worlds. Nevertheless, robots rudimentarily similar to those envisioned in R.U.R. have become a reality! Today robots are routinely used to substitute for manual labor in an effort to minimize human error and injury. Robots are also seen as a way to increase the production efficiency.

The transition from science fiction to reality has its genesis in a chance meeting in 1956 between entrepreneur Joseph Engelberger and inventor George Devol.

Engelberger was excited about George's latest invention, a Programmed Article Transfer device that led to formation of Unimation Inc., and building of the Unimate—the very first industrial robot. General Motors was the first to introduce the Unimate to assist in automobile production. Starting with Unimate way back in 1961 on the assembly line, application of robots in the automobile industry was explosive! Yet, General Motors' usage of Unimate in a die-casting plant remains the most significant moment in the history of robots and robotics [3].

Even though Devol and Engelberger were the most successful in transforming wishful thinking to an industrial revolution, they may not have been the first to create such machines [4]. Way back in 1948, William Grey Walter constructed some of the first electronic autonomous robots. Grey conducted experiments on mobile, autonomous robot. He was interested if the robots could model brain functions. Grey Walter's three-wheel machine was crude by today's standards but a marvel of the day. It had basic analog circuits but could even recharge their own batteries.

One of the most significant moments in robotics is when robots moved from the factory floor and made inroads into our living spaces! Pioneering work in this area was done at Stanford between 1966 and 1972, where Shakey a general-purpose mobile robot that was smart and could reason about its own actions was developed. Life Magazine referred to Shakey as the first electronic man. This definitely has been a significant moment in the history of robotics.

3   What Is a Robot?

Even though robots have penetrated into a number of domains and have taken various forms and functions, basic understanding of robots remains the same. A robot can be seen as a physical agent that manipulates the physical world; the physical agent is artificial rather than natural. In order to manipulate the physical world, a robot needs to work in a Sense–Plan–Act cycle. Each of the above, in turn, demands different capabilities from the robot. "Sense" or sensing requires the robot to take in information about its environment; a robot has to "Plan," i.e., use that information to make a decision, and finally, a robot needs moving parts to carry out commands or "Act."

3.1   RIA Definition

The Robotics Institute of America (RIA) defines a robot as follows:

> A robot is a reprogrammable, multifunctional manipulator designed to move material, parts, tools or specialized devices through variable programmed motions for the performance of a variety of tasks. The robot is automatically operating equipment, adaptable to complex conditions of the environment in which it operates, by means of reprogramming, managing to prolong, amplify and replace one or more human functions in its interactions with the environment.

> Robotics Institute of America [4]

This definition emphasizes the expected characteristics of a robot in terms of its functionality. Robots started its foray into the human space being an important component of modern automobile manufacturing plants. Even today, industrial robots form the single largest group of robots. Nevertheless, robots are now being used almost everywhere! However, robots differ depending on the end use. The hand of a robot or the "end-effector" may be specialized tools, such as spot welders or spray guns, or more general-purpose grippers. This ensures "multi-functionality." A robot needs to be "reprogrammable," i.e., define operation through computer commands, with robotic actions based on tasks or objectives. The robot needs to be capable of undertaking a "variety of tasks."

3.2 JIRA Classification

Japanese Industrial Robot Association (JIRA) has more detailed classification of the varieties of robots. JIRA classifies robots into the following categories (a) manipulators, (b) numerically controlled, (c) sensate, (d) adaptive, (e) smart, and (f) Intelligent Mechatronic System.

1. Manipulators—A manipulator is physically anchored to their workplace, usually with an entire chain of controllable joints, enabling placement of end-effectors in any position within the workplace. Manipulators are by far the most common type of industrial robots. Manipulators are categorized as (a) manual—machines slaved to a human operator; (b) sequential—device that performs a series of tasks in the same sequence every time they are activated; and (c) programmable—define operation through computer commands based on tasks or objectives. Note that Japan calls certain machines robots that would only best be termed "factory machines" in other parts of the world. For the three categories of manipulator highlighted here, only a programmable manipulator would qualify to be a robot outside of Japan.

2. Numerically controlled—These are industrial automation usually referred to as numerical control (NC) machines. These robots are a form of programmable automation instructed to perform tasks through information on sequences and positions in the form of alphanumeric data. The data represent relative positions between a tool and other processing element often referred to as a work-head and the work-part, i.e., object being processed. The operating principle of numerically controlled automation is to control the sequence of motion of the work-head relative to the work-part, leading to required machining. Three important components merge to create a numerical control system: (a) part program, (b) machine control unit, and (c) processing equipment. Part program refers to the detailed set of commands (in the form of alphanumeric data) to be followed by the processing equipment. The machine control unit is usually a microcomputer that stores and executes the program. This is accomplished by

converting each command into actions by the processing equipment; operation is sequential with one command being processed at a time.

3. Sensate—The word sensate is derived from the Latin sensatus 'having senses' and refers to perceiving or perceived by the senses. Therefore, robots that incorporate senor feedback into their circuitry could be referred to as sensate. However, by "sensate robots" we usually mean embodied machines with the unique capability to sense human body language, thus enabling these machines to better comprehend and respond to their human companions in a natural way. The family of robots that incorporate touch sensors, proximity sensors, vision systems, and so forth predominantly for human–machine societal interaction is referred to as sensate robots.

4. Adaptive—Advances in sensor technology coupled with artificial intelligence have infused new directions to robotics. Robots are slowly transforming from single-purpose, preprogrammed machines to multi-purpose, adaptive workers. Artificial intelligence is increasingly being used to make robots figure out on their own actions for a given goal, instead of being given explicit instructions. Robots that can change the way they function in response to their environment are termed as adaptive robots.

5. Smart—Robots that are considered to possess artificial intelligence leading to cognitive capabilities are smart!

6. Intelligent Mechatronic System—Tetsuro Mori from the Yaskawa Electric Cooperation coined the term "Mechatronics"—a fancy word to mean the intersection of mechanical/electrical and computer control systems [5]. Mechatronics refers to embedment of smart devices into systems already in place leading to Intelligent Mechatronic System.

4 Aspects of Robotics

Driven through science fiction and the proliferation of humanoids therein, robotics has always been concerned with design and synthesis of certain human functions by the use of mechanisms coupled with sensors, actuators, and information processing unit usually a microcontroller or a computer. This is a huge challenge and demands a multitude of concepts from a number of classical areas of study. There are different aspects of robotics research, each carried out by experts in various fields. Nonetheless, synergy between a number of areas including mechanics, control, and cognition plays an important role.

The field of robotics can be partitioned at a high level of abstraction into four major areas: (a) mechanical manipulation, (b) locomotion, (c) computer vision, and (d) artificial intelligence. The discussion in this section is broadly set along these lines.

4.1 Joints and Links

Robotic manipulators comprise nearly rigid links, which are connected by joints that allow relative motion of neighboring links. A serial manipulator is a set of bodies connected in a chain by joints. These bodies are referred to as links. Joints form a connection between a neighboring pair of links, each link maintaining a fixed relationship between the joints at its ends. Even though the serial manipulator is often compared with the human arm, unlike the joints in the human arm, the joints in a robotic manipulator are restricted to one degree of freedom. This is to simplify the mechanics, kinematics, and control of the manipulator.

Two types of joints are commonly found in robots: (a) revolute joints and (b) prismatic joints. A revolute or a rotary joint is a one-degree-of-freedom kinematic pair used in mechanisms. Displacement is referred to as joint angle. Manipulators also at times contain sliding or prismatic joints. The relative displacement between links in a prismatic joint is a translation. The usual convention is to number the immobile base of a serial manipulator as link 0. Links are thereafter marked with the first moving link as link 1, and so on, till one reached the free end of the arm, which is link n [6].

Degrees of freedom (DoF) are defined as the number of independent movements an object can have in 3-D space. A rigid body free in space can have six independent movements—three translations and three rotations—leading to six DoFs. These six independent variables completely specify the position and orientation of the object in 3-D space. Consequently, in order to position an end-effector arbitrarily in 3-D space a manipulator with six DoF is required—three joints for position and three joints for orientation of the end-effector. These are referred to as *spatial* manipulators. Serial manipulators are the simplest of the industrial robots. Every link in a serial manipulator except the last has a joint at each end. The last link has only one joint. This is located at the end closest to the base, referred to as the proximal end. The end-effector is at the free end, i.e., the end farthest away from the base, referred to as the distal end. The end-effector could be a gripper, a welding torch, an electromagnet, or another device depending on the intended application [6].

Not all robots are as simple; these robots have parallelogram linkages or other closed kinematic structures. Parallel manipulators have one or more loops. None of the link is designated as first or last! Consequently, there does not exist a natural choice of end-effector or output link. The output link must be chosen. More often than not, the number of joints is more than the degree of freedom, with several joints not being actuated [7].

4.2 Drives, Actuators, and Sensors

Actuators are akin to muscles of a robot and play a vital role in "Act" stage of a Sense–Plan–Act cycle. Sensors measure different stimuli that the robot can effectively use. Actuators and sensors together with a feedback control system are the most basic requirements for operation of a robot.

4.2.1 Drives and Actuators

Actuators convert energy to mechanical form; any device that accomplishes this conversion is an actuator. Different types of actuators are available; choice of one over the other is driven by a number of factors including force, torque, and speed of operation, accuracy, precision, and power consumption. Based on the three different modes of energy transfer, actuators are classified as (a) electrical, (b) pneumatic fluid power, and c. hydraulic fluid power.

Electrical Actuation: Electrical motors form the major chunk of electrical actuators. Nevertheless, there are other types of electrical actuators such as solenoids. Typical electrical motor types include (a) DC motors, (b) AC motors, and (c) stepper motors.

Small DC motors have stator magnetic poles produced by a permanent magnet. For large motors, the magnetic field is produced via a stator winding. Voltage supplied to the armature in turn controls the speed of the DC motor. For motors with stator windings, speed control is also possible through varying the current to the stator. AC motors are driven by an alternating current supply to the stator windings. The frequency of the input signal determines the speed of AC motors. Consequently, speed control is done through variation of the signal frequency. Stepper motors possess the ability to rotate a specific number of revolutions or fractions of a revolution. This in turn ensures specific angular displacement.

Pneumatic Power Actuation: Pneumatic actuation is possibly the most widely used in the manufacturing industry, for example in automated assembly, including jig and robot end-effector operation. Pressurized air is used as the power transfer medium. Pneumatic cylinder, a closed cylindrical barrel with a piston, is the primary type of energy transfer device for a linear motion. The piston is made to move by applying the pressurized air at one of two ports found at the ends of the cylinder. Limited rotation rotary actuators that provide back and forth rotation through a fixed angle are the other type of pneumatic energy transfer device.

Hydraulic Power Actuation: Hydraulic actuators use oil instead of air and give a very high power output. Hydraulic actuators are available as linear or rotary actuators. Linear actuators are akin to pneumatic actuators. Rotary actuators are either of limited rotation or of continuous rotation. The continuous rotation hydraulic actuators are referred to as hydraulic motors.

Motion Convertors: Mechanical power transmission systems required to convert actuator outputs to type of motions required by the system are referred to as motion convertors. For example, convertors are required for speed reduction or conversion from rotary to linear motion.

4.2.2 Sensors

A sensor is a transducer that converts a physical stimulus from one form into a more useful form to measure the stimulus. Sensors are a very critical part of any robot, whether autonomous or tele-operated. There was a time when "sensors" for an experimenter's robot were too rudimentary. It would be just a few whiskers connected to microswitches to sense walls and other obstacles. When the robot banged into a wall or obstacle, the switches were tripped and its simple logic steered it in another direction. But after the invention of microcontrollers, a new era of sensors has begun. We have numerous sensors such as active and passive IR sensors; sound and voice sensors; ultrasonic range sensors, positional encoders on arm joints, head and wheels; compasses, navigational and GPS sensors; active and passive light and laser sensors; a number of bumper switches; and sensors to detect acceleration, turning, tilt, odor detection, magnetic fields, ionizing radiation, temperature, tactile, force, torque, RF, UV, and visual sensors such as CCD cameras.

Sensors are the robot's contact with the outside world or its own inner workings. Any robotic system has two distinct categories of sensors for measuring internal and environmental parameters referred to as proprioceptors and exteroceptors, respectively [8].

Proprioceptors: Sensors that measure the kinematic and dynamic parameters of the robot are referred to as proprioceptors. Proprioceptive sensors are responsible for controlling internal status and monitoring self-maintenance.

Position transducers such as potentiometers, synchros and resolvers, encoders, and rotary variable differential transformer (RVDT) are the common rotary joint proprioception sensors. Tachometer transducers are used for the measurement of angular velocity. Traditional DC motor tachometers are getting replaced by digital tachometers using magnetic pickup sensors, for the latter are compact and suitable for embedment within robotic manipulators. Acceleration sensors measure the force that produces the acceleration of a known mass. Acceleration transducers based on stress–strain gage, piezoelectric, capacitive, and inductive principles are available. Micromechanical accelerometers where force is estimated by measuring the strain in elastic cantilever beams formed from silicon dioxide are available. Flexibility of a robot's mechanical structure is also estimated by strain gages mounted on the manipulator's link. Further joint shaft torques can also be estimated using strain gages mounted on shafts.

Exteroceptors: Sensors that sense the environment in order to estimate the location/position as well as force interaction of the robot with the environment are referred to as exteroceptors. Exteroceptors are broadly categorized into: (a) contact sensors, (b) range sensors, and (c) vision sensors.

Contact Sensors: Contact sensors detect a physical contact using mechanical switches that trigger when contact is made. Contact sensors are typically used to detect contact between mating parts and/or to measure the interaction forces and torques during part mating operations using the robot manipulator. In addition to triggers on physical contact, sensors to continuously sense contact forces ranging over an area are also available. This type of contact sensors is termed as tactile sensors. It is worthwhile to mention that tactile sensing is far more complex than contact sensing. Tactile sensing is based on inductive elastomer, strain gage, piezoelectric, capacitive, and optoelectronic technologies. In terms of their operating principles, tactile sensing can be seen as either force-sensitive or displacement-sensitive. Conductive elastomer, strain gage, and piezoelectric are examples of force-sensitive tactile sensors that measure the contact forces. Displacement-sensitive tactile sensors include optoelectronic and capacitive sensors that measure the mechanical deformation of an elastic overlay.

Range Sensors: Range sensors estimate the distance to objects in their operation area. The most common uses of range sensors include robot navigation, obstacle avoidance, or recovery of the third dimension for monocular vision. Range sensors are based on either principle of time of flight or triangulation. Range can be estimated from the time elapsed between the transmission and return of a pulse in time-of-flight sensors such as laser range finders and sonar. Estimation of range can also be accomplished by the principle of triangulation, i.e., detecting a given point on the object surface from two different points of view at a known distance from each other.

Vision Sensors: Robot vision is one of the most complex yet versatile sensing processes. Robotic vision is an involved process and includes extracting, characterizing, and interpreting information from images. This is to identify or describe objects in environment. A camera converts the visual information to electrical signals; these are in turn sampled and quantized yielding a digital image. CCD image sensors being lightweight, compact, and robust are preferred over conventional tube-type. Vision is largely dependent on illumination; for avoiding many problems associated with robotic vision, controlled illumination is preferred with structured lighting being widely used. Special light stripes, grids, or other patterns are projected on the scene with shape of the projected patterns on different objects offering valuable cues. This allows 3-D object parameters to be recovered from a 2-D image.

Digital image produced by a vision sensor is an array of pixel intensities and requires to be processed for an explicit and meaningful description. Digital image processing involves preprocessing, segmentation, description, recognition, and interpretation. Preprocessing techniques usually deal with noise reduction, and detailed enhancement is addressed in preprocessing. Objects are extracted from the scene using segmentation algorithms, like edge detection or region growing. Recognition involves classifying the objects in the feature space. Finally, interpretation assigns a meaning to the ensemble of recognized objects.

4.3   Kinematics and Dynamics

Robot kinematics is fundamental to describing an end-effector's position, orientation as well as motion of all the joints. Dynamic modeling lies at the heart of analyzing and synthesizing the dynamic behavior of a robot. In this section, a brief introduction to the kinematics and dynamics of manipulators is presented.

4.3.1   Kinematics

Kinematics is the study of position, velocity, acceleration, and other higher order derivatives of the position variables without considering forces causing these effects. Kinematics of manipulators involves study of the geometrical and time-based properties of motion of the manipulator.

Denavit–Hartenberg Notation: The Denavit–Hartenberg (D-H) notation is used for kinematic description of a robot. For such a description, the pose of each link in the manipulator relative to the pose of the preceding link is to be described. This should require six parameters in space. However, using a pair of constraints between the coordinate frames, the D-H formalism describes the spatial relation between coordinate frames attached to successive links using a set of four parameters. For a pair of consecutive links, the four parameters describe the links and the joint between them. Each link can be represented by two parameters: (1) the link length a_i and (2) the link twist angle α_i. The link twist angle α_i indicates the axis twist angle of two adjacent joints i and i − 1. Joints are also described by two parameters: (a) the link offset d_i, which indicates the distance from a link to next link along the axis of joint i and (b) the joint revolute angle θ_i, which is the rotation of one link with respect to the next about the joint axis [9]. Even though the choices of coordinate frames may not be unique, the D-H assignments always lead to the same expression for the pose of the end-effector with respect to the base [10].

Forward Kinematics: Forward kinematics is the problem of computing the position and orientation of the end-effector given the set of joint angles. Forward kinematics can be thought of as a mapping of the manipulator position from a joint space description into a Cartesian space description.

Inverse Kinematics: Inverse kinematics is the problem of computing all possible sets of joint angles to attain the given position and orientation. For any practical use of the manipulator such as a pick-and-place operation or line-following operation, inverse kinematics is the fundamental problem to be solved. Inverse kinematics can be thought of as a mapping of end-effector pose—position and orientation—in 3-D Cartesian space of the manipulator's internal joint space.

Although we, humans, solve the inverse kinematics problem routinely in our interaction with the physical world, it is not that simple to compute the joint angles from the Cartesian positions. Forward kinematics is simpler than inverse kinematics. This is compounded by the fact that the kinematic equations are nonlinear. Solution may not be easy or at times even possible in a closed form. Existence of multiple solutions or nonexistence of solutions is questions that need to be encountered. This in turn has effect on the workspace, the volume of space in which the end-effector of the manipulator can attain any desired position and orientation.

Velocities and Singularities: Apart from static positioning problems, kinematic analysis may involve manipulators in motion. A matrix called the Jacobian of the manipulator is defined to undertake velocity analysis. The Jacobian specifies a mapping from velocities in joint space to velocities in Cartesian space. Points at which this mapping is not invertible are called singularities. Knowledge of singularities is important as they lead to problems with motions of the manipulator arm in their neighborhood.

4.3.2 Dynamics

Motion is caused by forces acting on a body. Robot dynamics is devoted to studying the forces required at the joints to cause motion of the end-effector. To be more precise, dynamics includes kinematics as well, and the other part of it is called kinetics. Robot mechanism is modeled as a rigid body system. Consequently, robot dynamics involves the application of rigid body dynamics to robots. A manipulator at rest is accelerated and made to move at a constant end-effector velocity; thereafter, it needs to decelerate and stop. The joint actuators accomplish this feat through a complex set of joint torques. The actuator torque not only depends on path through which the end-effector moves but also on the mass properties of the links and payload.

Forward Dynamics: Forward dynamics involves finding end-effector motion for known joint torques/forces. An important use of dynamic equations involving forward dynamics is in simulation. Dynamic equations can be reformulated to express motion descriptors such as acceleration as a function of actuator torque. This makes it possible to simulate the motion of a manipulator under application of a set of actuator torques.

Inverse Dynamics: Finding joint torques/forces for given joint motions and end-effector moment/force is the problem of inverse dynamics. For a desired path of the end-effector, using the dynamic equations of motion of a manipulator, actuator torque can be estimated. This in turn can be used to control a manipulator.

Trajectory Generation: Path and trajectory describe the motion of an end-effector through space. The locus of points which the manipulator has to follow is the path. A path further qualified with specification of a timing law is referred to as trajectory. Often, a path is described not only by a desired destination but also by some intermediate locations, or via points, through which the manipulator must pass. Curves such as splines are used to approximate a function that passes through a set of via points. Trajectory generation involves computing the motion function of each joint of a manipulator so that motion appears coordinated. Further to force the end-effector to follow a desired path through space, motion in the Cartesian space is converted to an equivalent set of joint motions.

4.4 Planning and Control

Planning and control are fundamental components of robot systems. Motion planning is particularly important for autonomous robots. Over the years, considerable advances have been made in planning and control. Here, the traditional control paradigms with pointers to more advanced schemas are highlighted.

Forces or torques are usually supplied to actuators to drive the manipulators. Inverse dynamics computes the required torques that will cause the desired motion. Even though the problem of dynamics forms a basis of a framework for control of a manipulator, it in itself does not suffice.

Position Control: Control algorithms are based on linear approximations to the dynamics of a manipulator for linear position control. In nonlinear position control, control algorithms based on complete nonlinear dynamics of the manipulator are used. Nonlinear position control performs better than those based on the linear approximations.

Force Control: For many real-world applications, a robot needs to control contact forces between the end-effector and the object being manipulated.

Hybrid Control: Mixed or hybrid control involves some directions of the manipulator end-effector being controlled by a position control law and remaining directions controlled by a force control law.

4.5 Artificial Intelligence

For a robot to be able to perform at par with a human, only having a humanoid structure would not suffice. The robot needs to have the capability of "rational" decision making. Artificial embodiment of such intelligence in a robot is very involved and almost impossible. Nevertheless, intelligent robots form a vibrant and extremely exciting area within robotics.

Human sensory system assists human intelligence. Intelligent robots are equipped with a myriad of sensors, particularly for knowledge of the external world. In line with the use of visual and qualitative information for everyday commonsense reasoning, a robot in an unstructured environment (one that is not known a priori) makes use of "qualitative spatial reasoning" and "robotic vision." Interpretation of a scene and learning from vision are possibly the most difficult phase in the whole pipeline of visual image processing. This requires intelligence that in turn demands huge volume of knowledge. With maturity of learning techniques through evolving paradigms such as deep learning, artificial intelligence is expected to enhance decision-making capabilities of the robots.

5 Applications of Robotics

Starting from General Motors' use of Unimate, the first industrial robot in its assembly line, robots have come a long way. They are increasingly being applied in scenarios as diverse as space exploration, medical surgery, rescue, and rehabilitation. The predominant use still remains in the field of automation in manufacturing. Nevertheless, there has been an upsurge of robot activity in close interaction with human in applications as challenging as health service!

5.1 Robotics and Automation

Robotic mechanisms have been at the forefront of mechanization. There is a, however, subtle difference between mechanization and automation. Automation replaces the worker with intelligent control systems, thereby contributing to increase in productivity, speed, and repeatability. Automation using robotic technologies exploiting advances in computing, particularly machine learning and artificial intelligence, has been the trend. Improved flexibility is the hallmark of robotic technology to manufacturing processes [11] and invariably leads to an improvement in productivity.

Robotics and automation drive not only the manufacturing sector but also numerous other sectors such as construction, transportation, health and support, security. For developed countries particularly USA, Japan, Germany, and a few others, automation and robotics have become an integral part of their life. As the cost of human workers goes up, China is slowly exploring automation and robotics. In the near future, China is expected to overtake Japan as the largest operator of industrial robots [12].

5.1.1 Robotics in Manufacturing

Even though the first use of robots in manufacturing was in situations dangerous to a human user, motivation of using robots in manufacturing is driven more by economics rather than societal concern. Robotic manufacturing is preferred for high precision repetitive tasks. Such tasks are better performed by a robot rather than a human who is susceptible to error. Industrial robots which are programmable can be seen as advanced automation systems, control being executed using an onboard computer. Further with advances in sensing and control as well as computer vision, an industrial robot need not be restricted to environments completely known a priori. Traditionally, a stand-alone industrial robot is used for automation in manufacturing. More recently, another form of robotics in manufacturing is evolving through collaborative robots. Collaborative robots have human-in-the-loop, i.e., work in close coordination with human user. This makes it possible for humans to still have a role in automated manufacturing. Since its inception in 2014, collaborative robotics is poised to emerge as a key player in the automotive and electronics industries [12, 13].

Whether a traditional industrial robot or an adaptive collaborative robot, in an automated production line in manufacturing, its major roles include: (a) material handling, (b) welding, (c) assembly operations, (d) dispensing, and to a lesser extent (e) processing.

Robotic Material Handling: Material handling is possibly the most common application for industrial robots worldwide. Material-handling robots automate some of the most tedious, dull, and unsafe tasks in a production line, thus enhancing efficiency. Material handling encompasses part selection, transferring, packing, palletizing, loading, and unloading. Machine feeding or disengaging is also considered as material handling that can be automated using an industrial robot.

Robotic Welding: Robotic welding is the use of industrial robot which completely automates a welding process by both performing the weld and handling the part. Spot welding and arc welding mainly used by the automotive industry are done through robots. As a robotic application, robotic welding is relatively new, introduced only in the late 1970s [14].

Assembling Operations: Assembly operations include fixing, press-fitting, inserting, and disassembling. For each assembly robot, end-of-arm tooling is customized to cater to the specific requirements. Use of robots for assembly operations has undergone a paradigm shift. A traditional industrial robot is seldom used for assembly. Robotic assembly with sophisticated end-of-arm tooling is more commonly used. This shift can be attributed to the progress in terms of sensor-based technologies particularly force-sensing and machine vision [15].

Robot Dispensing Operations: Robots are the most suited for accurate repetitive processes. Therefore, industrial robots for dispensing jobs such as painting, gluing, applying adhesive sealing, spraying and filling are widely used.

Processing: Industrial robots are very rarely used for material processing. This is primarily due to availability of automated machines specifically for these applications. Nevertheless, industrial robots are used for application such as mechanical, laser, and water jet cutting.

5.1.2   Space Robotics

Like the dream of a mechanical man, outer space has also been a great fascination. Design and development of machines that could tolerate the extremes of yet-un-seen space environments and take forward space exploration have been a dream for science fiction writers and space explorers alike. Space robotics is an area concerned with design of machines capable of surviving in such extreme environment and performing maintenance and exploration tasks. Space robotics is a conglomeration of multiple disciplines grounded on well-founded knowledge of robotics, computer science, and space engineering. Starting in the 1950s, space exploration today not only involves manned space explorations but also robotic missions involving a range of mobility or locomotion systems.

Space robots can be either *orbital* robots or *planetary* robots. Robots that are conceived and designed for repairing satellites or assembling large space telescopes, etc., are usually deployed in orbits around the planetary bodies and are referred to as orbital robots. Planetary robots are deployed on to the planetary bodies. These robots such as the MARS Explorer undertake survey and observation, examination, and extraction of the extraterrestrial surfaces. The planetary robots are also engaged in preparing for subsequent human arrival on the planetary surface [16].

The four key issues in space robotics include a. mobility, b. manipulation, c. time delay, and d. extreme environments. Mobility and autonomy are essential for a space robot. Degree of mobility (i.e., locomotion capability) and manipulation capability are dictated by the application, i.e., whether it is designed to be an orbital or a planetary robot. Autonomy is decided based on the nature of the mission, particularly distance the robot is expected to travel from Earth. Autonomy could range from tele-operation to fully autonomous robots. Tele-operation is usually effected through master–slave control methodology, wherein the slave robot exactly replicates the movements of the master. Time delay is a particular challenge for manipulation in space robotics. Traditional master–slave approach would not work because of the considerable round trip time between the master and slave. Space manipulators are controlled from the earth station by way of supervisory control. Supervisory control involves human commands only for motion; contact forces are managed by the remote site electronics.

The extent of autonomy incorporated into a space robot leads to three distinct classes of space robots—(a) robotic agent, (b) robotic assistant, and (c) robotic explorer. One with the least autonomy is referred to as a robotic agent. It is in fact a human proxy performing various tasks using tele-operation. Fully autonomous space robots working in close coordination with human astronauts are referred to as robotic assistant. Fully autonomous robots capable of exploring unknown territory in space are space explorers. Largest proponent of space robotics, National Aeronautics and Space Administration (NASA), has to its credit a series of successful planetary rover missions. Possibly the most successful of these have been NASA's Mars rovers—sojourner, opportunity, and curiosity. The Japanese *Hayabusa Robotic Mission* to the near-Earth asteroid Itokawa and European Space Agency's Rosetta lander called *Philae* to the center of a comet are other noteworthy applications. Majority of the current space robots are in fact robotic agents. Increasingly challenging space missions necessitate higher level of autonomy for the robots, surging an evolution toward robotic assistants and robotic explorers [17].

5.1.3 Service Robots

Until recently, there was no strict technical definition of the term "service robot." The International Federation of Robotics (IFR) and the United Nations Economic Commission for Europe (UNECE) worked out a service robot definition and classification scheme resulting in ISO-Standard 8373 effective since 2012.

> A service robot is a robot that performs useful tasks for humans or equipment excluding industrial automation application. A personal service robot or a service robot for personal use is a service robot used for a non-commercial task, usually by lay persons. Examples are domestic servant robot, automated wheelchair, and personal mobility assist robot. A professional service robot or a service robot for professional use is a service robot used for a commercial task, usually operated by a properly trained operator. Examples are cleaning robot for public places, delivery robot in offices or hospitals, fire-fighting robot, rehabilitation robot and surgery robot in hospitals.
>
> Professional and Personal Service Robots [18]

Possible applications of robot outside of industrial automation to assist human in everyday chores are innumerable. Nevertheless, service robots at present are classified into three categories: (a) industrial service robots, (b) domestic service robots, and (c) scientific service robots.

1. Industrial service robots: Robots are used in the industry beyond industrial robots for automation. These assignments include routine tasks such as examination of welding, inspection of pipelines, and other services such as construction and demolition.
2. Domestic service robots: Domestic robots are substitute for workers in people's homes. Domestic robots address domestic chores including cleaning floors, mowing the lawn, and window cleaning. Domestic robots are aimed at substituting for the butler! Further, robots for the assistance of elderly and physically challenged such as robotized wheelchairs, personal aids, and assistive devices to carry out activities of daily living are classified as personal/domestic service robots.
3. Scientific service robots: Robotic systems either autonomous or collaborative are capable of performing a variety of repetitive functions required in research. For example, robots are regularly used for gene sampling and sequencing. Autonomous scientific robots are designed to perform difficult or impossible tasks, for example the Woods Hole Sentry for deep sea missions and the Mars rovers.

5.2 Medical Robotics

First used in 1985 for brain biopsy, robots are now regularly used in various medical areas including laparoscopy, neurosurgery, and orthopedic surgery [19]. Medical robotics is one of the fastest growing sectors within the medical devices industry. This unprecedented growth and acceptance of robotics within medical disciplines is largely due to advances in the areas of medical imaging and technological improvements in the area of actuators and control.

Robotic surgery and rehabilitation robotics have emerged as the largest use of medical robotics. Nevertheless, newer and novel uses for medical robots are regularly reported. Radiosurgery involves focused beams of ionizing radiation directed at the particular area. CyberKnife a robotic radiosurgery system ensures directing the beam through the tumor at various orientations for better outcome. There exist medical robot systems for disaster response and battlefield medicine.

5.2.1 Robotic Surgery

Surgical robots have made the greatest impact in medical robotics. Starting with ROBODOC [20] for orthopedic surgery, robots have made inroads into operating theaters in a big way. This is primarily because of the possibility to position, orient, and perform accurate motions of the scalpel in more ways than possible by a human expert. Robots are increasingly being used for a variety of malignant and benign conditions in urology, neurosurgery, and even cardiac operations demanding high order of precision.

Two pathbreaking surgical robots exploiting research on tele-manipulation and miniature tendon-driven wrists are the Zeus by Computer Motion [21] and the da Vinci from Intuitive Surgical [22]. Both feature a master–slave architecture, with the surgeon master console and a patient slave manipulator. The slave manipulator was with three arms—two for tissue manipulation and the third for positioning of the endoscopic camera. The Zeus system, however, is no longer in production. Today Intuitive Surgical's da Vinci system is undoubtedly the most widely accepted surgical robot in the world.

Following the success and popularity of da Vinci Surgical Robot, a number of research laboratories and companies intensified the quest of surgical robots. Amadeus, a four-armed laparoscopic surgical robot system, is being developed. This would be a direct competitor of Intuitive Surgical's da Vinci system. The Amadeus uses snakelike multi-articulating arms, expected to provide improved maneuverability. Improved MR-compatible robots are being investigated. Microrobotic and nano-robotic platforms are emerging. These would further improve surgical outcomes [23].

5.2.2 Rehabilitation and Assistive Robotics

Rehabilitation robotics refers to application of robotic technologies in therapeutic procedures for augmenting rehabilitation. Rehabilitation robotics is primarily concerned with restoration of sensorimotor functions for persons with impairment due to various disabling diseases such as stroke or brain and orthopedic trauma. Assistive robotics encompasses robotic assistive systems for persons with disability including design and development of personal aids and prosthetics for independently carrying out activities of daily living. For retention of residual skills, most advanced of the rehabilitation and assistive robots employ techniques to adapt its quantum of assistance based on the response and reaction of the patient.

Rehabilitation robotics has proved itself to be hugely effective, especially in motor impairments due to stroke. The MIT-MANUS, a robotic arm, is possibly the most widely used rehabilitation robot [24]. Other rehabilitation robotic devices include exoskeletons and improved treadmills. Another example of rehabilitation robot is the Hipbot used for people with limited mobility because of hip injury [25]. Even though rehabilitation using these robotic devices has been predominantly for restoration of motor impairment due to stroke, it is equally applicable to individuals with cerebral palsy or other disabling conditions.

Advances in machine learning and signal processing are allowing paradigm shift in medical rehabilitation and assistive systems, particularly through bio-signals and brain imaging. This has brought in newer and challenging role for robotics—robotic systems to be used as natural extensions of the human body—for restoration of neuromotor functions and motor capabilities. These include neurobionic hands and active legs. Use of robotics is not only limited to motor rehabilitation but has been effectively used for recovery of cognitive deficits [26].

5.3   Entertainment Robotics

Robots have not only been confined to utilitarian use. There is a huge segment of robotic industry engaged in select areas of culture and entertainment. Robots are designed and developed for the sole purpose of subjective pleasure of the human it serves. These robots vary in form and function, ranging from simple toy robots such as a pet dog to complex humanoids exploiting artificial intelligence. Entertainment robotics has come a long way. Starting with Sony's AIBO, a pet dog to its humanoid companion QRIO, we have now Pepper, an emotional humanoid. Designed and developed by Aldebaran Robotics, Pepper is a four-foot humanoid (on wheels) with cute features who have natural communication skills. The best part is that Pepper gives you all the attention you crave! Pepper not only offers advice but also "prattle on and on" making small talk. Even though we have not yet been able to have a slave robot in its truest sense, we are definitely one step closer to the type of artificial intelligence fantasized in science fiction movies.

Entertainment robotics also includes robotic technology used to create sophisticated narrative environments through preprogrammed movement patterns such as one used in Disneyland or the Cadbury World. Entertainment robots are also popular in trade shows and museums. Promotional robots to attract and lure show attendees toward particular products or services or guide them to particular booths are an established technology. For such guides, possibly Intel Museum's A.I.-driven interactive robot, ARTI, is one of the most well-known. ARTI can recognize faces, understand speech, and update museum guests on the history of the museum and its founders [27].

Perhaps one of the hugely successful endeavors of entertainment robotics is the RoboCup challenge. RoboCup is an international completion of robotic soccer with the ambitious official RoboCup motto: "By the year 2050, develop a team of fully autonomous humanoid robots that can win against the human world soccer champions." Robot soccer is possibly one of the toughest tests of building autonomous robots. The robots need to have active perception and real-time sensor-based planning. Apart from being a source of entertainment, the games have driven unprecedented advances in robotics research [28].

6   Current Trends

Starting from a modest industrial robot in the early 60s, robotics has achieved brilliant successes in manufacturing as well as personal and service robots. Robots have proliferated everywhere in society: in entertainment, health care, rehabilitation, and other applications. In this section, we survey the current trend toward social robotics, biomimetic robots, and the emergence of cloud robotics.

6.1 Social Robotics

A social robot is "a physically embodied, autonomous agent that communicates and interacts with humans on an emotional level" [29]. Interaction with human or other physical embodied agents is usually through well laid down social behavior associated with the role of the social robot. Personal and sociable robotics is a new entrant. Only by early 2000, robots were seen as teammates and researchers started to focus on human interactive robots leading to the birth of social robotics.

With requirement of close human interaction, physical form of the social robot is of fundamental importance. Most of the social robots are humanoids. Japanese companies focus on humanoids driven by the conviction human embodiment can ensure better human–robot communicative. This is based "on the human innate ability to recognize humans and prefer human interaction" [30].

6.2 Biomimetic Robots

In the late 80s, Brooks proposed a robot control architecture that couples sensory information to action selection without a symbolic representation of an external world [31]. He termed it the subsumption architecture. Following the subsumption architecture, there was a surge of embodiment-inspired robotics. An approach to a nonrepresentational approach to artificial intelligence was the area of "Artificial Life." It involved arriving at a myriad of complex behaviors based on simple rules. However, the major focus was on biomimetics—the mimicry of biological process or a system. Biomimicry does not only involve mimicking the geometry but also the functionality [32].

Living species have adapted over time to the environment leading to an optimal design. It is but natural to copy that design! It is not now that people have tried to copy nature. Based on anatomical studies of birds, Leanardo da Vinci had made detailed illustrations of a flying machine. Researchers are focussing on biological inspiration to design and develop flying, crawling, and swimming automatons. These are biomimetic robots—robots that are built based on principles extracted from biological systems.

The recent surge in biomimetic robotics lies in availability of (a) huge amount of biological process data and (b) low-cost, power-efficient computational resources.

6.3   Cloud Robotics

Networking of machines in manufacturing automation systems is not new! More than 30 years ago, General Motors delved on the idea of a network of machines [33]. By the late 90s, sporadic usage of Web interfaces to control and tele-operate robots led to a field of robotics christened as Networked Robotics [34]. Simultaneously within the realm of Internet Services, the concept of "cloud" took shape. Cloud is "a model for enabling ubiquitous, convenient, on-demand network access to a shared pool of configurable resources (e.g., servers, storage, networks, applications, and services) that can be rapidly provisioned and released with minimal management effort or service provider interaction" [35]. Networked robot and automation systems are greatly enhanced by the cloud.

Exploiting computational and storage resources offered through cloud and taking advantage of the rapid progress in communications, an area of robotics is slowly emerging. Christened cloud robotics, it minimizes "onboard" computation bringing down associated overheads and dependence on custom middleware. Cloud brings with it the possibility to store and handle large datasets facilitating machine learning. The RoboEarth and RoboBrain databases are laudable efforts in this direction. The RoboBrain project "learns from publicly available Internet resources, computer simulations, and real-life robot trials" [36]. The RoboEarth project "stores data related to objects and maps for applications ranging from object recognition to mobile navigation to grasping and manipulation" [37].

7   Future Directions

It is definitely hard to say what the future holds for robotics. With rapid pace of technological growth within as well as in allied field such as machine learning and artificial intelligence, robotics is poised for even greater influence on our life in general. In this section, we review three areas of research that is going to make considerable impact on how robotics evolves in the near future.

7.1   Embodied Cognition

Until now, most of AI within robotics stayed true to this symbolic tradition. Stages in the Sense–Plan–Act cycle were clearly separated. Sensing leads to a set of symbols corresponding to features based on state of the external world. These symbols with other symbolic knowledge are used by the Planner to choose appropriate action. Flow of information is unidirectional with processing at each stage clearly demarcated. It was Brooks, who with his subsumption architecture proposed a radical departure from this view. Recent finding within the psychology and neuroscience community emphasizes on embodiment of intelligence. Intelligence including cognitive functions such as decision making, perception, and language is grounded in our physical presence rather than on abstract symbolic models [38]. Further, grasping and walking dynamics have been explored as embodied, non-symbolic intelligence.

Rather than sensor-driven smart robots, paradigms of embodied cognition could define the design of robots. Recent models from cognitive sciences and neuropsychology could bias human interactive robotics. A new dream is being woven —models of embodied cognition from different fields driving robotic developments in that area. For example, cognitive insights into an industrial expert could drive industrial robotics, the robot learning from the human expert, leading to a novel human–robot apprenticeship [39]. Such developments in embodied cognition with growing embodiment of AI and the advances in human interactive robotics would churn a paradigm shift in adoption of embodied cognition models for autonomous interactive robots.

7.2 Robotics and Internet of Things

Internet of Things (IoT) connects different entities (living or non-living), using different but interoperable communication protocols. IoT envisages "to connect everything and everyone everywhere to everything and everyone else" [40]. Robots are slowly becoming ubiquitous! It is but natural to expect a robot to be connected as a thing and establish connections with other things over the Internet leading to a robot-incorporated IoT. This could be either in the form of Internet of Robots or IoT-aided robotics applications. IoT technology would manage and magnify the already existing sensing, computing, and communication capabilities of modern robots enabling them able to execute complex and coordinated operations. IoT would bring in several players to complement the work primarily undertaken by a robot. Sensing, planning, and intelligent computational goals of robots would be better achieved through a dense IoT network [41].

Smart objects within technological environments such as shop floors in manufacturing would enable more detailed information to be captured and processed. Pervasive and ubiquitous smart objects integrated with a network of robots would lead to development of a number of novel services and applications.

Major challenges that need to be met for interaction among varied devices over a dense IoT network include (a) secure and stable communication paradigm and (b) unique yet sharable data representation. IoT would lay down the foundation of full decentralization of control [41]. Technology is in place for IoT and networking of robots. A careful investigation of requirements thereby working out means to put the two together is all that is required for IoT-enhanced robotics applications.

7.3   Robotics Through Synthetic Biology

Self-replicating machines, machines those are capable of producing a detached, functional copy of itself have been a long-cherished dream. John Von Neumann was the first to put forward a detailed scientific proposal for a non-biological self-replicating system [42]. There has been work on self-reproduction of physical machine [43] demonstrated construction of arbitrarily complex self-reproducing machines [44]. Work on self-assembly and self-replication has got a fillip with emergence of synthetic biology. Synthetic biology can broadly be described as the "design and construction of novel artificial biological pathways, organisms or devices, or the redesign of existing natural biological systems" [45]. Synthetic biology has made it possible to generate "diagnostic tools that improve the care of patients with infectious diseases, as well as devices that oscillate, creep, and play tic-tac-toe" [46].

Design of "custom-specified" cells has been possible because of rapid advancements in synthetic biology. These cells are most preferred in microscale systems because a. these are self-contained systems and b. have natural ability to respond to environmental cues. For example, synthetic gene networks have been designed to create cell-based memory units, data loggers, counters, edge detectors, and multi-input logic circuits as well as analog processing functions such as filtering and timing [47]. Striving for advancements in microrobotics exploiting progress in synthetic biology has significant potential [48]. A case in point is the synthetic biology project of Cyberplasm, a microscale robot. Robotics through synthetic biology is achieved "interfacing multiple cellular devices together, connecting to an electronic brain and in effect creating a multi-cellular bio-hybrid microrobot" [49].

8   Conclusion

A brief overview of the history, trends, and future direction is covered in this chapter. The developments in robotics starting from its conceptualization in science fiction to its current craze of IoT-based robotics are discussed. Many important developments have been highlighted, whereas many more have been ignored in view of space limitations. Robotics based on evolutionary computation, swarm robotics, and other such ideas driven through progress in computing has been ignored. Discussion on specialized robots such as for process industries and robots primarily composed of easily deformable matter is not included. Further, trend toward creation of human-like robots such as Nadine, the robot receptionist [50], is not highlighted. Nevertheless, a clear and concise picture of the interdisciplinary field of robotics has been presented. It is clear that the field of robotics requires collaborative efforts of various disciplines of engineering.

True and False Questions

1. Issac Asimov coined the term robot.
2. Unimate was the first industrial robot.
3. D-H notation is used for kinematic description of a robot.
4. Surgical robots were first used for brain biopsies.
5. MIT-MANUS is an entertainment robot.
6. ARTI is a well-known entertainment robot.
7. One of the most popular surgical robots is da Vinci.
8. Biomimicry involves mimicking only the geometry.
9. Networked robots are greatly enhanced by the cloud.
10. IoT would achieve sensing needs of robots better.

(Keys 1-F, 2-T, 3-T, 4-T, 5-F, 6-T, 7-T, 8-F, 9-T, 10-T)

Glossary

Actuators Actuators are the devices that convert some form of energy to mechanical energy.

Automaton It is a technique to automatically perform a number of functions following a predetermined sequence of operations.

Biomimetics Biomimetics is the creation of new materials, devices, and systems based on the concepts and principles derived from the nature.

Degrees of freedom Degrees of freedom are the number of independent movements of an object in a 3-D space.

End-effector Robot end-effector is a generic term to refer to the device attached at the distal end of a robotic manipulator. It is with the end-effector that the robot interacts with the environment.

Exteroceptor Exteroceptors are sensors that provide information on the conditions of a body vis-à-vis the environment.

Forward Kinematics Forward kinematics is the problem of computing the position and orientation of the end-effector given the set of joint angles.

Humanoid A Humanoid is a machine that resembles and possesses the characteristics of a human.

Internet of Things It is a network of Internet-connected physical entities such as objects and devices.

Inverse kinematics Inverse kinematics deals with computing all possible sets of joint angles to attain a prescribed position and orientation of the end-effector.

Kinematics Kinematics is the branch of dynamics that studies position, velocity, acceleration, and other higher order derivatives of the position variables without considering forces causing these effects.

Manipulator A robotic manipulator is a chain of links with controllable joints that enables placement of end-effectors in any position and orientation within the workspace.

Path Path is a sequence of robot configurations in a particular order without regard to the timing of these configurations.

Proprioceptor Proprioceptors are sensors that provide information on the internal conditions of a body.

Rehabilitation robotics It refers to application of robotic technologies in therapeutic procedures for augmenting rehabilitation.

Robot A robot is a reprogrammable, multifunctional manipulator designed to move material, parts, tools, or specialized devices for the performance of a variety of tasks.

Singularities A matrix called the Jacobian of the manipulator specifies a mapping from velocities in joint space to velocities in Cartesian space. Points at which this mapping is not invertible are called singularities.

Synthetic biology Synthetic biology is the design and construction of new biological entities such as enzymes, genetic circuits, and cells or the redesign of existing biological systems.

Trajectory It is a path that a moving object follows as a function of time.

Transducer Transducer is a device that transforms a physical quantity into the form of an electrical signal or vice versa.

Workspace It is the region of space in which the end-effector can attain any desired position and orientation.

References

1. Lord, C. (1984). *The Politics* (Aristotle, Trans.). Chicago: University of Chicago Press.
2. Hobbes, T., & Gaskin, J. C. A. (1998). *Leviathan*. Oxford: Oxford University Press.
3. Nocks, L. (2007). *The robot: The life story of a technology*. Greenwood Technographies: Greenwood Press, London.
4. Jazar, R. N. (2010). *Theory of applied robotics: Kinematics, dynamics, and control* (2nd ed). Springer.
5. Dixit, U. S., Hazarika, M., Davim, J. P. (2016). *A brief history of mechanical engineering*. Springer.
6. Craig, John J. (1989). *Introduction to robotics: Mechanics and control* (2nd ed.). Inc, Boston, MA, USA: Addison-Wesley Longman Publishing Co.

7. Tsai, L. W. (1999). *Robot analysis: The mechanics of serial and parallel manipulators*. New York: Wiley.
8. Lewis, F. L., Dawson, D. M., & Abdallah, C. T. (2003). *Robot manipulator control: Theory and practice*. New York: CRC Press.
9. Corke, P. (2007). A simple and systematic approach to assigning Denavit-Hartenberg parameters. *IEEE Transactions on Robotics, 23*(3), 590–594.
10. Denavit, Jacques, & Hartenberg, R. S. (1955). A kinematic notation for lower-pair mechanisms based on matries. *Transactions on ASME Journal of Applied Mechanics, 23*, 215–221.
11. Laurent P., (2013). *Advanced manufacturing measurement technologies and robotics*. Business Innovation Observatory Contract No. 190/PP/ENT/CIP/12/C/N03C01.
12. Friis, D. (Ed.). (2016*). World robotics report 2016*. International Federation of Robotics.
13. Tobe, F. (2015). *Why Co-Bots Will be a huge innovation and growth driver for robotics industry*. IEEE Spectrum.
14. Wilson, M. (2002). Robots at the heart of the welding system. *Industrial Robot: An International Journal, 29*(2), 91–92.
15. Anandan, T. (2014). *Robotic assembly: Shrinking footprint, expanding market*. Robotics Industry Insights: RIA Robotics Online.
16. Yoshida, K., Nenchev, D., Ishigami, G., & Tsumaki, Y. (2014). Space robotics. In *International Handbook of Space Technology* (pp. 541–573).
17. Gao, Y., & Chien, S. (2017). Review on space robotics: Toward top-level science through space exploration. *Science Robotics, 2*(7).
18. Zielinska, T. (2016). Professional and personal service robots. *International Journal of Robotics Applications and Technologies (IJRAT), 4*(1), 63–82.
19. Lanfranco, A. R., Castellanos, A. E., Desai, J. P., & Meyers, W. C. (2004). Robotic surgery: A current perspective. *Annals of Surgery, 239*(1), 14–21.
20. Kazanzides P., Zuhars, J., Mittelstadt, B., Williamson, B., Cain, P., Smith, F., et al. (1992). Architecture of a surgical robot. In *IEEE International Conference on Systems, Man and Cybernetics* (pp. 1624–1629).
21. Ghodoussi, M., S. E. Butner, Y. Wang (2002). Robotic surgery—the transatlantic case. In *IEEE International Conference on Robotics and Automation* (pp. 1882–1888).
22. Guthart G. S., & Salisbury J. K., Jr. (2000). The IntuitiveTMtelesurgery system: Overview and application. In *IEEE International Conference on Robotics and Automation*, 2000 (pp. 618–621).
23. Bergeles, C., & Yang, G. Z. (2014). From passive tool holders to microsurgeons: Safer, smaller, smarter surgical robots. *IEEE Transactions on Biomedical Engineering, 61*(5), 1565–1576.
24. Hogan, N., Krebs, H. I., Charnnarong, J., Srikrishna, P., & Sharon, A. (1992). MIT-MANUS: A workstation for manual therapy and training. I. In *IEEE International Workshop on Robot and Human Communication*, 1992 (pp. 161–165).
25. Guzmán-Valdivia, C. H., Blanco-Ortega, A., Oliver-Salazar, M. A., Gómez-Becerra, F. A., & Carrera-Escobedo, J. L. (2015). HipBot—The design, development and control of a therapeutic robot for hip rehabilitation. *Mechatronics, 30*, 55–64.
26. Johnson, M., Micera, S., Shibata, T., & Guglielmelli, E. (2008). Rehabilitation and assistive robotics. *IEEE Robotics and Automation Magazine, 15*(3), 16–110.
27. Teo, G., & Reinerman-Jones, L. (2014). Robot behavior for enhanced human performance and workload. In *International Conference on Virtual, Augmented and Mixed Reality* (pp. 117–128). Springer International Publishing.
28. Veloso, M. M. (2002). Entertainment robotics. *Communications of the ACM, 45*(3), 59–63.
29. Darling, K. (2016). Extending legal protections to social robots: The effects of anthropo-morphism, empathy, and violent behavior towards robotic objects. In M. Froomkin, R. Calo, & I. Kerr (Eds.), *Robot law*. Edward Elgar.
30. Kanda, T., & Ishiguro, H. (2016). *Human-robot interaction in social robotics*. New York: CRC Press.

31. Brooks, R. A. (1991). Intelligence without representation. *Artificial Intelligence, 47,* 139–159.
32. Lepora, N. F., Verschure, P., & Prescott, T. J. (2013). The state of the art in biomimetics. *Bioinspiration & Biomimetics, 8*(1).
33. Irwin, J. D. (1997). *The industrial electronics handbook.* Boca Raton, FL, USA: CRC Press.
34. Goldberg, K., & Siegwart, R. (Eds.). (2002). *Beyond webcams: An introduction to online robots Cambridge.* MA, USA: MIT Press.
35. Mell, P., & Grance, T. (2009). *The NIST definition of cloud computing.* National Institute of Standards and Technology.
36. Saxena, A., Jain, A., Sener, O., Jami, A., Misra, D. K., & Koppula, H. S. (2014). Robobrain: Large-scale knowledge engine for robots. *arXiv preprint arXiv:1412.0691.*
37. Waibel, M., Beetz, M., Civera, J., D'Andrea, R., Elfring, J., Gálvez-López, D., et al. (2011). RoboEarth. *IEEE Robotics and Automation Magazine, 18*(2), 69–82.
38. Wilson, M. (2002). Six views of embodied cognition. *Psychonomic Bulletin & Review, 9*(4), 625–636.
39. Hoffman, G. (2012). Embodied cognition for autonomous interactive robots. *Topics in Cognitive Science, 2012,* 1–14.
40. Atzori, L., Iera, A., & Morabito, G. (2010). The internet of things: A survey. *Computer Network, 54*(15), 2787–2805.
41. Grieco, L. A., Rizzo, A., Colucci, S., Sicari, S., Piro, G., Paola, D. Di, et al. (2014). IoT-aided robotics applications: Technological implications, target domains and open issues. *Computer Communications, 54,* 32–47.
42. Von Neumann, J. (1966). The theory of self-reproducing automata. In Burke, A. W. (Ed.), *Essays on cellular automata* (pp. 4–65). Illinois: University of Illinois.
43. Chirikjian, G. S., Zhou, Y., & Suthakorn, J. (2002). Self-replicating robots for lunar development. *IEEE/ASME Transactions on Mechatronics, 7*(4), 462–472.
44. Zykov, V., Efstathios, M., Bryant, A., & Lipson, H. (2005). Robotics: Self-reproducing machines. *Nature, 435*(7039), 163–164.
45. Hu, X., & Rousseau, R. (2015). From a word to a world: The current situation in the interdisciplinary field of synthetic biology. *PeerJ, 3,* e728.
46. Benner, S. A., & Sismour, A. M. (2005). Synthetic biology. *Nature Reviews Genetics, 6*(7), 533.
47. Steager, E. B., Wong, D., Mishra, D., Weiss, R., & Kumar, V. (2014). Sensors for micro bio robots via synthetic biology. In *IEEE International Conference on Robotics & Automation (ICRA),* Hong Kong (pp. 3783–3788).
48. Sakar, M. S., Steagar, E. B., Kim, D. H., Julius, A. A., Kim, M. J., & Kumar, V. (2010). Biosensing and actuation for microbiorobots. In *IEEE International Conference on Robotics and Automation, Anchorage, AL* (pp. 3141–3146).
49. Lu, J., Yang, J., Kim, Y. B., & Ayers, J. (2012). Low power, high PVT variation tolerant central pattern generator design for a bio-hybrid micro robot. In *Circuits and systems (MWSCAS 2012), IEEE 55th International Midwest Symposium* (pp. 782–785).
50. Thalmann, N. M., Tian, Li, & Yao, F. (2017). Nadine: A social robot that can localize objects and grasp them in a human way. In *Frontiers in electronic technologies: Trends and challenges.* Singapore: Springer.

Chapter 8
Computer Vision in Industrial Automation and Mobile Robots

Frederico Grilo and Joao Figueiredo

Abstract Computer vision is presently a very relevant and important tool in both industrial manufacturing and mobile robots. As human vision is the most relevant sense to feed the brain with environmental information for decision making, computer vision is nowadays becoming the main artificial sensor in the domains of industrial quality assurance and trajectory control of mobile robots.

Keywords Computer vision · Industrial automation · Mobile robots

1 Computer Vision in Industrial Automation

1.1 Artificial Vision in Automation

Distributed systems began to be used in the telecommunication sector, pushed by the generalized spread of computers and their need to be interconnected. This context originated the rise of several network topologies. Distributed strategies were rapidly extended to other domains, as they bring several advantages. The distributed strategy has very interesting characteristics as it shares several resources, allowing a more rational distribution among the users. This philosophy presents enormous economical benefits in comparison to traditional centralized systems, where the same resources have to be multiplied.

Decentralized management of systems is nowadays an important development tool. This strategy reaches different fields, from agriculture to industry, building automation, etc. [1, 2].

Artificial vision and image processing have already an important role in several technical domains as: (i) pattern recognition, with multiple applications in industrial

F. Grilo · J. Figueiredo (✉)
CEM-IDMEC, University of Évora, R. Romão Ramalho, 59, 7000-671 Évora,
Portugal
e-mail: jfig@uevora.pt

© Springer International Publishing AG, part of Springer Nature 2018
J. P. Davim (ed.), *Introduction to Mechanical Engineering*, Materials Forming,
Machining and Tribology, https://doi.org/10.1007/978-3-319-78488-5_8

quality control [3], medical diagnosis [4], geometry restoration [5]; (ii) remote
displacement sensing [6]; (iii) visual servoing [7]; (iv) object tracking [8]; etc.

Despite the broad application of artificial vision systems in different domains,
their usual topology is as a centralized autonomous system. Here also, the advantages of distributed management systems can be applied with great advantages.

In this chapter, a quality control system implemented through artificial vision
technology is shown. The novelty of this system is its ability to be managed by a
completely decentralized environment and shared among several distributed users.

Experimental tests performed on a laboratory prototype are presented.

1.2 System Model

In advanced industrial systems, each automatic production unit is connected to a
global management platform through a typical PLC-master–slave network. The
global management platform—usually a SCADA system (supervisory control and
data acquisition) enlarges the system communication capabilities, allowing on-line
monitoring and control, events recording, alarm management, etc. [9].

The set of machines in one production unit constitutes an autonomous assembly
line.

In each assembly line, there exists an industrial master–slave network connecting
all PLCs from the several automatic machines. The EOL (end-of-line machine),
where the final tests are performed, is usually configured as the master PLC, for
each assembly unit. All other machines in the same production line are configured
as Slave-PLCs.

Figure 1 shows a typical industrial communication network implemented
through Profibus DP [10].

The typical use of automatic vision systems in industrial automation is as centralized autonomous systems. This autonomous centralized topology is illustrated in

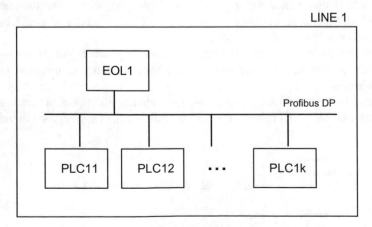

Fig. 1 Industrial Profibus-DP master–slave network in an automatic production line

Fig. 2. In this topology, only one user can access the vision system, and this system is completely managed by this system, usually configured as PLC-master in relation to its vision system.

1.3   Vision System Strategy

1.3.1   Control Strategy

We developed a distributed strategy that implements the vision system as a common intelligent sensor in an industrial network, in order that all network PLC-users can access this vision sensor, allowing the upload of different inspection programs in a decentralized environment. In this context, the vision system performs itself as any usual sensor/actuator inserted in an industrial network.

Figure 3 shows the communication strategy developed, where each vision system is accessible by all users in the network.

1.3.2   Implementation of the Vision System

The distributed vision system is implemented through a Profibus-DP industrial network, where the vision system is configured as a typical slave. However, this slave is not allocated to a single master but to several intelligent slaves. The vision system is composed of a Siemens VS710 vision sensor with 256 gray levels. This vision

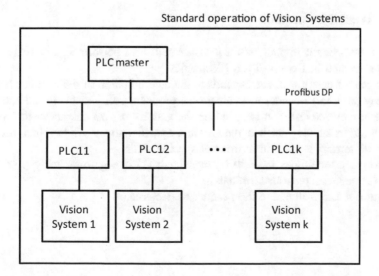

Fig. 2 Standard operation of vision systems in a centralized automatic production line

Vision Systems in a distributed industrial network

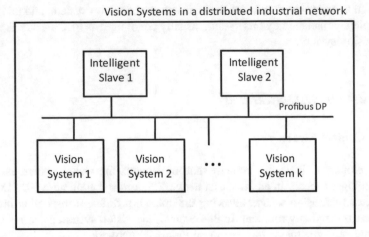

Fig. 3 Vision system in a distributed industrial network

sensor is an integrated system with a similar PC architecture that loads a similar DOS operative system, with the minimal requirement needs. This vision sensor has a solid state disk where the several user-inspection programs are stored. The selection of the running program is made by any of the intelligent slave-PLCs through a specific command, using the Profibus industrial network.

1.4 Laboratory Prototype

1.4.1 Overall Description

The prototype built implements a measurement task performed at a relay manufacturer, located in Évora—Tyco Electronics.

The measurement task corresponds to the determination of the distance between the relay terminals in order to evaluate the product quality status (good/bad part).

The acceptance criterion is based on the validity of two characteristics: (i) The number of terminals matches the product specifications; (ii) terminal distances among all terminals are within interval specifications.

Two relay topologies have to be evaluated: (1) change-over relay (five terminals); (2) make relay (four terminals).

Figures 4 and 5 illustrate the product characteristics.

Fig. 4 Overview of a power relay 12 V/20 A, for automotive industry

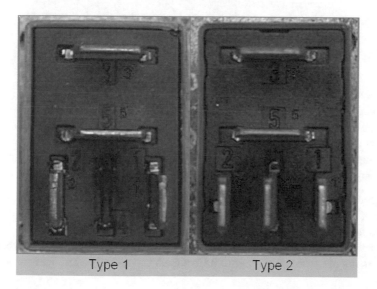

Fig. 5 Bottom view of two different types of power relay: Type 1 = make (four terminals); Type 2 = change-over (five terminals)

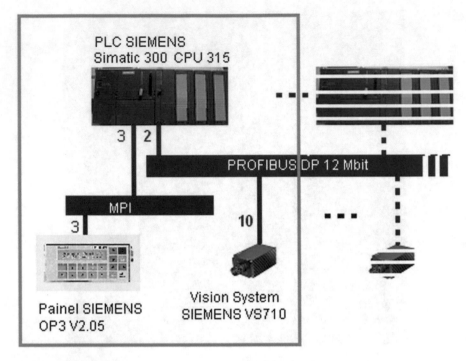

Fig. 6 Measurement application configured for a distributed industrial network

1.4.2 Hardware Requirements

The developed system for the measurement of the relay characteristics is composed of:

(1) a Siemens PLC S7-315, configured as an intelligent slave system;
(2) a vision sensor Siemens VS-710, configured as a slave system;
(3) a human machine interface Siemens OP3, connected with the PLC with a multi-point interface;
(4) a digital input operator panel to control the main system actuations (start/stop/emergency).

Figure 6 shows the implemented application where the enclosed rectangle identifies the distributed strategy described in Sect. 1.3.2. This distributed control application runs over a Profibus industrial network [10].

Figure 7 shows the actual appearance of the complete system, where the several sub-systems are referenced. Figure 8 illustrates the vision system (sensor and light source).

Fig. 7 Vision system—complete prototype: (1) intelligent slave-PLC Siemens S7-315; (2) vision sensor Siemens VS710; (3) Profibus network; (4) human machine interface Siemens Op3 connected to PLC CPU through MPI interface; (5) digital input operator panel to control the main system actuations (start/stop/emergency)

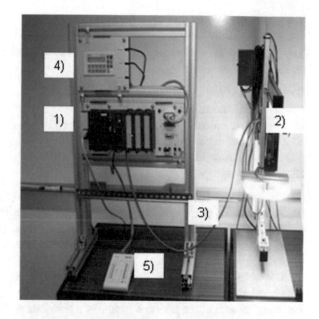

Fig. 8 Vision system: (a) vision sensor Siemens VS710; (b) light system with a fluorescent light source controlled by an electronic coil to avoid the flik effect; (c) monitor that displays the processing image results; (d) relay support

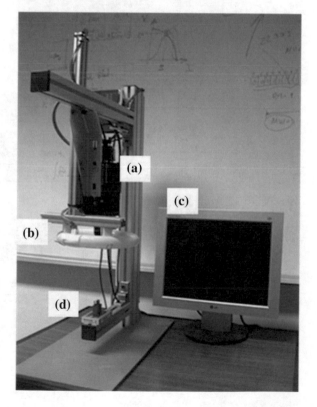

1.4.3 PLC Application Development

The PLC used to implement this vision system was the Siemens S7-300, composed of the following slots:

Slot1 = power supply module
Slot2 = CPU module 315-2DP
Slot4 = communication module (Profibus net)
Slot5 = digital input/output module
Slot6 = digital input/output module
Slot7 = analog input/output module

We used the Siemens software Simatic Step 7 [11] to implement the PLC application. The programming language used was the ladder diagram [12].

Figure 9 shows the control algorithm developed using the Grafcet methodology —sequential function chart. This Grafcet shows two areas: gray background and white background.

The operations done in the gray area refer to the control authorization and synchronism mode to be accomplished by any intelligent PLC-slave in the Profibus network, to access the vision system, in a distributed environment.

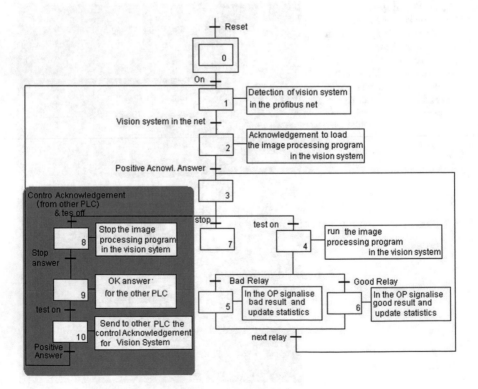

Fig. 9 Grafcet for the PLC application

The operations done in the white area refer to the standard sequence of specific tasks for the image processing application.

1.4.4   Image Processing Algorithm

The VS710 vision sensor [13] has the capability to store different image processing programs. According to the intelligent slave request, one of these programs is loaded into the sensor CPU. The program loading is similar to the usual process of uploading new device firmware.

Two image processing applications were developed. The first application verifies the number of terminals existing in the tested relay.

The second application measures the distances among all the existing terminals. Figure 10 shows the developed Grafcet for the image processing application concerning the relay type 2 (five terminals)—see Sect. 1.4.1.

Both image applications were developed using the Siemens ProVision software [13].

Figure 11 shows the work environment of the developed application.

1.5   Experimental Results

The laboratory prototype confirmed the correct performance of the developed image processing application identifying the actual relay type (correct evaluation of the number of terminals) and measuring the distances among terminals, with a good accuracy.

Referring the sharing of the vision system among different intelligent slave-PLCs, the prototype demonstrated also the correct commutation between different testing programs previously stored in the vision sensor.

The developed image processing application, which is a medium industrial complexity algorithm, had shown a total processing time of about 1.5 s for both the measurement tasks and for the upload of the corresponding processing program, implemented on the Siemens VS710 vision sensor.

Extrapolating this result for N intelligent slave-PLCs, sharing the same vision sensor, on the Profibus network, the correspondent time needed for the exchange of one PLC_i client, T_{vcs_i}, is calculated in Eq. (1):

$$T_{vcs_i} = t_{ps} + t_{cA_i} + \sum_{n=1}^{N} t_{s_n} - t_{s_i} \qquad (1)$$

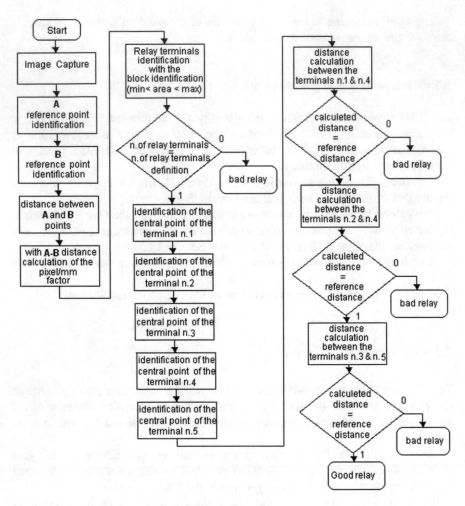

Fig. 10 Image processing algorithm—relay type 2

where:

t_{ps} time needed for the upload of a new program in the vision sensor;

t_{cA_i} time needed for the request of the vision sensor from a new intelligent PLC$_i$-slave;

t_{S_i} time for synchronizing with the new intelligent PLC$_i$-slave.

Considering now that the complete system integrates N intelligent PLC-slaves, the necessary time for changing the control of the vision sensor among the different PLCs, $T_{VCS_{total}}$, is given by:

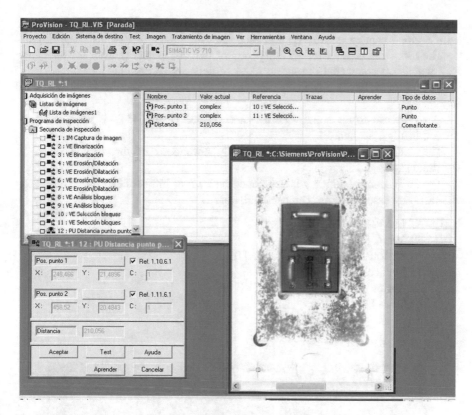

Fig. 11 Work environment of the developed application

$$T_{\mathrm{vcs}_{\mathrm{total}}} = t_{\mathrm{ps}} + t_{\mathrm{cA}_i} + \sum_{n=1}^{N-1} t_{\mathrm{s}_n} \qquad (2)$$

As the synchronizing time proved to be very small in comparison to the total processing time, it is verified that the sharing of the vision sensor among several intelligent PLC-slaves is very advantageous. In fact, the additional required computational work needed to implement the distributed strategy is largely compensating.

2   Computer Vision in Mobile Robots

2.1   Characterization of the AGV

The automatic guided vehicle—AGV—presented in this chapter is composed of two main systems: (1) the motor system and the (2) task-processor system.

The motor system is the AGV motorized platform that carries on a computer with a video camera (slave-PC) that processes the image and controls the motors.

The task-processor system is composed of two computers (master PC and slave PC) which are connected through a wireless network (XBee) [14].

The motor system is physically composed of five main components, as illustrated in Fig. 12.

One of these components—the motor controller board [15]—is deeper characterized in Figs. 13 and 14.

The task-processor system is physically composed of three main parts, as illustrated in Fig. 15:

The AGV control strategy is developed on a set of industrial software hosted on the task-processor system. The used software is:

  (i)   NI LabVIEW [16];
 (ii)   NI VISA [16];
(iii)   NI Vision Acquisition Software [16];
 (iv)   NI Device Drivers [16];
  (v)   NI Vision Development Module [16];
 (vi)   Arduino Uno [17].

Fig. 12 Moto-system characterization: (1) metallic structure; (2) motor controller board; (3) DC motor; (4) 24VDC battery; (5) Arduino controller

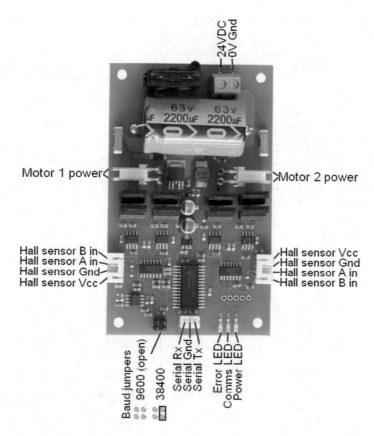

Fig. 13 Motor controller board [13]

2.2   *The Control Strategy of the AGV—Master/Slave*

The developed strategy for the AGV motion control is based on two main modes:

(i) Manual mode—which allows the user to control the AGV, through the PC-master, using four main commands (move forward—move backward—turn left—turn right);

(ii) Automatic mode—which shifts the complete movement control to the PC-slave, and the AGV performs automatically the tracking of a black stripe marked on the floor, by image processing technology.

Figure 16 illustrates the control strategy of the AGV.

Figure 17 characterizes the communication flow between PC-master and PC-slave.

command	Name	Bytes sent to MD49	Bytes returned by MD49	Description
0x21	GET SPEED 1	2	1	returns the current requested speed of motor 1
0x22	GET SPEED 2	2	1	returns the current requested speed of motor 2
0x23	GET ENCODER 1	2	4	motor 1 encoder count, 4 bytes returned high byte first (signed)
0x24	GET ENCODER 2	2	4	motor 2 encoder count, 4 bytes returned high byte first (signed)
0x25	GET ENCODERS	2	8	returns 8 bytes - encoder1 count, encoder2 count
0x26	GET VOLTS	2	1	returns the input battery voltage level
0x27	GET CURRENT 1	2	1	returns the current drawn by motor 1
0x28	GET CURRENT 2	2	1	returns the current drawn by motor 1
0x29	GET VERSION	2	1	returns the MD49 software version
0x2A	GET ACCELERATION	2	1	returns the current acceleration level
0x2B	GET MODE	2	1	returns the currently selected mode
0x2C	GET VI	2	3	returns battery volts, motor1 current and then motor2 current
0x2D	GET ERROR	2	1	returns a byte within which the bits indicate errors on the MD49
0x31	SET SPEED 1	3	0	set new speed1
0x32	SET SPEED 2 / TURN	3	0	set new speed2 or turn
0x33	SET ACCELERATION	3	0	set new acceleration
0x34	SET MODE	3	0	set the mode
0x35	RESET ENCODERS	2	0	zero both of the encoder counts
0x36	DISABLE REGULATOR	2	0	power output not changed by encoder feedback
0x37	ENABLE REGULATOR	2	0	power output is regulated by encoder feedback
0x38	DISABLE TIMEOUT	2	0	MD49 will continuously output with no regular commands
0x39	ENABLE TIMEOUT	2	0	MD49 output will stop after 2 seconds without communication

Fig. 14 Set of commands for the motor controller board [13]

2.2.1 Master LabView Program

This program runs in the master-PC, that stays on a desk, and it is responsible for the main control of the AGV. Basically, it has two command modes, as referred in 2.2.2: manual and automatic.

The GUI developed for the master-PC is shown in Fig. 18.

The base structure of the developed program is a "Flat Sequence Structure" composed of two sub-structures. This program initiates the wireless XBee communication, defines the main variables (default velocity, forward move, backward move), and selects the manual mode (0/1) or the automatic mode (0/1).

Figure 19 illustrates a small part of the developed LabView code for the mode selection: manual/automatic.

Fig. 15 Task-processor system characterization: (1) PC-master; (2) PC-slave with video camera; (3) XBee wireless network

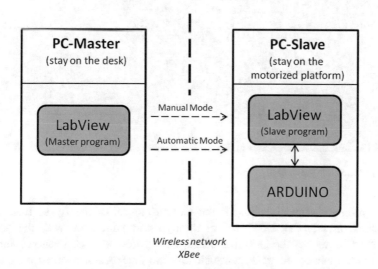

Fig. 16 Control strategy of the AGV

2.2.2   Slave LabView Program

This program runs in the slave-PC, that stays on the AGV platform, and it moves together with the AGV.

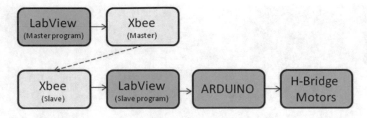

Fig. 17 Communication flow: PC-master/PC-slave

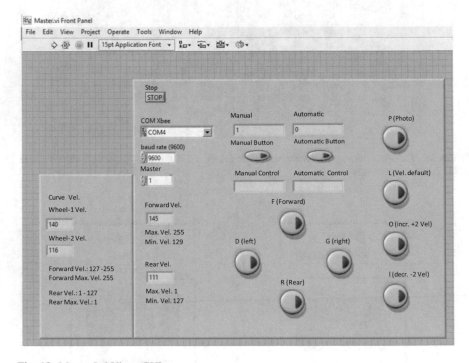

Fig. 18 Master LabView: GUI

This program is responsible for the motion control of the AGV. Basically, it commands directly the AGV motors through communication with the Arduino processor that attacks the H-bridge board, and it processes the image captured by the PC video camera to track automatically the black stripe marked on the floor (automatic mode).

The GUI developed for the slave-PC is shown in Fig. 20.

The base structure of the developed program is a "Flat Sequence Structure" composed of three sub-structures. This program initiates the wireless XBee communication, the Arduino communication, defines the main variables, and processes the captured images.

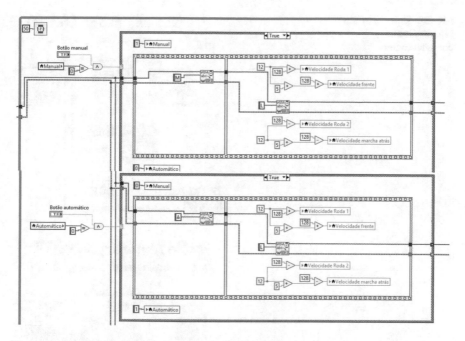

Fig. 19   Master LabView: mode selection manual/automatic

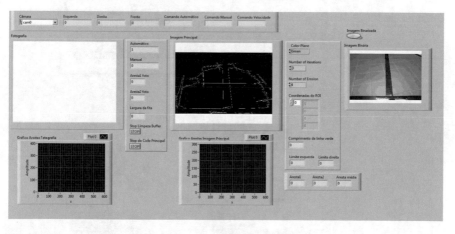

Fig. 20   Slave LabView: GUI

Figure 21 illustrates a small part of the developed LabView code for the initialization of the communications.

Fig. 21 Slave LabView:
initialization of
communications

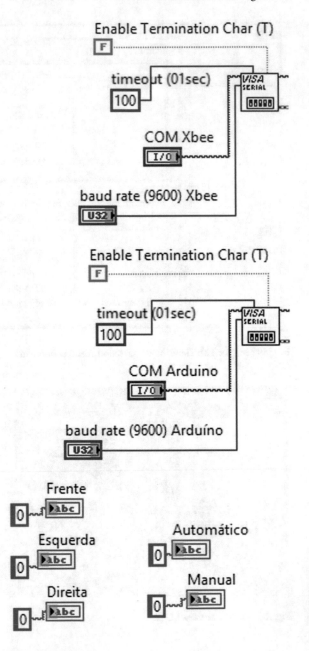

Fig. 21 Slave LabView: initialization of communications

2.2.3 Arduino Program

The Arduino processor is responsible for the serial communication between the
slave LabView program and the H-bridge board that controls the two DC motors.

According to the specifications of the H-bridge board (see Fig. 14), a program was developed to command the two DC motors in order to allow the AGV to perform its four main movements: forward, backward, turn left, turn right.

The initial part of this program, where the variables are defined according to the H-bridge specs (see Fig. 14), is shown in Fig. 22.

Figure 23 exemplifies a part of the program where the forward movement is processed.

Fig. 22 Arduino program: variables definition

```
#define CMD    (byte) 0x00
#define GET_SPEED1        0x21
#define GET_SPEED2        0x22
#define GET_ENC1          0x23
#define GET_ENC2          0x24
#define GET_ENCS          0x25
#define GET_VOLTS         0x26
#define GET_I1            0x27
#define GET_I2            0x28
#define GET_VER           0x29
#define GET_ACCEL         0x2A
#define GET_MODE          0x2B
#define GET_VI            0x2C
#define GET_ERROR         0x2D
#define SET_SPEED1        0x31
#define SET_SPEED2        0x32
#define SET_ACCEL         0x33
#define SET_MODE          0x34
#define RESET_ENCS        0x35
#define DISABLE_REG       0x36
#define ENABLE_REG        0x37
#define DISABLE_TIME      0x38
#define ENABLE_TIME       0x39
```

Fig. 23 Arduino program: forward movement

```
// Controlo e condução
else if (letra=='F' )
{
            Serial.println("Em frente");
            mySerial.write(CMD);
            mySerial.write(SET_ACCEL);
            mySerial.write(10);
            mySerial.write(CMD);
            mySerial.write(SET_SPEED1);
            mySerial.write(128+velocidade+5);
            mySerial.write(CMD);
            mySerial.write(SET_SPEED2);
            mySerial.write(128+velocidade+5);



}
```

2.3 Image Processing for Automatic Path Tracking

The image processing is responsible for the automatic path tracking. When the AGV is running in the automatic mode, the movement control of the AGV is switched to the slave LabView program. The aim of this image processing is to generate the appropriate movements to command the two DC motors in order to follow a black stripe marked on the floor (Fig. 24).

Fig. 24 Path marked on the floor with a black stripe

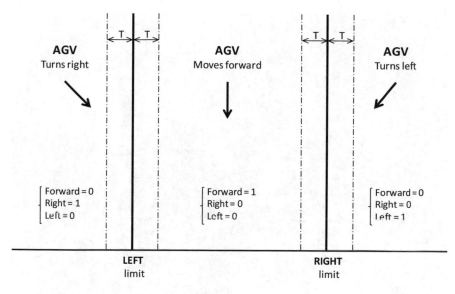

Fig. 25 Border lines for AGV path tracking (T = Tolerance)

For the image processing algorithm, it used a set of functions that are available at the *NI Vision Development Module* [16].

Basically, the main sequential operations performed on each captured image are: (i) image capture, (ii) conversion to 8-bit image, (iii) gray image processing, (iv) image erosion, (v) image edge detection.

These main transformations are used to identify the black stripe marked on the floor. The criterion to guide the vehicle along this black line is to maintain the relative location of this line between two defined border lines—left and right lines. The allowable distance between this two border lines is defined by the user (see Fig. 25).

Fig. 26 Image gray morphology processing

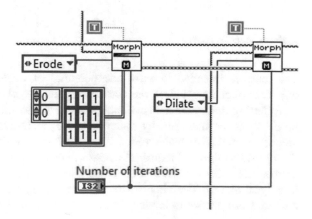

Fig. 27 AGV performing path tracking (in-door test)

In Fig. 26, it is shown a part of the developed program, where the image gray morphology is modified.

Several in-door tests were performed with different parameter variables: room luminosity, floor roughness, different paths, average speed.

The good performance of the AGV tracking the black line under the above mentioned different conditions proved the robustness of the developed program.

Figure 27 shows the AGV performing an in-door test.

2.4 Conclusions

This chapter intended to show the potential of image processing on industrial quality applications and on the automatic guidance of autonomous vehicles.

Referring first to the industrial automation, the novelty of the presented system is its ability to be managed by a completely decentralized environment and shared among several distributed users. The distributed vision system is implemented through a Profibus-DP industrial network, where the vision system is configured as a typical slave. However, this slave is not allocated to a single master but to several intelligent slaves. The obtained results show that the additional required computational work needed to implement the distributed strategy is largely compensating.

This distributed strategy permits a significant economic reduction in the necessary vision hardware, as the same system can be accessed by multiple users, configured in the same industrial communications network. It is important to define the sharing topology as well as the characteristics of the network where the distributed management is implemented. The selection of the network characteristics, among the several available options: Industrial Ethernet, Profibus; CAN bus, etc., is mainly dependent on the application specific requirements as well as on the economic factors.

Finally, referring to the guidance of AGVs, this study shows a simple way to implement an AGV.

The developed vehicle is composed of two main systems: the motor system and the task-processor system. The motor system is the AGV motorized platform that carries on a computer with a video camera (slave-PC) that processes the image and controls the motors. The task-processor system is composed of two computers (master-PC and slave-PC) which are connected through a wireless network (XBee).

The developed strategy for the AGV motion control is based on two main modes: (i) manual mode—which allows the user to control the AGV, through the PC-master, (ii) automatic mode—which shifts the complete movement control to the PC-slave, and the AGV performs automatically the tracking of a black stripe marked on the floor, by image processing technology.

The good performance of the AGV under the several performed in-door tests proved the robustness of the developed control strategy.

Questions

1. **Why is computer vision a key factor in automatic assembly?**

   (a) Because there is not enough personal.
   (b) Because it is a very robust and fast method that can be used for parameter control in mass production.
   (c) Because some operators can have vision problems.

   Answer: (b)

2. **Why are distributed systems gaining major influence in today management?**

   (a) The distributed strategy has very interesting characteristics as it shares several resources, allowing a more rational distribution among the users. This philosophy presents enormous economical benefits in comparison to traditional centralized systems, where the same resources have to be multiplied.
   (b) Because communication companies are pushing their services.
   (c) Although centralized management strategies are more profitable, the market prefers distributed systems.

   Answer: (a)

3. **What is an industrial vision sensor?**

   (a) It is an analog sensor which voltage output changes for different images.
   (b) It is a sensor that blinks for different images.
   (c) It is a digital sensor mainly composed of an image frame with n-columns and m-rows, corresponding to $n \times m$ pixels. For the monochromatic frames, each pixel has usually 8-bits in gray scale; for the polychromatic frames each pixel has usually 3×8-bits in RGB system (red–green–blue).

   Answer: (c)

4. **The acceptance criterion for products inspected by image processing methods is based on …**

   (a) the product's image similarity with an image pattern.
   (b) probability models.
   (c) mathematical laws.

   Answer: (a)

5. **What is a mobile robot?**

   (a) It is a robot that manipulates products while is driven by an operator.
   (b) It is an automated guided vehicle commonly known as an AGV.
   (c) It's a new trend of robotics.

   Answer: (b)

6. **What is the typical topology used for automatic vision systems in industrial automation?**
   Answer: The typical use of automatic vision systems in industrial automation is as centralized autonomous systems. In this topology, only one user can access the vision system, and this image sensor is completely managed by this system, usually configured as PLC-master in relation to its vision sensor.

7. **In which practical fields is pattern recognition processed by computer vision a common automatic methodology?**
   Answer: Image processing is a common tool used for pattern recognition, in several domains such as industrial quality control, medical diagnosis, geometry restoration.

8. **What are some typical filter/operators used in image processing algorithms?**
   Answer: Some of the common filter/operators used in image processing algorithms are: conversion to 8-bit image, gray image processing, image erosion, image edge detection.

9. **How is usually identified a pre-defined path to be tracked by an AGV?**
   Answer: Usually, a black stripe is marked on the plant floor.

10. **What are the usual available modes to control an AGV?**
    Answer: The two usual modes to control an AGV are: (i) *manual mode*—which allows the user to control the AGV, through the system-master, (ii) *automatic mode*—which shifts the complete movement control to the AGV, and the vehicle performs autonomously the path tracking by image processing technology.

Glossary

Computer vision Computer vision is an interdisciplinary field that deals with how computers can be made for gaining high-level understanding from digital images or videos. From the perspective of engineering, it seeks to automate tasks that the human visual system can do.

Pattern recognition Pattern recognition is a branch of machine learning that focuses on the recognition of patterns and regularities in data. This scientific and technological domain has great applicability in computer vision.

Industrial automation Industrial automation deals primarily with the automation of manufacturing, quality control, and material handling processes. Industrial automation is to replace the decision making of humans and manual command-response activities with the use of mechanized equipment and logical programming commands. One trend is the increased use of machine vision to provide automatic inspection and robot guidance functions.

AGV An automated guided vehicle or automatic guided vehicle (AGV) is a portable robot that follows markers or wires in the floor, or uses vision, magnets, or lasers for navigation. They are most often used in industrial applications to move materials around a manufacturing facility or warehouse.

Master–Slave control Master–slave is a model of communication where one device or process has unidirectional control over one or more other devices. In some systems, a master is selected from a group of eligible devices, with the other devices acting in the role of slaves.

References

1. Figueiredo, J., Ayala Botto, M., & Rijo, M. (2013). SCADA system with predictive controller applied to irrigation canals. *Control Engineering Practice, 21*(2013), 870–886. http://dx.doi.org/10.1016/j.conengprac.2013.01.008. (Elsevier).
2. Figueiredo, J., & Sá da Costa J. (2012). A SCADA system for energy management in intelligent buildings. *Energy and Buildings, 49*(2012), 85–98. http://dx.doi.org/10.1016/j.enbuild.2012.01.041). (Elsevier).
3. Du, C., & Sun, D. (2008). Multi-classification of pizza using computer vision and support vector machine. *Journal of Food Engineering, 86*(2008), 234–242.
4. Gouttière, C., & Coninck, J. (2007). Detection of synthetic singularities in digital mammographies using spherical filters. In *Proceedings of the ECCOMAS Thematic Conference On Computational Vision and Medical Image Processing*. Porto, Portugal.
5. Hlou, L., Lichioui, A., & Guennoun, Z. (2003). Degraded 3D-objects restoration and their envelope extraction. *International Journal of Robotics and Automation, 18*(2).
6. Lee, J., & Shinozuka, M. (2006). A vision-based system for remote sensing of bridge displacement. *NDT&E International, 39*(2006), 425–431. (Elsevier).

7. Nelson, B. (2003). A distributed framework for visually servoed manipulation using an active camera. *International Journal of Robotics and Automation, 18*(2).
8. Cui, J., Zha, H., Zhao, H., & Shibasaki, R. (2008). Multi-modal tracking of people using laser scanners and video camera. *Image and Vision Computing, 26*(2008), 240–252.
9. Figueiredo, J., & Martins, J. (2010). Energy production system management—renewable energy power supply integration with building automation system. *Energy Conversion and Management, 51*(2010), 1120–1126. http://dx.doi.org/10.1016/j.enconman.2009.12.020. (Elsevier).
10. SIEMENS. (2001). Simatic Net—NCM S7 for Profibus/FMS. Siemens 12/2001.
11. SIEMENS. (2000). System software for S7-300 and S7-400—Reference manual, Siemens 08/2000; A5E00069892-02.
12. SIEMENS. (2001). Simatic S7-300—Ladder logic (LAD) for S7-300, Simens, 2001.
13. SIEMENS. (2000). Provision and profibus-DP, vision sensor simatic VS710—Manual, Siemens 03/2000; A5E00063481.
14. https://www.digi.com/.
15. http://www.robot-electronics.co.uk/htm/md49tech.htm.
16. http://www.ni.com.
17. https://www.arduino.cc.

Part IV
Advanced Machining

Chapter 9
Advanced Machining Processes

Manas Das and Uday S. Dixit

Abstract Advanced machining processes are the material-removing processes different from conventional machining processes, in which a well-guided wedge-shaped tool removes the material in the form of chips by producing contact stresses. There are a variety of ways in which material is removed using these processes. One method is producing stresses in the workpiece by different means but not with a well-guided wedge-shaped tool. There are several processes in this category, e.g., ultrasonic machining, water jet machining, and abrasive jet machining. Another method is utilizing the thermal effect to melt or vaporize the material. This is accomplished by laser beam machining, electron beam machining, and electrical discharge machining. Ion beam machining bombards ions instead of electrons as in electron beam machining. It is principally different from electron beam machining in the sense that in the former, the material removal mainly takes place by sputtering and not by melting or vaporization. Chemical and electrochemical machining processes provide very good surface finish by making use of principles of chemistry. Combination of two or more processes is also in vogue. One such process is chemical–mechanical polishing, which removes the material by the combined action of chemical process and stresses caused by polishing. Advanced machining processes have become popular and economical and are finding their use in industries.

Keywords Advanced machining · Unconventional processes · Non-traditional processes · Ultrasonic machining · Water jet machining · Abrasive water jet machining · Abrasive jet machining · Chemical machining · Electrical discharge machining · Electrochemical machining · Chemical–mechanical polishing Laser beam machining · Ion beam machining · Electron beam machining Photochemical machining

M. Das · U. S. Dixit (✉)
Department of Mechanical Engineering, Indian Institute of Technology Guwahati,
Guwahati, India
e-mail: uday@iitg.ernet.in

© Springer International Publishing AG, part of Springer Nature 2018
J. P. Davim (ed.), *Introduction to Mechanical Engineering*, Materials Forming,
Machining and Tribology, https://doi.org/10.1007/978-3-319-78488-5_9

1 Introduction

Machining is a process of removing the material from a workpiece to create a part of desired shape, size, and surface finish. When the primary objective is only to get a good surface finish, the process is called finishing process. Traditional way of removing the material is to employ a wedge-shaped tool, which is harder than the workpiece. The tool makes physical contact with the workpiece and produces contact stresses, which cause the fracture on the surface of the workpiece. The tools were used by human being since about 10 million years ago [1]. These tools were operated by hand. In fact, the word manufacture originated from the Latin words "manu factum," which means made by hand. In the first millennium, human being started to use manually driven machine tools. Subsequently, the animal power was also used to operate the machine tools. In the sixteenth century AD, water-powered machine tools were developed. In the nineteenth century, steam engines were used as power sources of machine tools and electrically operated machine tools emerged since early twentieth century. During all these developments, the basic principle of material removal remained the same; i.e., a single wedge-shaped tool or a combination of such tools removed the material by having a guided relative motion against the workpiece.

Conventional machining was found to be unsuitable for machining of very hard material, for producing complicated and smaller shapes and for obtaining high accuracy, precision, and surface integrity. It prompted researchers to invent novel ways of removing the material. The newly invented processes were called unconventional or non-traditional machining processes. Most of these processes are now routinely employed by industries, and hence, the unconventional and non-traditional terms appear to be misnomers. A more appropriate name for these processes is advanced machining as a retronym of machining. A distinguishing feature of advanced machining processes is that they do not use a well-guided wedge-shaped tool that removes the material by its relative motion relative to workpiece. For example, ultrasonic machining process uses an oscillatory tool that moves a number of abrasive particles in somewhat random way on the surface of the workpiece. Similarly, laser beam machining employs a well-guided laser beam to remove the material by melting and/or evaporation. There are several advanced manufacturing processes like these.

One can classify the advanced machining processes based on the material removal mechanism. In the mechanical processes, there is a physical contact between the workpiece and another material, which generates contact stresses that eventually lead to the removal of material. Examples of such processes are ultrasonic machining, abrasive jet machining, water jet machining, abrasive flow machining, and magnetorheological finishing. In the chemical processes, chemical reactions take place on the surface of the workpiece that cause the removal of the material. Electrochemical machining process is governed by the Faraday's laws of electrochemistry. Thermal-based machining processes remove the material by melting and/or vaporization. In ion beam machining, the material is removed by sputtering of workpiece atoms due to bombardment of a focused ion beam.

In this chapter, a brief description of several advanced machining processes is provided. The processes have been grouped according to material removal mechanisms. The emphasis here is on the basic fundamentals related to the processes.

2 Processes Causing Material Removal Due to Stresses

The forces applied on the workpiece cause stresses. Stresses measure the intensity of the force. From a typical point in the workpiece, one can pass a plane that hypothetically divides the workpiece in two parts. The atoms of one part apply the force on the other part. The force per unit area provides the traction vector on that plane. If the traction vectors are known in three orthogonal planes, the traction vector on any plane can be obtained. Hence, once the components of the traction vectors are specified on three orthogonal planes, the state of stress at that point gets specified. As there are three components of a traction vector, stress at a point has nine components. There are a number of failure theories which predict the occurrence of failure or otherwise as a function of 9 components of the stress. In the material removal processes, the stresses are produced in order to cause local failure of the material leading to separation of some material from the workpiece. In conventional machining, stresses are produced by a wedge-shaped tool, which generates contact stresses in the workpiece. In advanced manufacturing processes, the local stresses are produced in a variety of ways, providing different types of material removal processes. Some popular processes are described in this section.

2.1 Ultrasonic Machining (USM)

In ultrasonic machining (USM) process, abrasives (of size 15–150 μm) are driven at high velocity against the workpiece by using a tool oscillating at a high frequency (about 19–25 kHz) and low amplitude (15–50 μm) [2–4]. The tool oscillates perpendicular to the surface to be machined. In USM, a tool oscillating at ultrasonic frequency throws the particles of different sizes on the workpiece. Sometimes, the abrasive particles are also hammered by the tool. The abrasive particles are suspended in a medium such as water, thus making a slurry. Typical abrasive particles are cubic boron nitride, aluminum oxide, silicon carbide, and boron carbide.

The process requires electrical energy. First, the low-frequency electrical supply signal is transformed to a high-frequency signal using an ultrasonic wave generator. The high-frequency electrical signal is converted to high-frequency linear mechanical motion by using a transducer that transfers its motion to the tool through a toolholder often called horn and made of Monel, titanium, and stainless steel. Often the horn is designed with a gradually reducing cross section toward the tool end in order to amplify the oscillations. Transducer can be piezoelectric or magnetostrictive type. In piezoelectric transducer, when the current passes through a piezoelectric crystal such as quartz, its length changes. This principle is used to convert electrical energy into mechanical energy with a conversion efficiency up to 95%. A typical magnetostrictive transducer comprises a laminated stack of nickel alloy sheets, which changes its length in a magnetic field. Electrical energy-to-mechanical energy conversion efficiency is 20–35%; hence, some cooling arrangement is required. Traditionally magnetostrictive transducers were more rugged and reliable and therefore found more application in ultrasonic machines in spite of their poor conversion efficiency. For achieving optimal material removal rate (MRR), tool and toolholder are designed to cause resonance (high amplitude of vibration). This happens when frequency of electrical signal matches with the natural frequency of the moving parts. A schematic of USM setup is depicted in Fig. 1. Cutting takes place due to brittle fracture of the workpiece by the hard abrasive particles. USM can machine hard (more than 40 HRC) and brittle materials, e.g., glass and ceramics. Recently, it has been found suitable for machining of titanium. It is uneconomical to machine ductile materials using USM. Abrasives are mixed with the water in a ratio of 1:1 by weight [4, 5]. The water helps in avoiding thermal damage to workpiece. Generally, the process generates negligible residual stresses [6].

USM has been applied for making holes in silicon nitride turbine blades for cooling purpose. It has also been used for cutting of semiconductor parts. It is also used for engraving on the glass, sintered carbide, and diamond. It can be used by dentists for drilling holes in teeth; skin and flesh being soft and ductile would not get cut.

2.2 Abrasive Jet Machining (AJM)

In abrasive jet machining (AJM) process, material removal takes place due to hitting of the workpiece with a jet of gas containing abrasive particles. A review on AJM is presented in [7]. Typical parameters of AJM are as follows: pressure 0.2–5 MPa, jet

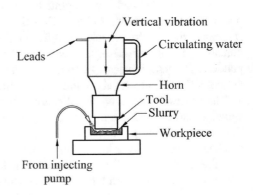

Fig. 1 A schematic of ultrasonic machining (USM) experimental setup

Vertical vibration

Circulating water

Leads

Horn

Tool

Slurry

Workpiece

From injecting pump

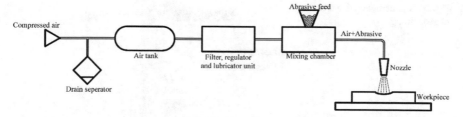

Fig. 2 A schematic of abrasive jet machining system

velocity 150–300 m/s, and the diameter of the nozzle 0.075–1.0 mm. Apart from the dry air, the process can use nitrogen, carbon dioxide, and helium. Erosive action of the abrasive particles is responsible for the removal of material from the workpiece surface. Aluminum oxide, silicon carbide, crushed glass, cubic boron nitride, glass beads, and sodium bicarbonate are the popularly used abrasive particles [8].

Elements of AJM are abrasive feeder, gas propulsion system, AJM nozzle, machining chamber, and abrasives. Gas propulsion system generally supplies gas to propel abrasive particles. Required quantity of abrasive particles is supplied by abrasive feeder, and quantity is controlled by inducing vibration to the feeder. The machining chamber is closed from outside environment to avoid the pollution of atmosphere. AJM nozzle is usually made of high-wear-resistant tungsten carbide or sapphire. Standoff distance, i.e., the distance between the nozzle tip to workpiece surface, is maintained at 0.75–1.0 mm for getting the maximum material removal rate. A schematic of AJM system is shown in Fig. 2.

AJM is generally used for trimming, deburring, cleaning, and polishing. It is suitable for machining very hard and brittle materials. It can produce fine and complicated shape. The abrasive jet can access the internal portions. The process can be used to manufacture electronic devices.

2.3 Water Jet Machining (WJM)

Water jet machining (WJM) process, also called hydrodynamic machining, works on the principle of the effect of erosion of a high velocity and small diameter jet of water on the workpiece surface. The process does not generate air-born dust and thus is suitable for machining of composites and asbestos. It uses a high-pressure and high-velocity water stream against the workpiece for cutting operation [9]. The fine stream of water can be obtained from a small nozzle with a diameter of 0.1–0.4 mm. The typical pressure in the nozzle is 400 MPa that can produce a jet of velocity up to 900 m/s. Such a high pressure is generated through a combination of oil pump and intensifier. Intensifier amplifies the oil pressure by about 40 times. To minimize pulsation in water pressure, an accumulator is used. Autofrettaged and shrink fitted tubes are used to handle high pressure of water. Generally, the nozzles

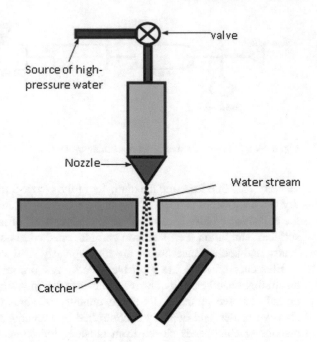

Fig. 3 A schematic of water
jet machining

are made of stainless steel, sapphire, ruby or in rare cases diamond. A good filtration system is essential to remove the small particles in water during operation. A noise catcher is also needed. A schematic of WJM setup is depicted in Fig. 3.

Process parameters for WJM are standoff distance, nozzle diameter, pressure, feed rate. Standoff distance should be 1–3 mm so that fluid should not disperse before reaching to the work surface. Thick materials can be machined at low feed rate with high pressure. This process can be used to cut floor tile, carpet, textiles, leather, plastic, composites, and cardboard [10]. WJM is also used for cutting composites, wire stripping, and deburring process. Water jet at lower pressure (60–200 MPa) has been used to cut insulation of the cable without damaging the metallic cable.

2.4 Abrasive Water Jet Machining (AWJM)

When there is need of cutting very hard and thick metallic parts, the abrasive water jet machining (AWJM) process is more appropriate than WJM. In AWJM process, the abrasive particles are homogeneously mixed with water stream that impinges on the workpiece surface with a very high kinetic energy. Water and abrasive streams come from two different rigid pipes, and these two streams mix in the mixing chamber before passing through the nozzle. Generally, abrasives used in this process are aluminum oxide, silicon dioxide, and garnet with grit sizes ranging between 60 and 120 [9]. For shallow cutting, fine particles and for high depth of cut coarse particles are used. The abrasive particles are continuously added to the water stream. The diameter of nozzle is 0.25–0.65 mm. Nozzle orifice diameter in this process is larger than WJM process so that high flow rate and energy of the jet can be achieved. Water pressures are almost same as in WJM [11]. AWJM process is used for machining hard materials [12]. In AWJM process, removal of material at upper position of kerf is due to erosion wear and at the lower portion of the kerf at higher depth is due to deformation wear.

AWJM consists of four important subsystems, i.e., pumping unit, abrasive water jet nozzle, abrasive feeding system, and catcher. The functions of abrasive water jet nozzle are mixing of abrasive and water and to form a high-velocity water abrasive jet. When workpiece is moving and nozzle is stationary, a catcher is generally used. A long tube-shaped catcher can be placed just below the jet at the lower part of the workpiece for capturing used jet. The important factors affecting the process are water pressure, nozzle diameter, size and concentrations of the abrasives, feed rate, standoff distance, workpiece material, mixing tube (diameter and length), angle of cutting, traverse speed, and number of passes.

AWJM is used for cutting metals (e.g., copper and its alloy, tungsten carbide, lead, and aluminum) and nonmetal (e.g., concrete, graphite, silica, glass, and acrylic). It can also be used for cutting sandwiched honeycomb-structure material used in aerospace industries. Other industries using this process are nuclear, oil, automotive, construction, etc. Slotting is a major application of this process. It can be done in any material like stainless steel, mild steel, and alloy steel in the form of plate, tube, and corrugated structure. Slot width is generally 0.75–2.25 mm, and there are no embedded abrasives in the machined surfaces [12].

2.5 Abrasive Flow Machining (AFM)

AFM is used for polishing and deburring workpiece with the use of abrasive particles mixed in a viscoelastic polymeric medium. The polymeric medium consists of silly putty, and generally, silicon carbide particles are used as abrasives. It is suitable for finishing of complicated shapes on internal and external surfaces. Mixture of abrasive and polymer is called medium, which flows onto the work surface under pressure ranging between 0.7 and 20 MPa [13].

The schematic of AFM experimental setup is shown in Fig. 4. AFM setup consists of two opposing cylinders. The workpiece is fixed through a workpiece fixture in between them, and the medium flows through the workpiece from one cylinder to the another cylinder. This process is repeated as many times as necessary to get the desired accuracy and finishing in workpiece. To maintain a constant viscosity of medium, an external cooling device is used to maintain the upper temperature. Both manual and automatically controlled AFM machines are available in the market. Tooling is used to confine and direct the flow of medium inside the workpiece for selective polishing of workpiece.

Fig. 4 A schematic of
abrasive flow machining setup

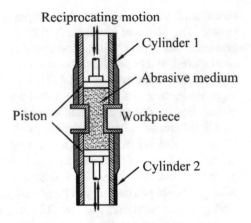

AFM medium is pliable; when the medium is forced through a passageway, it can act as a self-deforming grinding stone [13]. It comprises base material and abrasive grits. Base material consists of organic polymer and hydrocarbon gel. Degree of stiffness of the AFM medium can be determined by its base material. Soft material is used for small diameter holes, and stiff medium is used for large hole. It is reported that a stiff medium finishes the workpiece more efficiently than a soft medium. Process variables are extrusion pressure, number of cycles, stiffness of the medium, workpiece fixture configuration, abrasive size, and concentration.

AFM applications include rounding the sharp edges of workpiece and polishing the rough surface of a casting. It is mostly useful in industries like automobile, aerospace, and die making. It is used for uniform finishing of both metals and nonmetals for internal and external surfaces.

2.6 *Magnetic Abrasive Finishing (MAF)*

In this type of non-traditional machining process, finishing forces are controlled by an externally applied magnetic field [14]. The schematic of magnetic abrasive finishing (MAF) experimental setup is depicted in Fig. 5. This type of finishing process uses a brush comprising ferromagnetic particles, abrasive particles, and a binder that gets formed due to magnetic field. MAF brush contacts and acts upon the surface irregularities of the workpiece. Finishing depends upon workpiece traverse rate, type of work material, size of magnetic abrasive particles, magnetic flux density, pressure on the workpiece through magnetic abrasive brush, abrasive type and grain size as well as working gap between MAF brush and workpiece surface. Magnetic flux density is a function of type of magnetic pole (shape and size) and workpiece material. On increasing the value of magnetic flux density, surface finish improves and rate of material removal also increases. The material of the workpiece can be magnetic or non-magnetic in nature. Vibratory motion of the

Fig. 5 Schematic of magnetic abrasive finishing

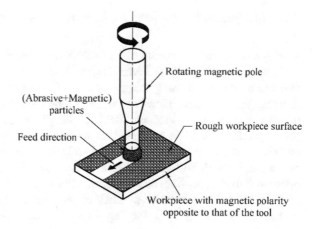

tool can be obtained by oscillating the magnetic poles relative to the workpiece [15]. Surface finish can be improved by increased flux density, increased machining time, higher workpiece traverse speed, and smaller working gap.

MAF is used for polishing of balls and roller, finishing of inner tube surface, the removal of oxide layer, polishing of fine component like printed circuit board, removing the burr in gear, and polishing of complicate shapes [12]. It is also used for finishing the internal and external surfaces of cylindrical workpiece. Types of operation that can be performed on the workpiece are finishing, rounding the edges, deburring, and introducing residual compressive stress.

Recently, nanoscale finish was achieved with the use of ultrasonic-assisted MAF [16]. An schematic of the setup is depicted in Fig. 6. The worktable was imparted ultrasonic vibrations at a frequency of 20 kHz and amplitude of 20 μm. A surface roughness of the order 50 nm could be achieved.

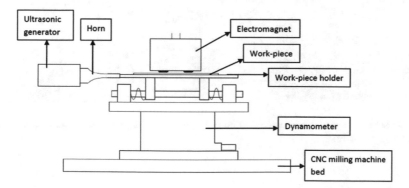

Fig. 6 Schematic of ultrasonic-assisted magnetic abrasive finishing process [16]. Copyright (2017) Authors. Open access under a Creative Common License

2.7 Magnetorheological Finishing (MRF) Process

Magnetorheological finishing (MRF) is a nano-finishing process developed by
William Kordonski at QED Technologies, Rochester, USA [17]. Surface roughness
that can be produced using this process is of the order of 10 Å. In MRF process,
finishing force by each abrasive particle in the medium is of the order of 10^{-9} N,
which is controlled by magnetic field. In this process, a smart magnetorheological
(MR) polishing fluid is utilized that usually comprises carbonyl iron particles,
abrasive particles, carrier medium, and additives. MR fluid changes its property in
the magnetic field. Stiffened MR fluid acts as a polishing tool. The relative motion
between finishing medium and workpiece can be achieved by rotating the polishing
medium [17]. The schematic of MRF experimental setup is shown in Fig. 7 [18].
In MRF, a nozzle is used for depositing MR fluid on a rotating wheel having
spherical shape. The rotating wheel transports fluid, and a MR fluid ribbon is
formed on the wheel surface. The workpiece surface to be polished and MR fluid
ribbon on rotating wheel rim form converging gap, which is exposed to magnetic
field by two poles of an electromagnet. Moving wall generates flow of MR fluid
through converging gap. MR fluid behaves as a Bingham plastic fluid where yield
stress varies with applied magnetic field [17]. Pressure is generated between converging gaps by the flow of stiffened MR fluid and workpiece surface [19].
Quasi-solid moving boundary is formed generating high stresses on the workpiece
surface that cause removal of material. This region is known as polishing spot.
Magnetorheological finishing process is generally used in finishing of lenses made
of calcium fluoride and fused silica.

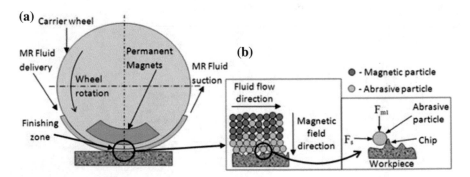

Fig. 7 Schematic of magnetorheological finishing process. With permission from [18]. Copyright
(2009) Talyor & Francis

2.8 Magnetorheological Abrasive Flow Finishing (MRAFF) Process

In MRF process, the finishing forces are controlled by external magnetic field. The process is not suitable for machining the internal surfaces of the workpiece. Although AFM process can polish both internal and external complex freeform surfaces, in this process, the viscosity of the rheological polishing medium cannot be controlled and the cost of polishing medium is very high. Magnetorheological abrasive flow finishing (MRAFF) process was developed by taking advantages of both the processes. The schematic of MRAFF setup is shown in Fig. 8 [20].

In MRAFF process, magnetorheological polishing fluid is transferred through workpiece using two opposing cylinders in the presence of a magnetic field in the polishing zone. MR fluid viscosity is a function of the magnetic field. Rheological behavior of MR fluid changes during entering and exiting the finishing zone from Newtonian to Bingham plastic and again changes to Newtonian fluid. Selective abrasion takes place due to applied magnetic field. The workpiece fixture is used to flow the MR fluid from one cylinder to another cylinder. Material of workpiece fixture should be non-magnetic, e.g., stainless steel. Around 0.6 T magnetic flux density is generated on the workpiece surface using two electromagnets. Hydraulic system is required for reciprocating motion of the pistons to simultaneously flow the polishing medium in two cylinders. A constant pressure throughout the finishing operation is maintained.

Rotational magnetorheological abrasive flow finishing (R-MRAFF) has working principle similar to MRAFF. The difference is that in this process the magnets that are just fixed in case of MRAFF are made to rotate with the help of a motor having

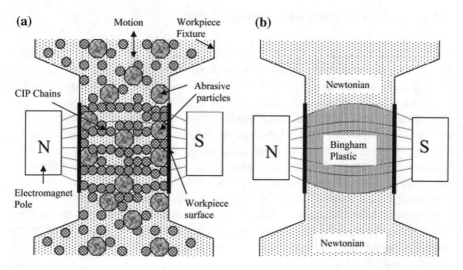

Fig. 8 A schematic diagram of MRAFF experimental setup. With permission from [20]. Copyright (2004) Elsevier

Fig. 9 A schematic of
R-MRAFF finishing process.
With permission from [21].
Copyright (2011) Taylor &
Francis

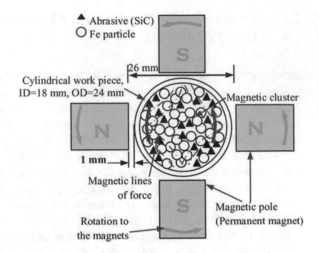

positive belt drive. On controlling the rpm of the motor, the finishing efficiency is also controlled as finishing forces get varied. A schematic of R-MRAFF is shown in Fig. 9 [21]. In R-MRAFF process, the abrasive particles follow a helical path due to reciprocating motion generated by the hydraulic unit and the rotational motion of the magnets by the motor surrounding the workpiece fixture. R-MRAFF not only reduces the surface roughness value but also improves textures of the internal surface of hollow cylindrical workpiece by generating cross-hatched pattern. This helps in enhancing oil retention capability of piston-cylinder assembly [22].

2.9   Elastic Emission Machining (EEM)

Elastic emission machining (EEM) removes materials with atomic level of accuracy (0.2–1 nm) by giving the workpiece a mirror-type finish. The schematic of EEM process is shown in Fig. 10 [23]. In EEM process, a polyurethane material ball is used and is mounted on a shaft which is driven by a variable speed motor. The rotational axis is oriented at an angle of 45° with the polishing surface. The workpiece is immersed in a mixture of zirconium dioxide or aluminum dioxide abrasive particles and water [24]. In some cases, the rate of material removal is nonlinear with respect to the concentration of the additives in the abrasive slurry. In this process, the material removal happens at atomic level due to the powerful interaction of the abrasive particles with the atoms of the workpiece without the introduction of the dislocations. The process has similarity with chemical etching and removes the material mainly by elastic fracture at atomic level. In EEM, surface roughness value as low as 0.5 nm is obtained in glass. Surface roughness value in silicon is reported to be less than 1 nm.

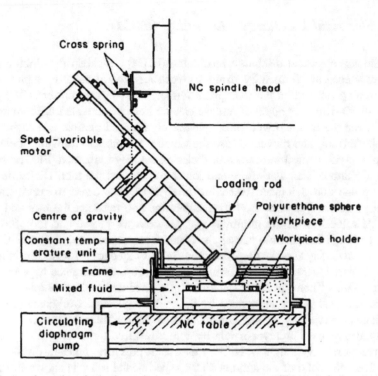

Fig. 10 A schematic of elastic emission machining process. With permission from [23]. Copyright (1987) Elsevier

This process is used in producing atomically stress-free and smooth surfaces of optical materials such as 4H-SiC, adaptive bimorph mirror, and silicon carbide. Ellipsoidal mirror in X-ray microscopy, which is a ring-shaped effective focusing device, can be polished by this process. Mirrors used for synchrotron radiation (SR) beam are fabricated using this process because focusing of SR beams requires atomically flat and perfect mirrors.

3   Processes Causing Material Removal Due to Melting/ Vaporization

There are some advanced machining processes in which material is removed by melting and/or vaporization. These processes are suited for machining of very hard materials. In the following subsections, some popular processes are described.

3.1 Electrical Discharge Machining (EDM)

In traditional electrical discharge machining (EDM), so-called die-sinking EDM, different shapes are formed by using an electrode having negative replica of the shape to be generated on the workpiece surface. Pulsed direct current (DC) power supply of 80–100 V is applied. Intense electric field is generated at the narrowest gap between the tool and workpiece. Negatively charged particle is detached from the cathode (tool) and moves toward the anode (workpiece). Collision of electrons takes place with neutral molecules of dielectric causing ionization. Electrons move toward the anode; i.e., workpiece and ions move toward the tool. Dielectric fluids like kerosene and deionized water are used in EDM process to create path for discharge. Servo system is used to maintain the gap between the tool and workpiece. At a location of the minimum gap between workpiece and tool, discharge occurs and high temperature causes the material to get melted or vaporized, which is flushed away by the flowing dielectric fluid. Frequency of sparking is around 5 kHz. Material is removed by forming craters into the workpiece by each spark, and thousands of hemispherical sparks are generated across the workpiece surface. Material removal rate is influenced by the melting point of workpiece [12]. Tool gets hit by positive ions causing unwanted wear. To minimize the tool wear, reverse polarity may be used. In reverse polarity, tool is positive and workpiece is negative. However, normally straight polarity is used as it provides more heat at the workpiece. The schematic of die-sinking EDM experimental setup is shown in Fig. 11.

EDM machine comprises power supply, dielectric system, cathode (tool) and anode (workpiece), and servo system. Power supply contains solid-state rectifier

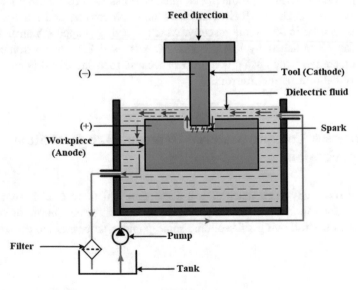

Fig. 11 A schematic of electric discharge machining process

that converts AC to DC. EDM machine is equipped with cutoff protection circuit, and in case of over voltage or over current, power cuts off. Dielectric system consists of dielectric fluid, reservoir, pump, and delivery device. Generally used dielectric fluids are lubricating oil, kerosene, transformer oil, deionized water, and hydrocarbon oils. The dielectric fluid functions as an insulating agent in the inter-electrode gap except at the time when ionization occurs in the presence of a spark. It flushes debris out from the gap between workpiece and tool. It also removes heat from tool and workpiece. Dielectric fluid should have high dielectric strength, minimum ignition delay time, effective coolant, and high degree of fluidity.

A tool material should have good machinability, conductivity, and wear resistance. At the same time, it should be inexpensive. Generally copper, graphite, brass, and tungsten are used as tool materials, the graphite being the most preferred. The important process parameters are current, voltage, frequency of current, inter-electrode gap, and duty cycle. With the increase in current or spark voltage, rate of material removal as well as surface roughness increases. With the increase in spark frequency, surface finish improves.

EDM process can be used for tool fabrication as well as for part production. It can be employed to machine any electrically conductive material. It is used for making dies for molding, casting, extrusion, wire drawing, stamping, forming, etc. It is also used in making tooling for plastic injection molding [12]. It can be used to machine thin and fragile components because in EDM there is no mechanical force.

A variant of EDM is wire EDM process, in which a wire of small diameter is used to cut narrow kerf in workpiece. Thermal energy is generated between the wire and the workpiece. Cutting path is achieved by feeding the wire to workpiece like a band saw. During cutting operation, wire is advanced between supply pool and take-up pool to get fresh electrode for better accuracy and constant narrow kerf. In wire EDM also there is a need of dielectric fluid. Wire diameters range from 0.15 to 0.30 mm [12]. Brass, copper, tungsten, and molybdenum can be used as wire material, and generally, deionized water is used as dielectric fluid because of its high cooling effect and fire hazard.

Wire EDM system consists of power supply, dielectric system, wire positioning system, and drive system. Pulse frequency of 1 MHz is used in wire EDM process. High frequency of power supply gives better surface finish and reduced crater size. Current-carrying capacity of wire is less than 20 A. For positioning system, 2-axis table CNC machine is used. In case of short circuit, servo system senses and moves back the wire to re-establish the gap between the work and tool. The function of wire drive system is to deliver fresh wire and keep wire always under tension. Wire is guided by guides generally made of sapphire and diamond. Wire is discarded after one-time use.

Wire EDM is used for making stamping dies. Kerf generated in this process is so narrow that it can be used to machine punch and die. Also, tools and parts with complex shapes, like extrusion dies, can be made. Tolerances as small as 0.07 μm can be achieved. A schematic of wire EDM process is shown in Fig. 12.

Fig. 12 A schematic of wire
EDM process

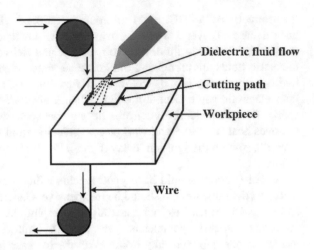

Dielectric fluid flow

Cutting path

Workpiece

Wire

3.2 Electron Beam Machining

In electron beam machining (EBM), a stream of high-velocity electrons, is focused
on the workpiece to remove the material by melting and vaporization. An electron
beam gun is used to generate high-velocity stream at around 75% of the speed of
light, and this beam is focused on workpiece by using a lens [12]. The lens is used
to reduce the diameter of beam up to 0.025 mm. As soon as beam is focused on
workpiece, the kinetic energy of the beam is converted into high-density thermal
energy that melts and vaporizes the workpiece material. EBM process needs a
vacuum chamber in order to avoid collision of electrons with the gas molecules.
The schematic of EBM setup is shown in Fig. 13. EBM is generally used for

Fig. 13 A schematic of
electron beam machining

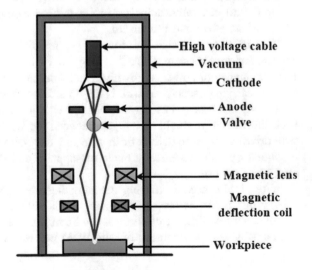

High voltage cable

Vacuum

Cathode

Anode

Valve

Magnetic lens

Magnetic
deflection coil

Workpiece

drilling of holes of high aspect ratio, down to 0.05 mm diameter, and cutting of slots. One drawback of the process is that it needs vacuum.

3.3 Laser Beam Machining

LASER (light amplification by stimulated emission of radiation) is a coherent beam of light, which is monochromatic in nature. A lens can focus the laser beam in a very small diameter with much higher density than that obtained from any other type of light. Material is removed in this process by using light energy of beam which has the potential to remove the material by melting and vaporization of material. Two types of laser heat sources are popularly used in machining process: gas lasers and solid-state laser. Solid-state lasers (e.g., ruby laser and Nd:YAG) have low efficiencies [25]. However, they are effectively used for drilling in pulsed mode. CO_2 laser, a typical gas laser, has good efficiency and is used for machining. Nowadays, fiber lasers are becoming popular due to their high efficiencies. Wavelength of commonly used lasers lies between 0.21 and 11 μm (Ruby: 0.7 μm, Nd:YAG: 1.0 μm, CO: 2.7 μm, and CO_2: 10.6 μm). Figure 14 shows the schematic of an LBM system [26].

Three important elements of laser device are lasing medium (a collection of atoms or molecules in gas or solids), pumping energy source required to uphold these atoms to higher energy level, and optical feedback system. For feedback mechanism, parallel mirrors at the two ends are used—one fully reflecting and another partially reflecting mirror. Feedback mechanism captures and redirects a

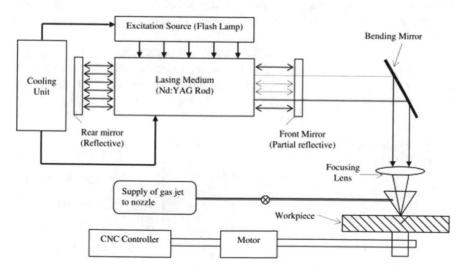

Fig. 14 A schematic of laser beam machining setup. With permission from [26]. Copyright (2008) Elsevier

few coherent photons back into active lasing medium. These photons further stimulate the emission of some more photons of the same frequency and phase. This mechanism also permits a small percentage of coherent photons to exit the system in the form of laser light.

For LBM, workpiece material should have low heat of fusion, high absorption of light energy, low specific heat, poor reflectivity, low heat of vaporization, and good conductivity of heat. LBM is utilized for the materials having high strength and hardness like ceramics, glass and glass epoxy, cloth, plastics, rubber, wood. Applications of LBM are drilling, slotting, scribing, and marking operations and 3D micromachining. It can be effectively used for drilling holes of diameter 0.025– 0.50 mm [25].

3.4 Plasma Arc Cutting (PAC)

In this process, a stream of plasma is used for machining. In PAC, temperature (as high as 10,000–14,000 °C) is generated to cut the workpiece. Plasma cutting operation is carried out by directing high-velocity plasma stream toward workpiece that melts and vaporizes the metal. Water is used to clean the kerf formed by plasma arc. Feed rate depends upon material that is going to be cut and its thickness along the cutting path [12].

PAC system uses DC power supply, and plasma arc can be transferred or non-transferred type. Non-transferred type arc has lower efficiency (60–85%) than transferred arc type (80–95%). In non-transferred arc mode, arc is generated between electrode and nozzle. Hence, workpiece can be either electrically conductive or non-conductive. In the case of transferred arc mode, arc is generated between electrode and workpiece. A schematic of plasma arc cutting system is shown in Fig. 15.

Elements of PAC system are power supply, gas supply, cooling water system, control console, and plasma torch. There are several torch designs. Air plasma torch uses compressive air that ionizes and does cutting. Torch may fail because of double arcing (arcing between electrode and nozzle as well as arcing between

Fig. 15 A schematic of plasma arc cutting system

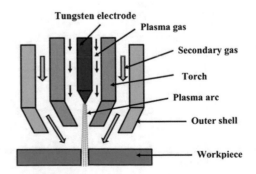

nozzle and workpiece. Generally used electrode materials are zirconium, hafnium (high resistance to oxidation), and tungsten (low resistance to oxidation). Zirconium has more life than tungsten electrode. Dual-gas torch uses primary and secondary gases. For example, nitrogen is used as a plasma gas, while oxygen or carbon dioxide is used as secondary gas. To avoid oxidation of electrode, oxygen-injected torch is used. In this type of nozzle, nitrogen is used as plasma gas; oxygen is inducted in the downstream of nozzle. It is generally used for cutting mild steel. In water-injected torch, water is used to constrict the plasma. A small quantity of water vaporizes, and this thin film of steam constricts the plasma and also insulates the nozzle. To avoid double arcing, the lower part of the nozzle may be made of ceramics. PAC can be used for cutting of flat metal sheets and plates. It can also be used for piercing holes. PAC can be used to cut material like plain carbon steel, stainless steel, copper, and aluminum. By using plasma arc system, different shapes can be formed on the workpiece.

4   Focused Ion Beam (FIB) Machining

In focused ion beam (FIB) machining, high-velocity electrons are released from a cathode (high-voltage tungsten material) and as it moves toward the anode, it gets hit with the fluid in the plasma region of the FIB tube. The ions of the gas/liquid get accelerated and bombard the workpiece material. When the ion's energy is higher than that of the bonding energy of the atom, the workpiece atoms are removed [8]. The phenomenon of dislodging the atoms of the workpiece is called sputtering and is akin to a football transferring its momentum to another football. The precision of machining in this process can be up to 5 nm. If gas ions are fed, then normally hydrogen or inert gases are used, and for liquid metal ion, gallium is used. Figure 16 shows the schematic of FIB process highlighting the situation before and after machining.

Focused ion beam machining is used in electronic industries where precision plays a vital role. Nano-finishing can be done using this process for hard and brittle

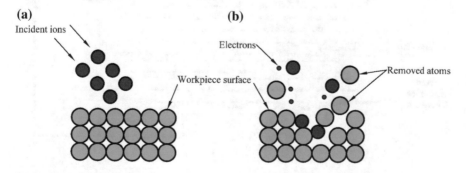

Fig. 16 Schematic diagram showing the process **a** before and **b** after FIB machining

materials like ceramics, semiconductors, and diamonds. Finishing of medical instruments is possible with very high precision in this process.

5 Electrochemical Machining (ECM)

Electrochemical machining (ECM) is also known as electrochemical forming process. In this process, material is removed from electrically conductive material with the help of anodic dissolution and shape of the workpiece is same as that of tool [27]. The workpiece is known as anode, and the tool as the cathode. The principle of this process is based on the Faraday's law of electrolysis. The electrolyte flow is used to remove depleted material so that it does not get deposited on the tool. Copper, brass, or stainless steel is generally used as tool material. To reduce electrolyte resistivity, salts such as NaCl or $NaNO_3$ are generally added in solution [27]. Debris is removed from the electrolyte by centrifuge, sedimentation, or other means. Reaction products are the barrier for flow of current. Hence, to minimize this effect, electrolyte velocity should be 20–30 m/s [28]. Flowing electrolytes limit the ion concentration, provide high MRR, dilute reaction products, and dissipate heat. Workpiece profile depends on feed rate, voltage, tool shape, and electrolyte properties. The schematic of ECM system is shown in Fig. 17.

Different elements of ECM are power supply, tool and tool feed system, electrolyte cleaning and supply system, work and work holding system. Rectifier is used to transform a high-voltage power supply into low-voltage and high-current power supply. Electrolyte supply and cleaning system comprises heating/cooling coil, control valves, pump, filter, piping, pressure gauge, and reservoir. Filter is

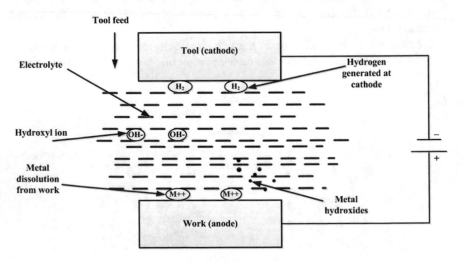

Fig. 17 A schematic of electrochemical machining

used for cleaning electrolytes and is made of anti-corrosive material, whereas piping system is made from stainless steel and anti-corrosive material. Fixtures should be made from insulating material, and there should be no vibration under hydraulic force. Tools are made from aluminum, brass, copper, Monel, and bronze. Tool material should have high machinability, conductivity, and corrosion resistance. Workpiece should be electrically conductive, but work holder should be insulator with good thermal stability and low moisture absorption. Most common work holder materials are plastic and Perspex.

ECM is used to machine work material of very complex geometry. It is very useful in die sinking, multiple hole drilling, and deburring. Typical products made from ECM are turbine blades, curvilinear slots and gears, etc. It provides very good surface finish.

Electrostream drilling (ESD) is a special variant of ECM, which is used for drilling high-aspect ratio (300:1) holes of less than 1 mm diameter. The use of acidic electrolyte ensures that metal sludge is dissolved and carried away as metal ion. In electrostream drilling, electrically negatively charged high-velocity acid electrolyte stream is used for the drilling operation [28]. As the stream strikes the positively charged workpiece, electrochemical dissolution occurs. Machining takes place in the same manner as a conventional ECM process. The dissolved material is flushed out from the machining zone in the form of metal ions in the solution. Two different ESD techniques are generally used for drilling, i.e., penetration drilling and dwell drilling. Penetration drilling is used for making deep and accurate hole, and dwell drilling is used for making shallow holes. In penetration drilling, cycle starts when nozzle starts to feed into workpiece and a gap monitoring device is used to maintain the gap, slow the feed rate, and trigger full power when accurate gap between nozzle and workpiece is achieved. During the drilling cycle, nozzle is fed at constant rate to maintain constant gap [12]. In dwell drilling, drilling operation is done without feeding the nozzle into workpiece and proper distance between workpiece and nozzle is maintained. In this technique, stream of electrolyte is responsible for the shape of workpiece as the nozzle does not penetrate into hole. This process can be used for drilling very small holes, curved holes, and holes at steep angle [28]. It also can be used for machining of fuel nozzles, machining of oil passage, machining of small hole, drilling regular array of hole in corrosion-resistant material of low machinability, and to make holes in the bearing for oil passage where EDM can crack the workpiece.

6 Chemical Machining (CM)

In chemical machining (CM), material is removed by using chemical etchant. There are several variants of the process, e.g., chemical blanking, chemical milling, photochemical machining, and chemical engraving. The schematic of chemical etching process is depicted in Fig. 18. The chemical machining process consists of following steps:

Fig. 18 A schematic diagram of chemical machining process

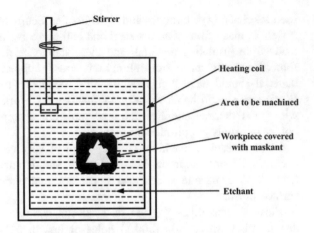

- **Cleaning**: Before chemical machining any workpiece, the surface of the workpiece should be cleaned of sand and grease.
- **Masking**: This is a protective coating, which should be applied on work surface. It should be applied at those surfaces that are not going to be etched. Masking material should be chemically resistant to the etchant.
- **Etching**: This is the step involved in removal of material. In this step, part is immersed inside etchant bath and etchant attacks the unmasked surface. Etchant chemically reacts with the material to remove it. After desired amount of material is removed, workpiece is withdrawn from the etchant bath.
- **Demasking**: In this step, mask is removed from the workpiece.

Masking materials include polyvinylchloride, neoprene, polyethylene, and a few polymers. Etchants are chemical electrolytic solutions. They etch the workpiece material when in contact. Some common etchants are $FeCl_3$, $FeNO_3$, HF, and HNO_3. High etch rate, minimum undercut, compatibility with the mask, high dissolved material capacity, and ease of control are the properties which an etchant should possess. Various chemical etching processes are as follows:

Chemical milling: It is used in aircraft industries by removing metal from aircraft components for weight reduction.

Chemical blanking: This process is conducted on thin material which has maximum thickness of 0.75 mm. Chemical blanking is suitable for hard and brittle materials on which any sort of mechanical methods of machining would cause fracture.

Chemical engraving: It is used for making nameplates or flat boards that have lettering on it or other artwork.

7 Photochemical Machining (PCM)

Photochemical machining (PCM) is just like chemical machining process [29]. The difference is that in this a photoresist mask is used instead of a normal masking material in PCM process. A photoresist is a UV photosensitive polymer supplied as a liquid or dry film, and it is used to produce an image on metal. Where the photoresist has been exposed to UV radiation, it becomes insoluble in a developer solution and adheres strongly to the metal surface to form an etchant-resistant coating. Once the photoresist has been imaged to produce an etchant-resistant patterned coating, chemical etching is used to transfer that pattern into the metal by selective removal as follows.

First, the removal of the dust, oil, debris, contaminants, and grease is done chemically. Next, the photoresist coating is deposited. The coating is deposited by hot rolling of the photoresist material against the metal. The metal sheets are processed in safe yellow light environment at this stage. Always a phototool is made for a new part. The phototool should suit the profile of the part. The profile of phototool is normally produced by Computer-Aided Design (CAD) modeling [30]. The drawing or the image that has been created is sent to laser photoplotter. Initially, it is created in a Mylar film, which is later developed. The polymer in the photoresist coating is activated in the printing section so that the phototool is transferred to the metal. A sandwich of metal coated with the phototool on the both sides is created. The process is done under vacuum condition in order to avoid air bubbles in between the metal and phototool layer. Now, the combination is exposed to ultraviolet rays [30]. An image gets formed in the photoresist material. Now, etching is carried out by spraying the etchants. After machining, the photoresist is removed using NaOH solution. After removing photoresist, the metal is rinsed and cleaned initially with normal water and then with deionized water. Table 1 shows the different workpiece metals with their suitable etchant. Integrated circuits' lead frame is manufactured using this method which is used in electronic industries.

Table 1 Workpiece material and etchants

Workpiece material	Etchant
Non-Fe alloys	$CuCl_2$
Mo and Ag alloys	$Fe(NO_3)_3$
Ti, Zn, and Ti–Zn-based alloys	Cu-based etchants
Ferrous alloys (stainless steel, cast iron, etc.)	Ferric chloride

8   Chemo-Mechanical Polishing (CMP)

Chemo-mechanical polishing (CMP), also known as chemical–mechanical polishing, has been adopted in semiconductor manufacturing industries for planarization of the Si wafer. The important material removal mechanism is the reaction occurring between the slurry (having chemical reactants) and the workpiece surface followed by smooth mechanical abrasive by the abrasive particles for the removal of reacted products [31]. The schematic diagram of CMP experimental setup is depicted in Fig. 19. The polishing pad (made of polyurethane cloth) is covered by SiO_2 slurry against which the Si wafers are pressed down by a rotating wafer chuck. The polishing pad is attached to a rotating plate, and polishing slurry is supplied on it by a slurry delivery pipe. A planetary motion is generated between the rotating polishing pad and Si wafer. The most widely used suspension used for polishing in CMP is aqueous colloidal silica. Initially, silicon wafer reacts with the aqueous solution forming a thin soft layer of silica which is then removed by mechanical action of the abrasives in polishing slurry. The small layer of silica helps in providing regular distribution of the slurry particles, reducing the friction, helping in removing the already eroded unwanted material off the surface of the material, and reducing the generation of heat. During CMP process, the erosion of the wafer layer happens due to the relative motion between the slurry particles and wafer. Silicon wafers had been manufactured for long using CMP process. CMP results in defect-free surfaces unlike any mechanical polishing process. Also, in thin film transistor technology, chemo-mechanical polishing substrate is used.

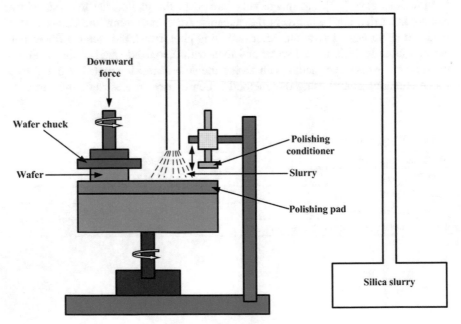

Fig. 19 A schematic of chemo-mechanical polishing experimental setup

9 Conclusions

In this chapter, a number of advanced manufacturing processes are described. Not a single process is suitable for machining all materials and/or manufacturing all features. Every process has a pros and cons. Sometimes, two are more processes can be combined to make a new process. At times, an advanced machining process is combined with conventional machining either in a parallel manner or sequentially. With the continuous development taking place, the advanced machining processes are becoming more economical day by day.

Review Questions
State True (T) or False (F):

1. AWJM cannot be used to cut composite materials.
2. Tool wear rate of a typical EDM tool is more than that of an ECM tool.
3. The surface achieved in ECM is of very high quality.
4. In USM, the tool vibrates at a frequency of around 19–25 kHz.
5. WJC is a well-suited cutting process for working in an explosive environment.
6. The AFM medium acts as a self-deforming grinding stone.
7. MRF process can polish lenses.
8. Ion beam machining is a thermal process that generates a lot of heat-affected zone.
9. Laser beam machining can be used to machine very hard and electrically non-conductive materials.
10. In PAC, a non-transferred arc has higher efficiency than a transferred arc.

Answers: 1. F 2. T 3. T 4. T 5. T 6. T 7. T 8. F 9. T 10. F

Glossary

Abrasives These are particles used to polish or clean hard surface by rubbing or grinding

Advanced machining processes These are the material removal processes different from the conventional machining in which a wedge-shaped tool removes the material by physical contact with the workpiece

Bingham plastic fluid Bingham plastic is a non-Newtonian viscoplastic fluid that behaves as a rigid body at low stresses but flows at higher stresses

Carbonyl iron particles (CIP) Carbonyl iron is 99.5% purity iron particles

Coherent beam A coherent beam is having wave sources with constant phase difference and same frequency

Cutting Tool Cutting tool in machining operation is used for removing material from the workpiece using shear deformation. Cutting tool may be single-point or multipoint based on the number of cutting edges

Deburring The unwanted piece of material that remains stuck to workpiece after grinding, drilling, milling, and turning processes, etc., is called burrs. The removal of burrs by some process is called deburring

Dielectric medium Dielectric materials are insulating material or a very poor conductor of electric current. While placed in an electric field, no current flows through them. Dielectric material can be polarized by an applied electric field

Electrical spark Electric spark is an abrupt electrical discharge that occurs when a sufficiently high electric field generates an ionized, electrically conductive channel through a normally insulating medium like air or gas

Electrolysis It is a process of chemical decomposition by passing an electric current through a liquid

Electrolyte An electrolyte is a solution that dissociates into ions and acquires electricity conducting capacity

Honeycomb structure *Honeycomb structures* are available in nature or can be fabricated having in the form of a *honeycomb*

Laser The full form of laser is light amplification by stimulated emission of radiation. Laser devices emit a coherent and monochromatic beam of light

Machining Machining refers to removal of material from workpiece for making a product

Magnetorheological fluid *MR fluid* is a smart material whose rheological behavior changes rapidly but it can be controlled easily in the presence of an applied magnetic field

Mesh size of abrasive Mesh is a measurement of particle size often used in determining the particle-size distribution of a granular material. Mesh size or sieve size is defined as the number of openings in one square inch of a screen. The higher the mesh number, the smaller the screen opening, and the smaller the particle that will pass through

Photoresist Photoresist is a light-sensitive material used in photolithography and photoengraving while making patterned coating on a surface

Piezoelectric Transducer *This type of transducer* works on the basis of *piezoelectric* effect, derived from Greek word piezen, i.e., to squeeze or press. When mechanical stress or forces are applied to a piezoelectric material like quartz crystal, it produces electrical charges on its surface

Plasma *Plasma* is an ionized gas consisting of positive ions and free electrons and is often called the fourth state of matter after solid, liquids, and gases

Rheology *Rheology* is the study of the deformation and flow of matter under applied forces

Sputtering During sputtering, atoms or molecules are ejected from a workpiece material due to the bombardment ion beam

Standoff distance (SOD) The distance between the nozzle tip and workpiece surface is called standoff distance in abrasive jet machining

Surface roughness Surface roughness refers to small local deviations of a real surface from an ideally perfect flat surface

Transducer A *transducer* is a device that converts one form of energy into another form

Ultrasonic frequency The term ultrasonic refers to frequency above audible sound frequency of human ear, generally over 20 kHz

Viscoelastic medium It is a material exhibiting both viscous and elastic property while undergoing deformation. Viscous materials generally resist shear flow and exhibit time-dependent strain. The elastic materials strain when stretched and quickly return to their original state once the stress is removed.

References

1. McNeil, I. (1990). Introduction: Basic tools, devices and mechanisms. In I. McNiel (Eds.), *An encyclopedia of the history of technology* (pp. 1–43). Routledge, London.
2. Jadoun, R. S., Kumar, P., Mishra, B. K., & Mehta, R. C. S. (2006). Manufacturing process optimisation for tool wear rate in ultrasonic drilling of engineering ceramics using the Taguchi method. *International Journal of Machining and Machinability of Materials, 1,* 94–114. https://doi.org/10.1504/IJMMM.2006.010660.
3. Kennedy, D. C., & Grieve, R. J. (1975). Ultrasonic machining—A review. *Production Engineer, 54,* 481–486. https://doi.org/10.1049/tpe:19750245.
4. Goetze, D. (1956). Effect of vibration amplitude, frequency, and composition of the abrasive slurry on the rate of ultrasonic machining in ketos tool steel. *The Journal of the Acoustical Society of America, 28,* 1033–1037. https://doi.org/10.1121/1.1908545.
5. Drozda, T. J. (1983). *Tool and manufacturing engineers handbook: Machining* (4th ed.). Dearborn, Michigan: Society of Manufacturing Engineers.
6. Lalchhuanvela, H., Doloi, B., & Bhattacharyya, B. (2012). Enabling and understanding ultrasonic machining of engineering ceramics using parametric analysis. *Materials and Manufacturing Processes, 27,* 443–448. https://doi.org/10.1080/10426914.2011.585497.
7. Ramachandran, N., & Ramakrishnan, N. (1993). A review of abrasive jet machining. *Journal of Materials Processing Technology, 39,* 21–31. https://doi.org/10.1016/0924-0136(93)90005-Q.
8. McGeough, J. A. (1988). *Advanced Methods of Machining.* London: Springer.
9. Miller, R. K. (1991). *Waterjet cutting: Technology and industrial applications.* Lilburn, GA: Fairmont Press.
10. Maid, M. (1988). Cutting with high-energy water-jet in industrial applications. *Materials and Design, 9,* 294–296. https://doi.org/10.1016/0261-3069(88)90007-6.

11. Momber, A. W., Kovacevic, R. (2012). *Principles of abrasive water jet machining*. Berlin: Springer Science & Business Media.
12. Benedict, G. (1987). *Nontraditional machining processes*. New York: Marcel Decker Inc.
13. Rhoades, L. (1991). Abrasive flow machining: A case study. *Journal of Materials Processing Technology, 28,* 107–116. https://doi.org/10.1016/0924-0136(91)90210-6.
14. Jha, S., & Jain, V. K. (2006). Nanofinishing techniques. *Micromanufacturing and nanotechnology* (pp. 171–195). Berlin, Heidelberg: Springer.
15. El-Hofy, H. A-G. (2013). *Fundamentals of machining processes: Conventional and nonconventional processes* (2nd ed.). USA: CRC Press.
16. Shukla, V. C., Pandey, P. M., Dixit, U. S., et al. (2017). Modeling of normal force and finishing torque considering shearing and ploughing effects in ultrasonic assisted magnetic abrasive finishing process with sintered magnetic abrasive powder. *Wear, 390,* 11–22. https://doi.org/10.1016/j.wear.2017.06.017.
17. Kordonsky, W. I. (1993). Magnetorheological effect as a base of new devices and technologies. *Journal of Magnetism and Magnetic Materials, 122,* 395–398. https://doi.org/10.1016/0304-8853(93)91117-P.
18. Sidpara, A., Das, M., & Jain, V. K. (2009). Rheological characterization of magnetorheological finishing fluid. *Materials and Manufacturing Processes, 24,* 1467–1478. https://doi.org/10.1080/10426910903367410.
19. Tichy, J. A. (1991). Hydrodynamic lubrication theory for the Bingham plastic flow model. *Journal of Rheology, 35,* 477–496. https://doi.org/10.1122/1.550231.
20. Jha, S., & Jain, V. K. (2004). Design and development of the magnetorheological abrasive flow finishing (MRAFF) process. *International Journal of Machine Tools and Manufacture, 44,* 1019–1029. https://doi.org/10.1016/j.ijmachtools.2004.03.007.
21. Das, M., Jain, V. K., & Ghoshdastidar, P. S. (2011). The out-of-roundness of the internal surfaces of stainless steel tubes finished by the rotational-magnetorheological abrasive flow finishing process. *Materials and Manufacturing Processes, 26,* 1073–1084. https://doi.org/10.1080/10426914.2010.537141.
22. Das, M., Jain, V. K., & Ghoshdastidar, P. S. (2012). Nanofinishing of flat workpieces using rotational–magnetorheological abrasive flow finishing (R-MRAFF) process. *International Journal of Advanced Manufacturing Technology, 62,* 405–420. https://doi.org/10.1007/s00170-011-3808-2.
23. Mori, Y., Yamauchi, K., & Endo, K. (1987). Elastic emission machining. *Precision Engineering, 9,* 123–128. https://doi.org/10.1016/0141-6359(87)90029-8.
24. Mori, Y., Ikawa, N., Okuda, T., & Yamagata, K. (1976). Numerically controlled elastic emission machining. *Osaka University, Technology Reports, 26,* 283–294.
25. Meijer, J. (2004). Laser beam machining (LBM), state of the art and new opportunities. *Journal of Materials Processing Technology, 149,* 2–17. https://doi.org/10.1016/j.jmatprotec.2004.02.003.
26. Dubey, A. K., & Yadava, V. (2008). Laser beam machining—A review. *International Journal of Machine Tools and Manufacture, 48,* 609–628. https://doi.org/10.1016/j.ijmachtools.2007.10.017.
27. Hewidy, M. S. (2005). Controlling of metal removal thickness in ECM process. *Journal of Materials Processing Technology, 160,* 348–353. https://doi.org/10.1016/j.jmatprotec.2003.08.007.
28. Ahmed, M., & Duffield, A. (1990). The drilling of small deep holes by acid ECM. In *Proceedings of Advanced Machining Technology III*.
29. Weller, E. (1984). *Nontraditional machining processes*. Dearborn, Michigan: Society of manufacturing engineers.
30. Allen, DM. (1986). *The principles and practice of photochemical machining and photoetching*. Adam Hilger, Techno House, Redcliffe Way, Bristol BS 1 6 NX, UK.
31. Ein-Eli, Y., & Starosvetsky, D. (2007). Review on copper chemical–Mechanical polishing (CMP) and post-CMP cleaning in ultra large system integrated (ULSI)—An electrochemical perspective. *Electrochimica Acta, 52,* 1825–1838. https://doi.org/10.1016/j.electacta.2006.07.039.

Chapter 10
Comparative Assessment and Merit Appraisal of Thermally Assisted Machining Techniques for Improving Machinability of Titanium Alloys

O. A. Shams, A. Pramanik, T. T. Chandratilleke and N. Nadim

Abstract Titanium-based alloys are highly recognised for the outstanding strength, lightweight, stable properties and exceptional resistance to corrosion which make them greatly suitable material for industry applications involving harsh environmental conditions such as elevated temperature. However, in product manufacture, machining of these alloys tends to produce poor surface quality with accelerated tool wear and low material removal rate, resulting from large cutting forces, excessive workpiece temperatures and chemical reactivity owing to high yield stress and low thermal conductivity. In overcoming these manufacturing challenges, industry practice supported by current research identifies the significantly useful potential for thermally assisted machining (TAM) techniques for improving machinability of titanium-based alloys whereby localised workpiece heating is applied to temporarily reduce metal hardness at the cutting point for lessening cutting forces. Reviewing current literature, this paper appraises various manufacturing issues related to the machining of these alloys and the potential improvement to machinability characteristics of these alloys from the application of TAM techniques. A detailed evaluation is presented on the alloy machinability influenced by workpiece heating prior to or during machining process using external heat source. This investigation recognises that the laser-assisted machining (LAM) reduces cutting forces by 30–60% and tool wear by 90%. It observes that the plasma-assisted machining (PAM) reduces cutting force by 20–40%, increases material removal rate by 200% and enhances the tool life by 150 times. Additionally, the induction-assisted machining (IAM) is noted to reduce cutting force by 36–54% while increasing the tool life and material removal rate by 206 and 214%, respectively.

O. A. Shams · A. Pramanik (✉) · T. T. Chandratilleke · N. Nadim
Department of Mechanical Engineering, Curtin University,
U1987, Perth, WA 6845, Australia
e-mail: alokesh.pramanik@curtin.edu.au

O. A. Shams
Department of Machinery and Equipment, Al Anbar Technical Institute,
Middle Technical University, Baghdad, Iraq

© Springer International Publishing AG, part of Springer Nature 2018
J. P. Davim (ed.), *Introduction to Mechanical Engineering*, Materials Forming,
Machining and Tribology, https://doi.org/10.1007/978-3-319-78488-5_10

Keywords Machinability · Preheating temperature · Thermally assisted machining · Titanium alloys

1 Introduction

A tremendous growth has been witnessed in the use of advanced materials for improving technical efficiency and cost-effectiveness in engineering applications recently. In complying with industry needs, high-performance materials are constantly developed to satisfy superior products quality with extended dimension accuracy and surface finish. Among many advanced engineering materials used, titanium alloys are at the forefront of technical choice due to their extraordinary material properties that surpass traditional steel and aluminium alloys, particularly for applications where lightweight, higher strength and stable properties at elevated service temperatures and harsh environments are prime considerations [1–3]. Some such application examples are the product design in automotive, aerospace, marine and power generation industries, where the system efficiency, reliability and fuel economy largely dependent upon the strength-to-weight ratio of alloys used and their operating longevity.

Titanium alloys offer unique combinations of exceptional mechanical properties and superior physical characteristics of high resistance to corrosion, fatigue and oxidation, underpinning the alloys' popularity as highly sought-after lightweight alternative [4, 5]. In particular, Ti-6Al-4V alloy is widely utilised in important components in the aerospace sector for higher ratio of strength-to-weight and in heat exchangers [6–9] for high temperature and corrosion tolerance. Chemically inert and stable physical properties make titanium alloys highly attractive and biocompatible for producing biomedical implements for diseased bone replacement, cartilage and spinal fixation [10–14], and in food processing industries. In producing modern aircraft such as Boeing 787, titanium alloys have become even more attractive due to their similar thermal expansion coefficients to common aviation materials of aluminium and graphite [3, 8, 15] with the added benefits of superior galvanic resistance and excellent creep resistance.

Table 1 shows inclusive research studies involving the machining of titanium alloys under dry conditions for comparative assessment and merit appraisal of techniques for improving machinability. These studies aptly recognise titanium-based alloys to be extremely hard to machine, especially at higher surface speed, even by latest cutting tools and methods [16, 17]. It is observed that poor surface finish quality is evidently noticeable at rather low machining speeds between 60 and 120 m/min and beyond [18]. Such machining issues are attributed to excessive cutting forces generation at tool tip due to high material hardness that also leads to elevated workpiece temperatures contributed by lower thermal conductivity of these materials. Tool chatter arising due to large machining forces leads to rough surface finish which is further compounded by chemical reactivity due to high cutting temperatures [19, 20]. These effects are exaggerated with increased cutting speeds, making it very complex for machining titanium

Table 1 Summary of research work on dry machining different alloys using several types of cutting tools

Author/publication	Titanium alloy	Aim of investigation	Outcomes/findings
Komanduri et al. [44]	Ti-6Al-4V	Chip formation mechanism during turning.	–Rapid flank wear –Uninterrupted contact at/near the cutting tool tip –Higher reactivity to most tool materials –High temperature near shear bands
Hartung et al. [45]	Ti-6Al-4V	Determine the wear mechanism during titanium alloy turning machining	–Rapid desolation or chemical reaction occurs during machining process between titanium alloy and two conventional cutting tool materials including C2 grade (Carboloy 820) and C3 grade (Kennametal K68)
Dornfeld et al. [46]	Ti-6Al-4V	Examine the effects of tool dimensions and machining conditions on the burr cross-sectional shape during drilling	–During dry machining, four shapes of burr were formed, such as a uniform, roll back, lean back and roll back with expanded exit –Roll back phenomena happened because of the thermal expansion due to increasing local temperature at the inside burr face –During wet machining, there are three types of burr observed, namely standard even burr, burrs with a drill cap, burrs ring-shaped burr –Depending on the drill types, smaller burrs are produced by helical point drill –The height and the thickness of burr are reduced owing to enlarging helix angle with increasing point angle.
Jawaid et al. [47]	Ti-6246	Investigate the grooved carbide cutting tool wear during turning	–The rate of flank wear was increased rapidly when the feed and speed were increased –High cutting temperature facilitates diffusion wear which smoothens the worn area –Higher speed and feed cause chipping of cutting tool –A significant increasing in tool life achieved during dry machining when the feed and speed

(continued)

Table 1 (continued)

Author/publication	Titanium alloy	Aim of investigation	Outcomes/findings
			were 0.25 mm/rev and speed 60 m/min, respectively
López de lacalle et al. [8]	Ti64 & 718 Inconel	Study the influence of uncoated hard cutting tools [quality (K10) microgram] and coating tools [titanium carbide (TiC) and titanium-aluminium nitride (TiAlN)] with milling machining parameters on the machinability	–The chip thickness was enlarged with rising depth of radial cut and decreasing feed when using uncoated tool –The development in the cutting tool life was achieved with using uncoated tools at the range of clearance angle from 18° to 20° –The coating material caused a delay the wear in flank
Che-Haron [22]	Ti-6Al-2Sn-4Zr-6Mo	Investigate wear of tool and integrity of machined surface during turning	–Straight grade cemented carbides tool is more convenient to machine this type of alloys –Chipping and wear in flank face are key reasons for the failure of tools –Machined surface contains austere tearing and plastic deformation –Machined surface hardness is higher than that of original material
Nabhani [48]	TA48	Experimentally study the work quality of the polycrystalline diamond (SYNDITE) cutting tool in turning titanium alloy	–SYNDITE PCD is the most functional cutting tool due to its minimum wear rate and better surface finish
Wang et al. [49]	Ti-6Al-4V	Investigate the effect of binder of cubic boron nitride tools during high speed milling	–The wear in the flank is not even and it dominates the wear of binder-less tools. –The binder-less performed better during titanium alloys machining –The optimal machining with use BCBN tools at depth of cut 0.075 mm, speed 400 m/min, and feed 0.075 mm/rev

(continued)

Table 1 (continued)

Author/publication	Titanium alloy	Aim of investigation	Outcomes/findings
			–At higher feed and depth of cut, Ti-6Al-4V adhered to tool rake face
Wang et al. [50]	Ti-6Al-4V	Examine the wear characteristics of cubic boron nitride tool without binder while milling at high speed	–Grooves of microsize appear on the flank during the initial cutting stage owing to the abrasion –Workpiece material diffuses from the tool into chip
Che-Haron [51]	Ti-6Al-4V	Investigate the surface integrity during turning with uncoated carbide of ISO label cutting tools, CNMG 120408-883-MR4 and CNMG 120408-890-MR3	–Surface finish is slightly worse at slower speed, whereas surface finish improves when the speed rises. –Surface microstructure experiences plastic flow and tearing when the tool fails –Machined surface has higher hardness than that of original material
Nouari et al. [52]	Ti-6242S	Study the effect multi-layer CVD-coating during end milling	–The local wear dominates tool performance –Higher speed, lower feed and a smaller depth of cut give improved finish of machined surface
Che-Haron et al. [53]	Ti-6246S	Examine the influence of using the alloyed carbide tools on the end milling	–The carbide without coating and CVD-coated carbide tools are better for titanium alloys milling
Nurul Amin et al. [54]	Ti-6Al-4V	Compare the tool wear, surface finish while using polycrystalline diamond (PCD) and uncoated tungsten carbide–cobalt (WC-Co) during end milling	–Faster machining and longer tool life with uniform wear in PCD compare to that of uncoated WC-Co –To avoid tool wear, the allowed speed should be below 120 m/min for PCD, while for WC-Co is up to 40 m/min –Because of the better tool performance and lesser chatter in PCD, the surface roughness is lower.

(continued)

Table 1 (continued)

Author/publication	Titanium alloy	Aim of investigation	Outcomes/findings
Jianxin et al. [55]	Ti-6Al-4V	Investigate diffusion wear during turning	–The tool wear is accelerated due to the element diffusion to and from the chips and tool
Dargusch et al. [56]	Grade 2 Titanium (99.8% Ti)	Study the effect of end milling without and with lubricant on the deformation zone depth	–Residual compressive stress increases while the speed rises during dry machining –Deeper deformation layer and higher sub-surface twin density observed during dry machining than machining with lubricant at higher cutting speed (48.3 m/min) –Microhardness of the machined surface without lubricant increases slightly with the increase of compressive residual stresses
Ibrahim et al. [57]	Ti-6Al-4V	Analyse the surface integrity after turning with coated carbide cutting tools	–Better surface with acceptable, cracks and tears was achieved at 95 m/min speed, 0.35 mm/rev feed and 0.1 mm cutting depth –Higher surface roughness (4.31 μm) at initial machining then regularly decrease until the surface becomes smoother (2.92 μm) at the end.
Sun et al. [58]	Ti-6Al-4V	Understand the chip formation and machining force generation during turning	–At high feed rate and low machining speed, both the continuous and segmented chips are generated due to a static force and a recurring force, respectively –The slip angle is lower in continuous chips than that in the segmented chip
Abdel-Aal et al. [59]	Ti-6246S	Effect of thermal conductivity of cutting tool material on alloyed carbide (WC-Co) cutting tools during turning	–The cutting temperature and thermodynamic forces such as stress, strain rate, and temperature gradient, are led to anisotropy of the tool material's thermal conductivity

(continued)

Table 1 (continued)

Author/publication	Titanium alloy	Aim of investigation	Outcomes/findings
			–The high gradient of flank wear was noted at machining speed 150 m/min when cutting with tool without coating –The flank wear is noted at the edge of the tool during machining by coated cutting tool at 125 m/min and 0.2 mm/rev speed and feed respectively
Fang et al. [60]	Ti-64 & 718 Inconel	Study the machinability in term of force ratio (cutting force/thrust force) during high speed turning	–Machining forces reduce at the rise of speed, whereas the force ratio increases –Increasing feed caused increasing forces as well as force ratio
Armendia et al. [61]	Ti-6Al-4V & Ti54 M	Compare the Ti-6Al-4V machinability with Ti54 M machinability during the turning	–Ti54 M alloy machines better than Ti-6Al-4V –The lower crater and flank wear rates are noted easily during machining of Ti54 M owing to its microstructure –The machining forces of Ti-6Al-4V are larger than those of Ti54 M –In speeds 50–60 m/min, both the titanium alloys showed adiabatic shear bands during chip generation
Özel et al. [62]	Ti-6Al-4V	Investigate the effect of single and multi-layered coatings of cBN and TiAlN on tungsten carbide cutting tools during turning	–Both coatings show decreasing cutting forces at around 50 m/min speed, while, increased near to 100 m/min speed due to larger edge radius, especially in multi-layer coated tool
Dass et al. [63]	(Grade 5) alloy (89.75% Ti)	Explore the influence of cutting depth, speed, approach angle as well as feed on the surface finish and machining forces while turning	–The decreasing of feed affects more substantially than other Variables on the reduction of force and roughness –On the contrary, increasing approach angle has a slight effect on the improvement of surface finish

(continued)

Table 1 (continued)

Author/publication	Titanium alloy	Aim of investigation	Outcomes/findings
Honghua et al. [64]	TA15	Compare the performance of polycrystalline cubic boron nitride (PCBN) as well as polycrystalline diamond (PCD) in terms of surface finish, tool wear, wear morphology and tool life while milling	–At higher cutting speeds, PCBN shows short tool life compared to the PCD –Non-uniform flank wear occurred for both two types of tools
Mhamdi et al. [65]	Ti-6Al-4V	Investigate the influence of hemispherical shape of tool, speed and feed on the machined surface of during milling	–The significant improvement of surface roughness was achieved during cutting upward and downward surface with hemispherical tool compared when machining at the top of concave surface –The increment in speed feed (900 mm/min) leads to increase the roughness of the surface
Ugarte et al. [66]	Ti-6Al-4V, & Ti-5553	Study the machinability of two titanium alloys during three different operations, such as face milling, interrupted cutting and orthogonal milling	–Adhesion wear is appeared Very clearly in Ti-5553 machining compared with Ti-6Al-4V machining. –Machining forces while machining Ti-5553 alloy are larger than the forces of Ti-6Al-4V alloy –The cutting temperature for Ti-5553 alloy was greater than the temperature for Ti-6Al-4V alloy –Due to higher segmentation and higher temperature during machining, the Ti-6Al-4V alloy machinability is greater than those of Ti-5553 alloy
Nouari et al. [67]	Ti-6Al-4V & Ti-555	Study the effect of microstructure of workpiece material on tool wear as well as stability of the turning process	–The Ti-6Al-4V has higher thermal softening sensitivity than that of Ti-555 alloy –Due to a nodular structure and fine grains (1 μm) of Ti-555, it is harder to machine compare to Ti-6Al-4V alloy –Adhesive wear is higher during Ti-6Al-4V machining compare to that during machining of Ti-555

(continued)

Table 1 (continued)

Author/publication	Titanium alloy	Aim of investigation	Outcomes/findings
Balaji et al. [68]	Grade 2 unalloyed Titanium (99.8% Ti)	Examine the effects of rake angle, speed and feed on machining forces, temperature and chips generation in turning	–higher speed with positive rake angle leads to reduced forces, cutting tool wear and temperature
Cotterell et al. [69]	Ti-6Al-4V	Investigate the shear strain and cutting temperature during milling	–The shear strain increased with cutting speed increases, while, slight changes in shear strain noticec with a Variation of feed rate Value –Temperature at cutting zones increases with increasing cutting speed. –Increase cross section of the chip due to increase cutting feed rate chip section which in turn increases friction between chip and tool
Shetty et al. [70]	Ti-6Al-4V	Determine optimum drilling parameters for the better hole quality and chip type	–The shape of chip at low cutting speed is spiral cone with minimum thickness. –The size of formation burr was minimum at reduced cutting speed –Surface finish was better at high speed (20 m/min) because of increased heat generated which is lead to softening the material of workpiece at the top layer
Li et al. [71]	Ti-6Al-4V	Study the effect of deep submillimeter-scaled texture on the cutting forces and the friction properties	–Implanted three types of texture on rake faces of cutting tools caused a substantial influence on the distribution of stress and therefore affects tool life

to a precise size and shape at higher cutting speed [21, 22], hence imposing limits on production economy and efficiency. Therefore, the present industry trend is to search for effective approaches or methods for improving machinability of these alloys.

Some investigations are dedicated to the development and implementation of technological innovations [23, 24] in overcoming operational challenges and enhance performance capabilities [25]. At the forefront of such techniques is thermally assisted machining (TAM) that reduce ductile deformation in the cutting zone [26, 27]. In this, the workpiece surface is heated just before machining and the material temperature is raised to reduce yield strength locally within the deformation zone at the cutting tool edge, thus minimising cutting forces [28]. TAM process requires operating conditions to be managed inside an optimal domain for a certain workpiece material for obtaining the desired outcome, such as expected surface integrity and extended cutting tool life [27, 29].

For machining titanium alloys, TAM technique has been attempted with different heating sources and configurations, namely high-frequency induction coil [30–33], plasma [34–38], and laser beam [39–43]. These studies have appraised potential improvement in the machinability of titanium-based alloys. This paper provides an exhaustive compilation and a detailed review of these investigations, identifying specific issues related to titanium alloys machining and the merits of various TAM approaches, including workpiece preheating before/during the machining. Specific emphasis is placed on the reduction of machining forces, surface roughness and tool wear.

The following section presents a detailed discussion on the difficulties and issues associated with machining of titanium-based alloys.

2 Machining Difficulties and Issues with Titanium Alloys

In practical sense, titanium machining brings about operational challenges because of the manufacturing needs to achieve higher cutting speed and material removal rate with these alloys having extraordinarily hard mechanical properties and poor thermal characteristics in maintaining production efficiency and cost-effectiveness [72, 73].

During machining of titanium alloys, excessively large cutting forces are produced along with vigorous tool chatter due to very high yield stresses of these alloys, while highly elevated tool/workpiece temperatures are generated from unyielding plastic deformation under low ductility. These interactive effects lead to poor surface finish, which is adversely affected by increased cutting speeds and rate of material removal. Additionally, reduced ability to conduct heat hinders heat dispersion and creates highly concentrated temperature rise within the cutting zone and at the tool tip. Consequently, high-temperature chemical reactivity is triggered at the machining surface while the tool tip undergoes burnout reducing tool life [74].

Each property of titanium has different effects on the machinability [20, 74]. Past studies identify that, when machining titanium alloys, the workpiece surface undergoes work hardening process at the cutting edge, making the material harder than normal for the cutting tool. Increased cutting force arising from this effect is detrimental to the machining process and leads to very high cutting tool wear rate. Moreover, it has been reported that both compressive and tensile residual stresses may present at the cutting zone in titanium alloys machining [75] depending on the machining processes. The tensile stresses are not beneficial and should be minimised while compressive stresses are useful for fortifying fatigue strength [76]. TAM techniques enable to release these in varying degrees.

Upon applying cutting tool pressure, machining processes generate large cutting forces and strong friction at the tool tip from plastic deformation at the workpiece–tool contact interface [77]. Intense heat is also invariably produced at the cutting zone, largely from plastic deformation and to a lesser extent by tool tip friction [63, 78]. Cutting forces and heat generation are acutely stronger with titanium alloys having higher hardness and high-temperature strength [79]. It has been observed that owing to elevated cutting zone temperature, a built-up edge is formed on cutting tool as a protective layer of workpiece material [80]. Thickness of this layer tends to decrease with increased cutting speed, causing higher rate of tool wear. Additionally, acutely high temperature and resulting thermal stresses over workpiece–tool contact length is a known major reason for micro/macro-crack development along the cutting edge, which is designated as edge depression [81]. Figure 1 typically illustrates cracks developed at the tool's cutting edge during machining of Ti-555 alloy [67].

In machining, cutting zone temperature plays a crucial role in determining product surface quality, manufactured tolerances and tool wear [82]. Typically in machining, 90% of the work expended on plastic deformation is transformed to heat creating intense temperature rise within the cutting zone [28]. About 75–80% of this heat is captivated in cutting tool [19, 45], whereas the rest of heat is taken away by the chips and workpiece. In easy-to-machine materials, chips may remove up to 25% of the generated cutting heat [83].

Titanium alloys tend to have poor thermal conductivity values [56, 84] because of the constrained molecular slip in hexagonal crystal microstructure of these alloys. This naturally hinders the heat diffusion process within the workpiece and creates highly concentrated thermal stresses and steep temperature rise in the

Fig. 1 Cracks at cutting edge after machining Ti-555 alloy on a heavy-duty lathe machine [67]

cutting zone [56, 70, 85–87]. Affected by elevated temperatures and compressive stresses, the plastic deformation occurring within the primary cutting region produces adiabatic shear bands [78, 88–90].

At high cutting speeds, rapidly forming chips would have lesser contact length with the tool face and somewhat lower friction coefficient. However, resulting from reduced heat removal by chips [19, 45, 70], the cutting temperature is significantly elevated over the small interface between the tool face and chips [83, 91]. This tends to severely damage the tool cutting edge, as shown in Fig. 2 [47]. With titanium alloys machining, higher localised tool temperature is a serious and concerning issue, especially at large cutting speeds.

Retention of high yield strength at higher temperature with lower elastic modulus adversely affects the titanium alloys machinability [22, 73, 85]. With such conditions, cutting tool pressure would deflect the workpiece significantly leading to tool chatter/vibration and undercut [92, 93], hence creating detrimental effects of poor surface quality and tolerance. It has also been noted that, under the compressive force exerted by the tool, high cutting zone temperature would weld some workpiece material to the tool edge. This welded material is built-up edge (BUE) [32]. Figure 3 [61] shows the BUE which was generated during machining titanium alloys with cemented carbide (WC/Co) tools at machining speed of 60 m/min [80].

Fig. 2 Damaged cutting tool edge in machining Ti-6Al-4V alloy at speed and feed of 100 m/min and 0.25 mm/rev, respectively [47]

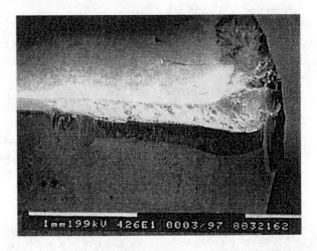

Fig. 3 Build-up edge (BUE) at cutting tool edge [61]

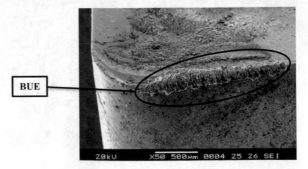

Chemical reaction initiates BUE formation at the tool-chip contact zone, which induces chamfering and chipping of the cutting edge [46, 94]. This increases the cutting forces eventually beginning to deflect the workpiece [94]. Therefore, BUE is regarded as a root cause for producing unacceptable surface finish and poor dimensional accuracy [61, 95, 96]. The development of BUE is influenced by speed, machining temperature, workpiece material and the tool. A stable BUE plays the significant role of acting as a protective layer for the cutting edge [97]. To the contrary, an unstable BUE adversely impacts on both the cutting tool and surface finish. Trent [98] has stated that the BUE reduces the contact length, implying high cutting temperature and high stress concentration simultaneously occur restricted to a minor region at the cutting edge (within 0.5 mm) [99]. This increases pressure load on the tool edge while machining of titanium alloys [18]. Under limited machining situations, machining of titanium-based alloys shows no tendency for BUE to form, depending on the titanium alloy type (Table 2).

It was noted that cutting tools are very rapidly consumed owing to cohesion between workpiece and the cutting tool while machining of titanium alloys [22, 84]. This is attributed to prevalent wear mechanisms, such as attrition, dissolution–diffusion and adhesion at contact length between workpiece and cutting tool face. Tool wearing triggers different types of tool failure [100, 101], as presented in Figs. (4, 5, 6 and 7) [67]. These wear mechanisms do occur due to microhardening and topography of material beneath the machined surface in machining of titanium alloys [102, 103].

Coated tools may get delaminated at the cutting edge making it blunt [64, 104], as illustrated in Fig. 5, because of incompatibility in thermal expansion coefficients of coating and tool materials [59]. Ezugwu et al. [105] have identified that poor thermal diffusion leading to higher machining zone temperature is the main reason of tool consumption during machining of Ti-6Al-4V. Amalgamation of increased machining load as well as elevated temperature at upper speeds causes rapid chipping on tool edge [21, 106, 107], as shown in Fig. 6 [49]. High cutting temperature also produces adhesion of small morsels of workpiece material on cutting tool [59, 85], such as depicted in Figs. 6 [49].

The adhesion erodes the tool flank material, and such effects are accelerated when the cutting depth and machining speed rise in high speed milling with PCD/PCBN tools [64]. Repetitive adhesion of workpiece material promotes BUE formation that gradually alters the cutting edge geometry [61]. For these reasons, it is hard to machine titanium alloys with HSS steel cutting tools at speed 30 m/min and by tungsten carbide tools at 60 m/min. Ceramic and cubic boron nitride tools are not appropriate for machining titanium alloys since these react with titanium alloys [64].

Titanium alloys react with many cutting tool materials, such as CBN and PCBN at elevated (cutting) temperatures typically above 500 °C [3, 83]. In this thermally triggered reactions, the cutting tool material is removed on an atomic scale [59, 111] producing a Tribo-chemical wear process. This mechanism can occur as molecular–mechanical wear or corrosive wear or as a combination of both [81, 84]. Such fast reactivity between workpiece and tool causes smearing, galling and

Table 2 Machining conditions for no built-up edge development

Author/year	Titanium alloy	Cutting tool	Speed (mm/min)	Feed (mm/tooth or mm/rev)	Depth of cut (mm)	Machining type
Amin [54]	Ti-6Al-4V	Uncoated cemented carbide WC/Co and polycrystalline diamond (PCD)	160	0.1	1.0	Milling
Armendia [61]	Ti-54 M & Ti-6Al-4V	Uncoated cemented carbide WC/Co	80	0.1	2.0	Turning
Rashid [142]	Ti-6Cr-5Mo-5V-4Al	Uncoated tungsten carbide	160	0.19	1.0	Turning
Jawaid [47]	Ti-6246	Tungsten carbide CNMG 120408-(890 & 883)	100	0.25	2.0	Turning
Ugarte [66]	Ti-6-4 MA	PVD coated (TiAlN-TiN)	40	0.15	2.0	Milling
	Ti-6-4 STA	cemented carbide	40			
	Ti-5553		25			
Wang [49]	Ti-6Al-4V	Cubic boron nitride (CBN) and Binder-less CBN (BCBN)	350	0.1	0.1	Milling

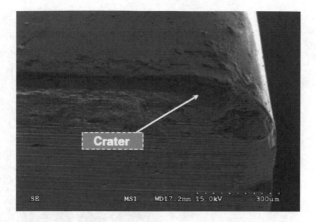

Fig. 4 Crater wear at cutting edge in machining titanium alloy (Ti-555) [67]

(a)   **(b)**

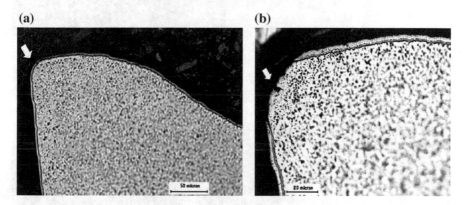

Fig. 5 Delamination of coated cutting tool (WC-Co) in machining Ti-6242S alloy [59]

chipping of the machined surfaces [59, 108] as well as fast cutting tool wear [74], as depicted in Fig. 7. Additionally at elevated cutting temperatures, titanium alloys undergo oxidation forming compounds [109], while chips may get welded on to the cutting tool face and machined surface [52, 57, 67, 110], causing machining issues.

Chip formation in machining is categorised to be (i) continuous chip, (ii) continuous chip with BUE and (iii) discontinuous chip [3, 112]. Chips produced during titanium machining are observed to be discontinuous or segmented [80], particularly at lesser speed around 40 m/min [112, 113], as illustrated by Fig. 8 [114]. This is because, with titanium alloys, the plastic deformation within the primary cutting zone is very difficult and said to undergo "Catastrophic thermos plastic shear" [61, 83], where the process is unstable and non-homogeneous [88] due to extreme material hardness of these alloys. Consequently, the chip formation occurs with cyclic frequent chip breakage [88] and increased flank wear.

Discontinuous chips are observed to have much shorter contact length with tool face compared to a continuous chip—approximately one-third length [18, 44, 74].

312 O. A. Shams et al.

Fig. 6 Adhesion at cutting edge during machining of Ti-6Al-4V alloy [49]

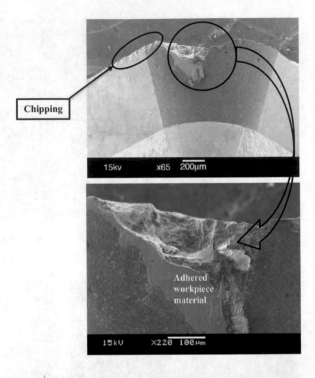

Fig. 7 Wear and adhering workpiece material at rake face while high speed milling of Ti-6Al-4V alloy [50]

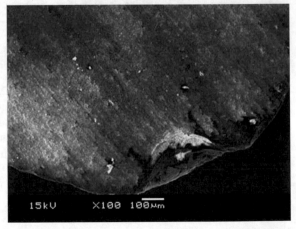

Pramanik and Littlefair [96] have stated that the microstructure of titanium alloy determines the cutting speed at which segmented chip formation would begin. For titanium alloys, Fig. 9 [113] illustrates various chip formation shapes/types associated with different cutting speeds [44, 61].

Discontinuous chips have tendency to induce fluctuations in cutting forces, especially when alpha–beta alloys are machined [99]. Shivpuri et al. [109] have

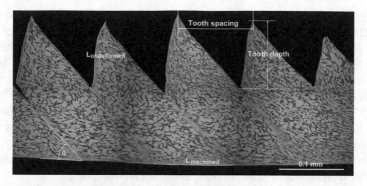

Fig. 8 Geometrical features of segmented chip formation [114]

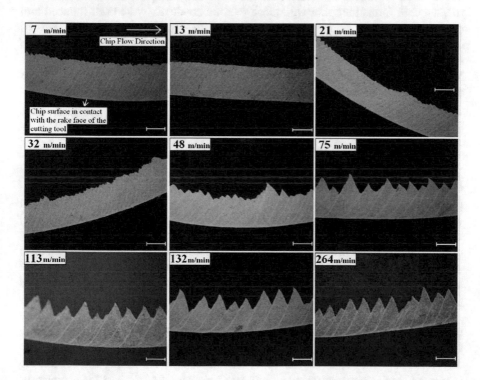

Fig. 9 Types of chip microstructures in conventional machining of titanium [113] (*Scale bar represents 100 µm*)

reported that the phenomenon of chip segmentation would limit the metal removal rates during cutting operation and causes cyclic vibration of forces.

The self-excited vibration significantly affects the formation of chips as well as machined surface, which often referred to as chatter [80, 115]. The changing frictional force due to chip flow on tool face generates shear localisation and also contributes to forced vibration and self-excited chatter [19, 116]. These vibrations can lead to catastrophic tool failure. It has been observed that the tool vibration and chatter from segmented chip are associated with the thrust components of the cutting forces and the periodic oscillation during machining [54, 91]. Also, the serration chip intensity increases the fluctuations of cutting forces and induces dynamic stresses at the cutting tool, triggering high tool wear [18, 117].

Chip formation instability induced by high cutting temperature and the micro-fatigue loading thus applied on the tool tip is accountable for austere wear in flank and decreased cutting tool life [118, 119] in titanium alloy machining. These invariably affect the stability of the cutting edge leading to sub-surface defects and inaccurate dimensions of the machined surface [92, 105, 120].

In machining hard titanium alloys, extraordinarily high cutting temperatures are encountered owing to the combined effects of difficult plastic deformation, segmented chip formation, poor workpiece thermal conductivity and self-induced tool chatter, resulting in unacceptable machined surface quality and various forms of cutting tool damage to shorten tool life. All these detrimental effects lead to machining process inefficiencies and higher manufacturing product cost. The following section examines the status-quo in methods to increase the machinability of titanium alloys, based on thermally assisted machining (TAM) approaches.

3   Thermally Assisted Machining (TAM)

Over the last four decades, different techniques have been considered and applied in overcoming problems associated with surface quality and tool wear during titanium alloy machining [104]. Among these methods, thermally assisted machining (TAM) has emerged as a forerunner, which involves workpiece preheating with precisely coordinated external heat sources. This technique was first devised by Tigham in 1889 and since then has evolved to use heating methods incorporating plasma, laser and induction coil systems with varying degrees of thermal generation intensities [121, 122].

Conceptually, TAM approach applies localised heating in advance of the cutting point to soften the workpiece surface layer that reaches cutting tool for machining [123]. The external heating locally improves ductility of workpiece and reduces forces required for plastic deformation, thus potentially alleviating the concerning issues of high temperature rise and tool chatter [91, 124, 125]. Figure 10 schematically shows a TAM configuration, where the heat source position and power are essentially dependent on machining variables and the distribution of temperature to be achieved in workpiece. In some cases, instead of the workpiece, the cutting tool is heated to manage the behaviour of work material [126, 127].

Due to low thermal conductivity and material hardness, in machining titanium alloys, TAM requires careful and appropriate selection of preheating method based on source heating power, source position, machining speed, depth of cut and feed. Also, the material of tool should be capable to endure more than 300 °C surface

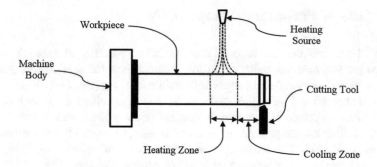

Fig. 10 Preheating workpiece during turning process

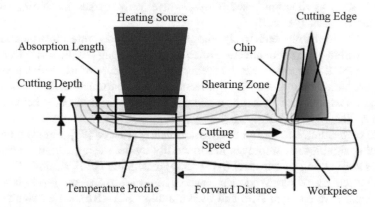

Fig. 11 Preheating workpiece during milling process [128]

temperatures and have steady chemical and thermal behaviours, and adequate resistance to wear.

It is noted that the heating strategy for turning process is slightly different to that of milling, due to rotational differences in the cutting configurations of the two methods. In turning, the cutting area is repeatedly heated [104] therefore would require less heating intensity compared to milling, where the cutting area does not move as fast as in turning. A typical TAM arrangement for milling is shown in Fig. 11 [128] along with workpiece temperature distribution.

In TAM, the effectiveness of the approach is critically dependent on the pre-heating temperature, which requires careful control to prevent overheating of the workpiece and material deformation [121, 129]. In selecting external heating, considerations should include not to over soften the surface layer in the primary cutting zone, which can lead to unwanted deformation that can bring about other machining complexities [3, 41].

4 Effects of Preheating Temperature

Figure 12 presents the influence of temperature on mechanical strength of some
selected hard-to-machine materials [128], where it is noted that heating generally
weakens materials by reducing stress and strain hardening rate. This effect, which is
known as thermal softening [96, 104], is the principle behind TAM technique.

In TAM, workpiece temperature within the primary shear zone is raised above
the recrystallization temperature by external heating to reduce the workpiece yield
strength [130, 131]. This assists plastic deformation and reduces work hardening
effects and cutting force requirements in machining processes [133, 134]. Thus,
power consumption is reduced while material removal rate and productivity [129,
135] are increased. This approach is particularly beneficial for hard-to-machine
materials such as titanium-based alloys, where TAM makes machining of these
alloys much easier [41, 132].

For TAM to operate effectively, the preheating temperature must be controlled
and kept within a range not affecting microstructural change of the workpiece [104].
In this regard, it is reported that the preheating temperature of titanium alloys needs
to be maintained below 882 °C to prevent phase transformation [83]. Also, it has
been observed that rate of strain hardening decreases significantly between temperatures 200–734 °C [128].

With preheating, cutting forces are dramatically reduced [96]. For example, laser
preheating significantly improves machinability by decreasing cutting forces by as
much as 20% during carbon steel (AISI 1536) machining [136]. Amin [137] studied
the influence of whole workpiece preheating in furnace on machinability of
high-temperature resistant steel and titanium alloy BT6 (Russian Standard). It has
been found that the preheating tends to stabilize chip formation process and
increases interaction length between cutting tool and chip.

Figure (13) [67] depicts force components in machining, namely cutting force
(F_c), feed force (F_a), normal force (F_n) and thrust force (F_t). Preheating reduces

Fig. 12 Relationship
between temperature and
mechanical strength of
selected hard-to-machine
alloys [128]

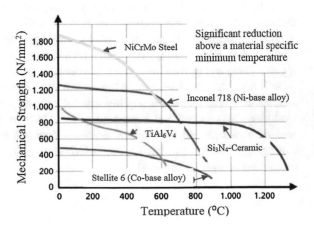

Fig. 13 Typical machining
forces while machining [67]

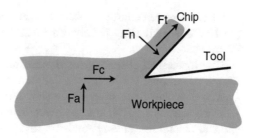

vertical component of the cutting force which reduces dynamic impact on the tool
and advances tool's mechanical and thermal fatigue characteristics, operational life
and machining cost-effectiveness [96, 138].

Sun et al. [139] explored the influence of laser preheat in milling of Ti-6Al-4V.
They noticed significant reduction of feed force within 200–450 °C temperature.
Ayed et al. [140] observed reduced machining forces with the laser power up to a
certain value. Heating beyond this adversely affected because of chip melting and
melted chips sticking to the tool rake face, especially when the laser-cutting tool
clearance was 3 mm.

Figure 14 [140] illustrates the cutting force reduction with the applied laser
power for preheating. It is noted that a more than 50% reduction in machining
forces was achieved when the laser beam and cutting tool gap was 5 mm. However,
increased feed and speed impart opposite effect on the reduction of machining
forces. It is concluded that a substantial decrease of forces is generally possible with
higher levels of external heating.

The machining force is linearly proportional to the variation of feed and/or depth
of cut [141] in the course of laser-assisted machining of titanium alloys
(Ti-6Cr-5Mo-5V-4Al) [142] and (Ti-10V-2Fe-3Al) [113]. During the turning of
(Ti-6Cr-5Mo-5V-4Al) alloy, the cutting forces decrease by about 15% at laser
power of 1200 W, feed 0.15–0.25 mm/rev and speeds within 25–100 m/min.

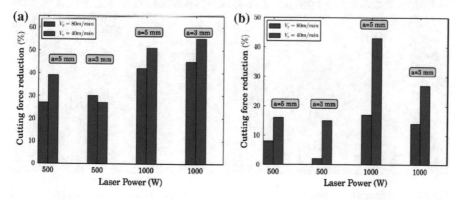

Fig. 14 Reduction rate of machining force while machining Ti-6Al-4V alloy at feed, **a** 0.1 mm/
rev, and **b** 0.2 mm/rev [140] with speed 40 and 80 m/min and cutting depth 1 mm

Whereas, the percentage of decrease in machining forces during machining of (Ti-10V-2Fe-3Al) alloy is between 10 and 13% in machining speed range of 45–70 m/min.

Due to thermal softening, workpiece preheating also reduces the amplitude of force fluctuations, which in turn dampens tool vibration [143, 144]. Hence, damages to rake and flank faces are lessened, as illustrated by Fig. 15 [142].

Dandekar et al. [138] studied machinability of Ti-6Al-4V alloy in terms of the specific machining energy, tool life and surface roughness with laser assistance. A marked decrease in the rate of tool wear was noted with increased material removal temperature (T_{mr}) up to (250 °C). At this temperature, cutting tool life increased by about 1.7 times at machining speed below 107 m/min compared to that of conventional machining. The wear rate of tool started to increase with the rise of workpiece temperature over 250 °C.

TAM offers further benefits in reducing the occurrence of adhesion of chip fragments and BUE at the tool cutting edge, which are direct implications of high temperature build up in the plastic deformation zone. With thermal softening induced by preheating, TAM improves surface finish with fewer surface cracks and less porosity than in traditional machining [145]. This is further evident from the fact that the microhardness at the surface layer is decreased with higher levels of

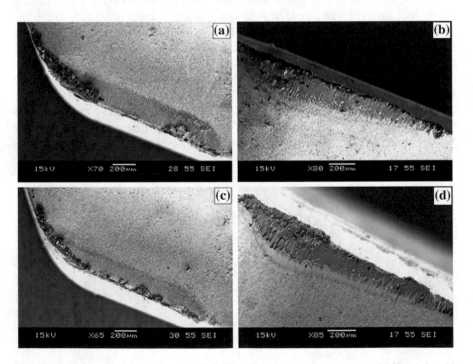

Fig. 15 Tool damage, (**a**) rake face (**b**) flank face of cutting tool inserts at speed 160 m/min with laser heating, and (**c**) rake face (**d**) flank face of cutting tool inserts under similar machining circumstances under without laser heating [142]

preheating temperature because of the reduced density of dislocation [146]. The increased preheating temperature stimulates grain growth and larger grains stay close to the heated surface of the workpiece. Increased grain size reduces strain hardening effect, and as a result, the hardness increases as the distance from the workpiece surface rises along the depth.

Several variables influence the yield strength of the workpiece during TAM and affect the machined surface. These variables are heat source power, the gap between heat source and tool, and the distance and angle between the workpiece surface and heat source. Germain et al. [147] observed a slight reduction in the surface roughness between 0.3 and 0.6 μm, when the laser power was between 500 and 1000 W while machining of Ti-6Al-4V alloy at speed 26 and 54 m/min.

Localised heating softens the surface and changes the material removal mechanism. These affect the morphology and microstructure of the chips in TAM where the chips are formed with clearer edges, uniform thickness and homogeneous segmentation [139]. Due to high temperature in TAM, the chip morphology changes from sharp saw-tooth (brittle fracture type) to a continuous chip (plastic flow type). This is represented by the significant difference in both the depth of sow-teeth and segment spacing, as shown in Fig. 16 [140]. This type of chips reduces the surface roughness and minimises self-induced vibration [120, 148]. The segmented chip is occasionally found at low cutting speed. With increased preheating temperature, the deformability of work material at heated surface improves. Hence, the continuous chips are produced at higher machining speeds. It has been noted that the localised melting in the chips occurs at less than 20 m/min speeds, as illustrated in Fig. 17 [113]. It is also reported that the increased surface temperature raises chip thickness and reduces the distance Dc (shown in Fig. 16) and shear angle [140]. All these increase the frequency of the chip segmentation.

Using laser of 1500 W, Braham-Bouchnak et al. [149] have observed semi-continuous and continuous chips owing to the increase of saw-tooth frequency. It is also reported that the chip shrinkage coefficients decreased with higher heating temperature that increases chip length [150], or the reduction in chip shrinkage

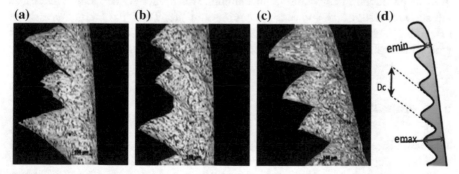

Fig. 16 SEM views of chip sections at varying laser powers at speed 40 m/min and feed 0.2 mm/rev, **a** no laser, **b** 500 W, **c** 1000 W, **d** schematic chip morphology [140]

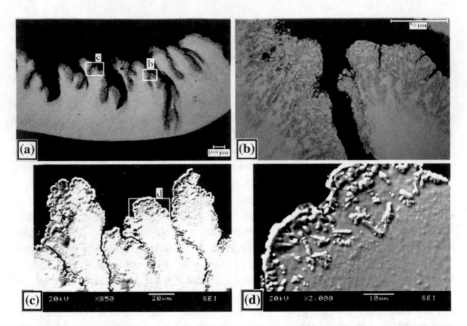

Fig. 17 a, b Chip microstructures formed in laser-assisted machining at speed 7 m/min, **c, d** SEM secondary electron images of dendrites at the same chip surface [113]

coefficient leads to stable and thinner chip. Using polycrystalline diamond (PCD) inserts for machining, Ginta et al. [150] have noted that TAM reduces vibration amplitude from 53 to 86% due to preheating at 650 °C.

Above review identifies that TAM decreases the flow stress and strain hardening of workpiece material, hence significantly suppressing tool chatter/vibration and improving tool life. Elevated temperature in TAM facilitates smoother material flow over the flank as well as rake faces of the cutting tool thereby improving the machined surface. Chips formed under proper thermal conditions tend to be thin, long and continuous. These would increase the chip–tool contact length compared to those produced in conventional machining. This longer contact length decreases the stresses at tool face and extends cutting tool life. Thus, the thermally assisted machining technique represents an effective solution to machining hard materials with an acceptable cost with improved productivity and quality.

5   Comparative Analysis

Titanium alloy machining is usually performed at lower speed. Therefore, product manufacturing cycle time and cost are major concerns and issues. Currently, many methods have been introduced and attempted to overcome these difficulties, where TAM is regarded as an emerging technology for advanced manufacturing.

Table 3 Summary of TAM techniques used in titanium-based alloys machining [122]

Features/TAM technique	Critical parameters	Controlling heating operation	Shape of heating area	Class of heat concentration	Advantages
LAM	–Laser beam incident angle –Spot size. –Cutting tool-laser beam distance –Laser power	Trouble-free operation	Spot area	High class	–Improving MRR –Decreasing 30–60% in cutting forces –Getting a better surface finish –Reducing tool wear around 90%
PAM	–Voltage electric arc. –Type of material –Surface temperature –Emissivity of surface	Difficult operation	Localised area	High class	–Decreasing 20–40% in cutting forces –Increasing MRR around 2 times –Enhancing cutting tool life around 1.5 times –Lower power requirement
IAM	–Frequency and intensity of the input electric current –Induction coil shape –Specific heat, magnetic permeability, and current flow resistance of the material	Simple operation	Surface area immediately adjacent to the coil shape	Low class	–Reduction of acceleration of amplitude to range 25% –80% –Reducing the cutting force in range 36–54% –Increasing by around 206% and 214% for tool life and volume of metal removal, respectively –Decreasing sharply in surface roughness up to certain heating temperature, then beyond this temperature slightly increases –Almost continuous chip produced with increasing heating temperature

Workpiece preheating in TAM, either immediately prior to or during cutting operation, improves machinability from assisted plastic deformation. In this, the optimum workpiece surface temperature delivers maximum benefits, where thermal assistance would vary from alloy to alloy.

Table 3 [122] summarises commonly used thermal assistance techniques to advance titanium alloy machinability. These are: laser-assisted machining (LAM), plasma-assisted machining (PAM) and induction-assisted machining (IAM). These TAM techniques are carried out using a laser beam, plasma torch and an induction coil, respectively. LAM requires complex system components. Nonetheless, controlled operation of heat source movement is much easier than the other two techniques. Due to high laser power, microhardness at cutting zone is minimised, which in turn reduces tool pressure on the machined surface. Therefore, tool life significantly extended as a consequence of reduced cutting forces and rate of tool wear. Similar to LAM, cutting forces are reduced by PAM and IAM techniques. However, achievable cutting tool life is shorter than with LAM. This is because, rate of flank wear tends to be more due to the high chip temperatures formed with these two techniques. Also, the power density of PAM is low; therefore, heating temperatures are difficult to control compared with LAM. Therefore, laser techniques are highly recognised for improved titanium alloy machinability in terms of lower machining forces, higher metal removal rate (MRR), lesser heat-affected zone, reduced tool wear and better integrity in machined surface.

The optimum benefits of TAM are achieved at certain ranges of feed, speed and the temperature from preheating, depending on the behaviour of titanium alloys, such as the ability to conduct heat and modulus of elasticity. Within these limits, thermal effects produced by surface heating leads to significant decrease in cutting forces at tool cutting edge. Although below these parametric ranges, a significant reduction of machining forces and machining energy is observed, and surface finish becomes unacceptable. In contrast beyond these ranges, tool wear is observed to be significantly higher.

6   Conclusions

This study provides a comprehensive appraisal of current state of development in thermally assisted machining (TAM) techniques for improving the machinability of titanium alloys. It is identified that the main strategy of TAM is to reduce yield strength and the work hardening effects of the workpiece. With appropriate workpiece heating in TAM, plastic deformation is promoted in the cutting zone to reduce shear resistance at the cutting edge thereby lessening the cutting force and specific cutting energy while improving surface finish and tool life. Prevention of workpiece overheating is critical for successful implementation of TAM that is to be ensured by estimating the peak preheating temperature. As such, careful analysis of surface temperature and workpiece temperatures distribution forms an essential research element in studies to determine optimum preheating conditions and machining parameters. In achieving these objectives, predictive thermal models are viewed to be essential for clear understanding of heat transfer processes associated with TAM, supported by further experimentation.

Among current TAM approaches, Laser-Assisted technique is the most efficient in improving the machinability of titanium alloys in terms of low cutting forces, increased metal removal rate (MRR), reduced heat-affected zone, increased tool life and superior workpiece surface integrity.

Questions of the Chapter

1. What is the role of cutting speed on BUE thickness?

Answer: Thickness of BUE layer tends to decrease with increased cutting speed, causing a higher rate of tool wear.

2. Explain briefly the effect of high machining speed on cutting tool?

Answer: At high machining, minimum contact length occurs with the tool face and has a somewhat lower friction coefficient. This leads to reduce heat removal by chips, thereby increasing the cutting temperature over the small contact area between the chip and the tool face. This tends to severely damage the tool cutting edge.

3. Provide a list of the factors that BUE development depends on?
Answer: The development of BUE depends on the following factors:

   1. Cutting speed.
   2. Machining temperature.
   3. Material of both workpiece and tool.

4. What are negative impacts of BUE formation?

   Answer: 1. Reducing contact length
             2. Implying high cutting temperature and high stress concentration at the cutting edge.

This increases pressure load on the cutting tool edge during machining of titanium alloys.

5. What are the mechanisms of wear that occurred in titanium alloys machining?

Answer: (i) adhesion, (ii) attrition and (iii) dissolution–diffusion at contact length between workpiece and cutting tool face.

6. What does Tribo-chemical wear process produce?

Answer: This process produces fast reactivity between workpiece and tool which causes galling, smearing and chipping of the workpiece surface, and rapid cutting tool wear on the rake face.

7. How can you categorise chip formation?

Answer: Chip formation can be categorised as (i) continuous chip, (ii) continuous chip with BUE and (iii) discontinuous chip.

8. Describe briefly TAM approach?

Answer: TAM approach applies localised heating in advance of the cutting point to soften the workpiece surface layer that reaches cutting tool for machining.

9. Explain the role of local heating on the material of the workpiece?

Answer: The heating locally improves the ductility of the workpiece and reduces forces required for plastic deformation, thus potentially alleviating the concerning issues of high temperature rise and tool chatter.

10. What are the main conditions of TAM preheating method?

Answer: This method based on source heating power, source position, cutting speed, feed rate and depth of cut.

Glossary

Machinability The term machinability refers to how easily materials can be removed to reshape the workpiece during machining, permitting the removal of the material with a satisfactory finish at low cost. A material is said to have good machinability if tool life is long, machining forces are low, and the surface roughness is low. The high chemical reactivity of the material with many tool materials and the low elastic modulus contribute to its difficult machinability.

Thermally assisted machining (TAM) Thermally assisted machining is a high speed machining method for materials with low machinability, such as titanium- and nickel-based alloys. In TAM, the surface of the workpiece just ahead of the machining point is preheating locally by a suitable heating source. This reduces the strength of the material and cutting force in the milling and turning process substantially which increase the machinability of the difficult-to-machine materials.

Built-up edge (BUE) A built-up edge (BUE) is an addition of workpiece material against the rake face that grasps to the tool tip, separating it from the chip. It is not tool wear though it decreases the efficiency of the cutting edge. BUE changes the geometry of the cutting tool which induces dimensional inaccuracy and poor surface finish on the part.

References

1. Ezugwu, E., & Wang, Z. (1997). Titanium alloys and their machinability—A review. *Journal of Materials Processing Technology, 68*(3), 262–274.
2. Gorynin, I. (1999). Titanium alloys for marine application. *Materials Science and Engineering A, 263*(2), 112–116.
3. Yang, X., & Richard Liu, C. (1999). Machining titanium and its alloys. *Machining Science and Technology, 3*(1), 107–139.
4. Shokrani, A., Dhokia, V., & Newman, S. T. (2012). Environmentally conscious machining of difficult-to-machine materials with regard to cutting fluids. *International Journal of Machine Tools and Manufacture, 57,* 83–101.
5. Gurrappa, I. (2003). Characterization of titanium alloy Ti-6Al-4V for chemical, marine and industrial applications. *Materials Characterization, 51*(2), 131–139.
6. Boyer, R. (1996). An overview on the use of titanium in the aerospace industry. *Materials Science and Engineering A, 213*(1), 103–114.
7. Zhou, Y., Zeng, W., & Yu, H. (2005). An investigation of a new near-beta forging process for titanium alloys and its application in aviation components. *Materials Science and Engineering A, 393*(1), 204–212.
8. Pérez, J., Llorente, J., & Sanchez, J. (2000). Advanced cutting conditions for the milling of aeronautical alloys. *Journal of Materials Processing Technology, 100*(1), 1–11.
9. Ezugwu, E. (2005). Key improvements in the machining of difficult-to-cut aerospace superalloys. *International Journal of Machine Tools and Manufacture, 45*(12), 1353–1367.
10. Rack, H. J., & Qazi, J. (2006). Titanium alloys for biomedical applications. *Materials Science and Engineering C, 26*(8), 1269–1277.
11. Niinomi, M. (1998). Mechanical properties of biomedical titanium alloys. *Materials Science and Engineering A, 243*(1), 231–236.
12. Niinomi, M. (2008). Mechanical biocompatibilities of titanium alloys for biomedical applications. *Journal of the Mechanical Behavior of Biomedical Materials, 1*(1), 30–42.
13. Liu, X., Chu, P. K., & Ding, C. (2004). Surface modification of titanium, titanium alloys, and related materials for biomedical applications. *Materials Science and Engineering: R: Reports, 47*(3), 49–121.
14. Murr, L., et al. (2009). Microstructure and mechanical behavior of Ti–6Al–4V produced by rapid-layer manufacturing, for biomedical applications. *Journal of the Mechanical Behavior of Biomedical Materials, 2*(1), 20–32.
15. Thomas, M., Turner, S., & Jackson, M. (2010). Microstructural damage during high-speed milling of titanium alloys. *Scripta Materialia, 62*(5), 250–253.
16. Mantle A., & Aspinwall, D. K. (1998). *Tool life and surface roughness when high speed machining a gamma titanium aluminide, progress of cutting and grinding*, in *Fourth International Conference on Progress of Cutting and Grinding*, U. A. Turpan (Ed.) (pp. 89–94). China: International Academic Publishers.
17. Abele, E., & Hölscher, R. (2014). New technology for high speed cutting of titanium alloys. In *New Production Technologies in Aerospace Industry* (pp. 75–81). Berlin: Springer.
18. Venugopal, K., Paul, S., & Chattopadhyay, A. (2007). Growth of tool wear in turning of Ti-6Al-4V alloy under cryogenic cooling. *Wear, 262*(9), 1071–1078.
19. Oosthuizen, G. A., et al. (2010). A review of the machinability of titanium alloys. *R&D Journal of the South African Institution of Mechanical Engineering, 26,* 43–52.
20. Veiga, C., Davim, J., & Loureiro, A. (2013). Review on machinability of titanium alloys: The process perspective. *Reviews on Advanced Materials Science, 34*(2), 148–164.
21. Bandapalli C., Sutaria B. M., & Bhatt, V. D. (2013). High speed machining of Ti-alloys—A critical review. In *Proceedings of the 1 International and 16 National Conference on Machines and Mechanisms (iNaCoMM2013)* (pp. 324–331). IIT Roorkee: India.
22. Che-Haron, C. (2001). Tool life and surface integrity in turning titanium alloy. *Journal of Materials Processing Technology, 118*(1), 231–237.

23. Lauwers, B., Klocke F., & Klink, A. (2010). Advanced manufacturing through the implementation of hybrid and media assisted processes. *International Chemnitz Manufacturing Colloquium, 54*, 205–220.
24. Gupta, K., & Laubscher, R. F. (2016). Sustainable machining of titanium alloys: A critical review. *Proceedings of the Institution of Mechanical Engineers, Part B: Journal of Engineering Manufacture*, 0954405416634278.
25. Warap, N., Mohid, Z., & Rahim, E. A. (2013). Laser assisted machining of titanium alloys. In *Materials Science Forum*. Trans Tech Publ.
26. Madhavulu, G., & Ahmed, B. (1994). Hot machining process for improved metal removal rates in turning operations. *Journal of Materials Processing Technology, 44*(3), 199–206.
27. Tosun, N., & Ozler, L. (2004). Optimisation for hot turning operations with multiple performance characteristics. *The International Journal of Advanced Manufacturing Technology, 23*(11–12), 777–782.
28. Radovanovic, M. R., & Dašić, P. V. (2006) LASER ASSISTED TURNING. In *Research and Development in Mechanical Industry* (pp. 312–316), RaDMI 2006, Budva, Montenegro..
29. Rebro P. A., et al. (2002). *Comparative assessment of laser-assisted machining for various ceramics* (Vol. 30, pp. 153–160). Transactions of North American Manufacturing Research Institution.
30. Amin, A., & Abdelgadir, M. (2003). The effect of preheating of work material on chatter during end milling of medium carbon steel performed on a vertical machining center (VMC). *Journal of Manufacturing Science and Engineering, 125*(4), 674–680.
31. Amin, A. N., et al. (2008). Effects of workpiece preheating on surface roughness, chatter and tool performance during end milling of hardened steel D2. *Journal of Materials Processing Technology, 201*(1), 466–470.
32. Ginta, T. L., et al. (2009). Improved tool life in end milling Ti-6Al-4V through workpiece preheating. *European Journal of Scientific Research, 27*(3), 384–391.
33. Lajis, M. A., et al. (2009). Hot machining of hardened steels with coated carbide inserts. *American Journal of Engineering and Applied Sceices, 2*(2), 421–427.
34. Kttagawa, T., & Maekawa, K. (1990). Plasma hot machining for new engineering materials. *Wear, 139*(2), 251–267.
35. Leshock, C. E., Kim, J.-N., & Shin, Y. C. (2001). Plasma enhanced machining of Inconel 718: Modeling of workpiece temperature with plasma heating and experimental results. *International Journal of Machine Tools and Manufacture, 41*(6), 877–897.
36. Novak, J., Shin, Y., & Incropera, F. (1997). Assessment of plasma enhanced machining for improved machinability of Inconel 718. *Journal of Manufacturing Science and Engineering, 119*(1), 125–129.
37. De Lacalle, L. N. L., et al. (2004). Plasma assisted milling of heat-resistant superalloys. *Journal of Manufacturing Science and Engineering, 126*(2), 274–285.
38. Popa, L. (2012). Complex study of plasma hot machining (PMP). *Revista de Tehnologii Neconventionale, 16*(1), 26.
39. Jau B. M., Copley S. M., & Bass, M. (1981). Laser assisted machining. In *Proceedings of the Ninth North American Manufacturing Research Conference* (pp. 12–15). Pennsylvania: University Park.
40. Rajagopal, S., Plankenhorn, D., & Hill, V. (1982). Machining aerospace alloys with the aid of a 15 kW laser. *Journal of Applied Metalworking, 2*(3), 170–184.
41. Chryssolouris, G., Anifantis, N., & Karagiannis, S. (1997). Laser assisted machining: an overview. *Journal of Manufacturing Science and Engineering, 119*(4B), 766–769.
42. Dumitrescu, P., et al. (2006). High-power diode laser assisted hard turning of AISI D2 tool steel. *International Journal of Machine Tools and Manufacture, 46*(15), 2009–2016.
43. Thomas, T., & Vigneau, J. O. (1999). *Laser-assisted milling process*, Google Patents.
44. Komanduri, R., & Von Turkovich, B. (1981). New observations on the mechanism of chip formation when machining titanium alloys. *Wear, 69*(2), 179–188.

45. Hartung, P. D., Kramer, B., & Von Turkovich, B. (1982). Tool wear in titanium machining. *CIRP Annals-Manufacturing Technology, 31*(1), 75–80.
46. Dornfeld, D., et al. (1999). Drilling burr formation in titanium alloy, Ti-6AI-4V. *CIRP Annals-Manufacturing Technology, 48*(1), 73–76.
47. Jawaid, A., Che-Haron, C., & Abdullah, A. (1999). Tool wear characteristics in turning of titanium alloy Ti-6246. *Journal of Materials Processing Technology, 92,* 329–334.
48. Nabhani, F. (2001). Machining of aerospace titanium alloys. *Robotics and Computer-Integrated Manufacturing, 17*(1), 99–106.
49. Wang, Z., Wong, Y., & Rahman, M. (2005). High-speed milling of titanium alloys using binderless CBN tools. *International Journal of Machine Tools and Manufacture, 45*(1), 105–114.
50. Wang, Z., Rahman, M., & Wong, Y. (2005). Tool wear characteristics of binderless CBN tools used in high-speed milling of titanium alloys. *Wear, 258*(5), 752–758.
51. Che-Haron C. H., & Jawaid. A. (2005). The effect of machining on surface integrity of titanium alloy Ti-6% Al-4%V. *Journal of Materials Processing Technology, 166,* 188–192.
52. Nouari, M., & Ginting, A. (2006). Wear characteristics and performance of multi-layer CVD-coated alloyed carbide tool in dry end milling of titanium alloy. *Surface & Coatings Technology, 200*(18), 5663–5676.
53. Haron, C. C., Ginting, A., & Arshad, H. (2007). Performance of alloyed uncoated and CVD-coated carbide tools in dry milling of titanium alloy Ti-6242S. *Journal of Materials Processing Technology, 185*(1), 77–82.
54. Amin, A. N., Ismail, A. F., & Khairusshima, M. N. (2007). Effectiveness of uncoated WC–Co and PCD inserts in end milling of titanium alloy—Ti–6Al–4V. *Journal of Materials Processing Technology, 192,* 147–158.
55. Jianxin, D., Yousheng, L., & Wenlong, S. (2008). Diffusion wear in dry cutting of Ti–6Al–4V with WC/Co carbide tools. *Wear, 265*(11), 1776–1783.
56. Dargusch, M. S., et al. (2008). Subsurface deformation after dry machining of grade 2 titanium. *Advanced Engineering Materials, 10*(1–2), 85–88.
57. Ibrahim, G., Haron, C. C., & Ghani, J. (2009). Surface integrity of Ti-6Al-4V ELI when machined using coated carbide tools under dry cutting condition. *The International Journal of Mechanical and Materials Engineering, 4*(2), 191–196.
58. Sun, S., Brandt, M., & Dargusch, M. (2009). Characteristics of cutting forces and chip formation in machining of titanium alloys. *International Journal of Machine Tools and Manufacture, 49*(7), 561–568.
59. Abdel-Aal, H., Nouari, M., & El Mansori, M. (2009). Tribo-energetic correlation of tool thermal properties to wear of WC-Co inserts in high speed dry machining of aeronautical grade titanium alloys. *Wear, 266*(3), 432–443.
60. Fang, N., & Wu, Q. (2009). A comparative study of the cutting forces in high speed machining of Ti–6Al–4V and Inconel 718 with a round cutting edge tool. *Journal of Materials Processing Technology, 209*(9), 4385–4389.
61. Armendia, M., et al. (2010). Comparison of the machinabilities of Ti6Al4V and TIMETAL® 54M using uncoated WC–Co tools. *Journal of Materials Processing Technology, 210*(2), 197–203.
62. Özel, T., et al. (2010). Investigations on the effects of multi-layered coated inserts in machining Ti–6Al–4V alloy with experiments and finite element simulations. *CIRP Annals-Manufacturing Technology, 59*(1), 77–82.
63. Kali, D., & Chauhan, S. R. (2011). Machinability study of titanium (Grade-5) alloy using design of experiment technique. *Engineering.*
64. Honghua, S., et al. (2012). Tool life and surface integrity in high-speed milling of titanium alloy TA15 with PCD/PCBN tools. *Chinese Journal of Aeronautics, 25*(5), 784–790.
65. Mhamdi, M., et al. (2012). Surface integrity of titanium alloy Ti-6Al-4V in ball end milling. *Physics Procedia, 25,* 355–362.
66. Ugarte, A., et al. (2012). Machining Behaviour of Ti-6Al-4V and Ti-5553 Alloys in interrupted cutting with PVD coated cemented carbide. *Procedia CIRP, 1,* 202–207.

67. Nouari, M., & Makich, H. (2013). Experimental investigation on the effect of the material microstructure on tool wear when machining hard titanium alloys: Ti–6Al–4V and Ti-555. *International Journal of Refractory Metals & Hard Materials, 41,* 259–269.
68. Balaji, J., Krishnaraj, V., & Yogeswaraj, S. (2013). Investigation on high speed turning of titanium alloys. *Procedia Engineering, 64,* 926–935.
69. Cotterell, M., et al. (2013). Temperature and strain measurement during chip formation in orthogonal cutting conditions applied to Ti-6Al-4V. *Procedia Engineering, 63,* 922–930.
70. Shetty, P. K., et al. (2014). Machinability study on dry drilling of titanium alloy Ti-6Al-4V using L 9 orthogonal array. *Procedia Materials Science, 5,* 2605–2614.
71. Li, N., et al. (2017). Experimental investigation with respect to the performance of deep submillimeter-scaled textured tools in dry turning titanium alloy Ti-6Al-4V. *Applied Surface Science, 403,* 187–199.
72. Nakayama, K., Arai, M., & Kanda, T. (1988). Machining characteristics of hard materials. *CIRP Annals-Manufacturing Technology, 37*(1), 89–92.
73. Darsin, M., & Basuki H. A. (2014). Development machining of titanium alloys: A review. In *Applied Mechanics and Materials.* Trans Tech Publ.
74. Lütjering, G., & Williams, J. C. (2003). Titanium (Vol. 2). Berlin: Springer.
75. Pramanik, A., et al. (2017). Fatigue life of machined components. *Advances in Manufacturing, 5*(1), 59–76.
76. Ulutan, D., & Ozel, T. (2011). Machining induced surface integrity in titanium and nickel alloys: A review. *International Journal of Machine Tools and Manufacture, 51*(3), 250–280.
77. Jaffery, S., & Mativenga, P. (2009). Assessment of the machinability of Ti-6Al-4V alloy using the wear map approach. *The International Journal of Advanced Manufacturing Technology, 40*(7), 687–696.
78. Arrazola, P.-J., et al. (2009). Machinability of titanium alloys (Ti6Al4V and Ti555. 3). *Journal of Materials Processing Technology, 209*(5), 2223–2230.
79. Isbilir, O., & Ghassemieh, E. (2013). Comparative study of tool life and hole quality in drilling of CFRP/titanium stack using coated carbide drill. *Machining Science and Technology, 17*(3), 380–409.
80. Odelros, S. (2012). *Tool wear in titanium machining.*
81. Balažic, M., & Kopač, J. (2010). Machining of Titanium Alloy Ti-6Al-4V for biomedical applications. *Strojniški vestnik-Journal of Mechanical Engineering, 56*(3), 202–206.
82. Ying-lin K., et al. (2009). *Use of nitrogen gas in high-speed milling of Ti-6Al-4V. Transactions of Nonferrous Metals Society of China, 19,* 530–534.
83. Machado, A., & Wallbank, J. (1990). Machining of titanium and its alloys—A review. *Proceedings of the Institution of Mechanical Engineers, Part B: Journal of Engineering Manufacture, 204*(1), 53–60.
84. Zhecheva, A., et al. (2005). Enhancing the microstructure and properties of titanium alloys through nitriding and other surface engineering methods. *Surface & Coatings Technology, 200*(7), 2192–2207.
85. Kuttolamadom, M., et al. (2010). *Investigation of the Machining of Titanium Components for Lightweight Vehicles.* 2010, SAE Technical Paper.
86. Kumar, J., & Kumar, V. (2011). Evaluating the tool wear rate in ultrasonic machining of titanium using design of experiments approach. *World Academy of Science, Engineering and Technology, 81,* 803–808.
87. Donachie, M. J. (2000). *Titanium: A technical guide.* ASM international.
88. Komanduri, R., & Hou, Z.-B. (2002). On thermoplastic shear instability in the machining of a titanium alloy (Ti-6Al-4V). *Metallurgical and Materials Transactions A, 33*(9), 2995–3010.
89. El-Wardany, T., Mohammed, E., & Elbestawi, M. (1996). Cutting temperature of ceramic tools in high speed machining of difficult-to-cut materials. *International Journal of Machine Tools and Manufacture, 36*(5), 611–634.
90. Ezugwu, E., Bonney, J., & Yamane, Y. (2003). An overview of the machinability of aeroengine alloys. *Journal of Materials Processing Technology, 134*(2), 233–253.

91. Pramanik, A. (2014). Problems and solutions in machining of titanium alloys. *The International Journal of Advanced Manufacturing Technology, 70*(5–8), 919–928.
92. Dearnley, P., & Grearson, A. (1986). Evaluation of principal wear mechanisms of cemented carbides and ceramics used for machining titanium alloy IMI 318. *Materials Science and Technology, 2*(1), 47–58.
93. Bridges, P., & Magnus, B. (2001). *Manufacture of Titanium Alloy Components for Aerospace and Military Applications*. France: Research and Technology Organization.
94. Shaw, M. C. (2005). *Metal cutting principles* (Vol. 2). New York: Oxford University Press.
95. Amin, A., Hossain M. I., & Patwari, A. U. (2011). Enhancement of machinability of Inconel 718 in end milling through online induction heating of workpiece. In *Advanced Materials Research*. Trans Tech Publ.
96. Pramanik, A., & Littlefair, G. (2015). Machining of titanium alloy (Ti-6Al-4V)—Theory to application. *Machining Science and Technology, 19*(1), 1–49.
97. Amin, A., & Ginta T. L. (2014). *Heat-assisted machining*. Netherlands: Elsevier.
98. Trent, E. M., & Wright P. K. (2000). *Metal cutting*. UK: Butterworth-Heinemann.
99. Kumar, J., & Khamba, J. (2008). An experimental study on ultrasonic machining of pure titanium using designed experiments. *Journal of the Brazilian Society of Mechanical Sciences and Engineering, 30*(3), 231–238.
100. Pramanik, A., et al. (2013). Machining and tool wear mechanisms during machining titanium alloys. In *Advanced Materials Research*. Trans Tech Publ.
101. Corduan, N., et al. (2003). Wear mechanisms of new tool materials for Ti-6Al-4V high performance machining. *CIRP Annals-Manufacturing Technology, 52*(1), 73–76.
102. Zanger, F., & Schulze, V. (2013). Investigations on mechanisms of tool wear in machining of Ti-6Al-4V using FEM simulation. *Procedia CIRP, 8*, 158–163.
103. Ginta, T. L., et al. (2007). Tool life prediction by response surface methodology for end milling titanium alloy Ti–6Al–4V using uncoated carbide inserts. In *International Conference on Mechanical Engineering*.
104. Abd Rahim, E., Warap, N., & Mohid, Z. (2015). *Thermal-assisted machining of nickel-based alloy*.
105. Ezugwu, E. O., et al. (2007). Surface integrity of finished turned Ti–6Al–4V alloy with PCD tools using conventional and high pressure coolant supplies. *International Journal of Machine Tools and Manufacture, 47*(6), 884–891.
106. Rahman, M., Wang, Z.-G., & Wong, Y.-S. (2006). A review on high-speed machining of titanium alloys. *JSME International Journal Series C: Mechanical Systems Machine Elements and Manufacturing, 49*(1), 11–20.
107. Chauhan, D. K., & Chauhan, K. K. (2013). Optimization of Machining Parameters of Titanium Alloy for Tool Life. *Journal of Engineering Computers & Applied Sciences, 2*(6), 57–65.
108. Uddin, M., et al. (2017). Comparative study between wear of uncoated and TiAlN-coated carbide tools in milling of Ti6Al4V. *Advances in Manufacturing, 5*(1), 83–91.
109. Shivpuri, R., et al. (2002). Microstructure-mechanics interactions in modeling chip segmentation during titanium machining. *CIRP Annals-Manufacturing Technology, 51*(1), 71–74.
110. Sun, J., & Guo, Y. (2009). A comprehensive experimental study on surface integrity by end milling Ti–6Al–4V. *Journal of Materials Processing Technology, 209*(8), 4036–4042.
111. Kato, K. (2000). Wear in relation to friction—A review. *Wear, 241*(2), 151–157.
112. Ke, Q., Xu, D., & Xiong, D. (2017). Cutting zone area and chip morphology in high-speed cutting of titanium alloy Ti-6Al-4V. *Journal of Mechanical Science and Technology, 31*(1), 309–316.
113. Rashid, R. R., et al. (2013). The response of the high strength Ti–10V–2Fe–3Al beta titanium alloy to laser assisted cutting. *Precision Engineering, 37*(2), 461–472.
114. Sun, S., Brandt, M., & Dargusch, M. (2010). The effect of a laser beam on chip formation during machining of Ti6Al4V alloy. *Metallurgical and Materials Transactions A, 41*(6), 1573–1581.

115. Palanisamy, S., et al. (2007). A rationale for the acoustic monitoring of surface deformation in Ti6Al4V alloys during machining. *Advanced Engineering Materials, 9*(11), 1000–1004.
116. Altintas, Y., & Weck, M. (2004). Chatter stability of metal cutting and grinding. *CIRP Annals-Manufacturing Technology, 53*(2), 619–642.
117. Dogra, M., et al. (2010). Tool wear, chip formation and workpiece surface issues in CBN hard turning: A review. *International Journal of Precision Engineering and Manufacturing, 11*(2), 341–358.
118. Hong S. Y., Markus I., & Jeong, W.-C. (2001). New cooling approach and tool life improvement in cryogenic machining of titanium alloy Ti-6Al-4V. *International Journal of Machine Tool & Manufacture, 41*, 2245–2260.
119. Amin, A. (1986). Influence of the instability of chip formation and preheating of work on tool life in machining high temperature resistant steel and titanium alloys. *Mechanical Engineering Research Bulletin, 9*, 52–62.
120. Namb, M., & Paulo, D. (2011). Influence of coolant in machinability of titanium alloy (Ti-6Al-4V). *Journal of Surface Engineered Materials and Advanced Technology, 1*(01), 9.
121. Lei, S., & Pfefferkorn, F. (2007). A review on thermally assisted machining. In *ASME 2007 International Manufacturing Science and Engineering Conference*. American Society of Mechanical Engineers.
122. Shams, O., Pramanik, A., & Chandratilleke, T. (2017). Thermal-Assisted Machining of Titanium Alloys. In *Advanced Manufacturing Technologies* (pp. 49–76). Berlin: Springer.
123. Maity, K., & Swain, P. (2008). An experimental investigation of hot-machining to predict tool life. *Journal of Materials Processing Technology, 198*(1), 344–349.
124. Klocke, F., & König, W. (2007). *Fertigungsverfahren 3: Abtragen, Generieren und Lasermaterialbearbeitung* (Vol. 3). Berlin: Springer.
125. Bermingham, M., Palanisamy, S., & Dargusch, M. (2012). Understanding the tool wear mechanism during thermally assisted machining Ti-6Al-4V. *International Journal of Machine Tools and Manufacture, 62*, 76–87.
126. Luo, J., Ding, H., & Shih, A. J. (2005). Induction-heated tool machining of elastomers—Part 1: Finite difference thermal modeling and experimental validation. *Machining science and technology, 9*(4), 547–565.
127. Luo, J., Ding, H., & Shih, A. J. (2005). Induction-heated tool machining of elastomers—Part 2: Chip morphology, cutting forces, and machined surfaces. *Machining science and technology, 9*(4), 567–588.
128. Brecher, C., Rosen, C.-J., & Emonts, M. (2010). Laser-assisted milling of advanced materials. *Physics Procedia, 5*, 259–272.
129. Kitagawa, T., Maekawa, K., & Kubo, A. (1988). Plasma hot machining for high hardness metals. *Bulletin of the Japan Society of Precision Engineering, 22*(2), 145–151.
130. Sun, S., Brandt, M., & Dargusch, M. (2010). Thermally enhanced machining of hard-to-machine materials—A review. *International Journal of Machine Tools and Manufacture, 50*(8), 663–680.
131. Dogra M., & Sharma, S. V. (2013). Techniques to improve the effectiveness in machining of hard to machine materials: A review. *International Journal of Research in Mechanical Engineering & Technology, 3*(2), 122–126.
132. Anderson, M., Patwa, R., & Shin, Y. C. (2006). Laser-assisted machining of Inconel 718 with an economic analysis. *International Journal of Machine Tools and Manufacture, 46*(14), 1879–1891.
133. Tian, Y., et al. (2008). Laser-assisted milling of silicon nitride ceramics and Inconel 718. *Journal of Manufacturing Science and Engineering, 130*(3), 031013.
134. Weck, M., Zeppelin, W., & Hermanns, C. (1994). Laser—A tool for turning centres. In *Proceedings of the LANE, Laser Assisted Net Shape Engineering*.
135. Tian, Y., & Shin, Y. C. (2007). Laser-assisted burnishing of metals. *International Journal of Machine Tools and Manufacture, 47*(1), 14–22.
136. Shi, B., & Attia, H. (2013). Integrated process of laser-assisted machining and laser surface heat treatment. *Journal of Manufacturing Science and Engineering, 135*(6), 061021.

137. Amin, A. (1983). Investigation of the mechanism of chatter formation during metal cutting process. *Mechanical Engineering Research Bulletin, 6*(1), 11–18.
138. Dandekar, C. R., Shin, Y. C., & Barnes, J. (2010). Machinability improvement of titanium alloy (Ti–6Al–4V) via LAM and hybrid machining. *International Journal of Machine Tools and Manufacture, 50*(2), 174–182.
139. Sun, S., et al. (2011). Experimental investigation of cutting forces and tool wear during laser-assisted milling of Ti-6Al-4V alloy. *Proceedings of the Institution of Mechanical Engineers, Part B: Journal of Engineering Manufacture, 225*(9), 1512–1527.
140. Ayed, Y., et al. (2014). Experimental and numerical study of laser-assisted machining of Ti6Al4V titanium alloy. *Finite Elements in Analysis and Design, 92,* 72–79.
141. Wu, Z., et al. (2009). Evaluation and optimization method of high speed cutting parameters based on cutting process simulation. *Intelligent Robotics and Applications,* pp. 317–325.
142. Rashid, R. R., et al. (2012). An investigation of cutting forces and cutting temperatures during laser-assisted machining of the Ti–6Cr–5Mo–5V–4Al beta titanium alloy. *International Journal of Machine Tools and Manufacture, 63,* 58–69.
143. Ginta, T. L., & Amin, A. N. (2010). *Machinability improvement in end milling titanium alloy TI-6AL-4V.* (vol. 3, pp. 25–33).
144. Rashid, R. R., et al. (2014). A study on laser assisted machining of Ti10V2Fe3Al alloy with varying laser power. *The International Journal of Advanced Manufacturing Technology, 74*(1–4), 219–224.
145. Lauwers, B. (2011). Surface integrity in hybrid machining processes. *Procedia Engineering, 19,* 241–251.
146. Ginta, T. L., & Amin, A. N. (2013). Surface integrity in end milling titanium alloy Ti-6Al-4V under heat assisted machining. *Asian Journal of Scientific Research, 6*(3), 609.
147. Germain, G., et al. (2006). Effect of laser assistance machining on residual stress and fatigue strength for a bearing steel (100Cr6) and a titanium alloy (Ti 6Al 4V). In *Materials Science Forum.* Trans Tech Publ.
148. Joshi, S., Tewari, A., & Joshi, S. (2013). Influence of preheating on chip segmentation and microstructure in orthogonal machining of Ti6Al4V. *Journal of Manufacturing Science and Engineering, 135*(6), 061017.
149. Braham-Bouchnak, T., et al. (2013). The influence of laser assistance on the machinability of the titanium alloy Ti555-3. *The International Journal of Advanced Manufacturing Technology, 68*(9–12), 2471–2481.
150. Ginta, T. L., Amin, A. N., & Lajis, M. (2012). Suppressed vibrations during thermal-assisted machining of titanium alloy Ti-6Al-4V using PCD inserts. *Journal of Applied Sciences, 12*(23), 2418.

Chapter 11
Smart Machining System Using Preprocessor, Postprocessor, and Interpolation Techniques

Fusaomi Nagata, Koga Toshihiro, Akimasa Otsuka, Yudai Okada, Tatsuhiko Sakamoto, Takamasa Kusano, Keigo Watanabe and Maki K. Habib

Abstract The authors have developed earlier an industrial machining robotic system for foamed polystyrene materials. The developed robotic CAM system provides a simple and effective interface without the need to use any robot language between operators and the machining robot. In this chapter, a preprocessor for generating cutter location source data (CLS data) from Stereolithography (STL data) is first proposed for smart robotic machining. The preprocessor enables to control the machining robot directly using STL data without using any commercially provided CAM system. The STL deals with triangular patches representation for a curved surface geometry. The preprocessor allows the machining robot to be controlled along a zigzag or spiral path directly calculated from STL data. Then, a smart spline interpolation method is proposed and implemented for smoothing coarse CLS data. The effectiveness and potential of the developed approaches are demonstrated through experiments on actual machining and interpolation.

Keywords Interpolation techniques · Robotic CAM system · Smart machining system

F. Nagata (✉) · K. Toshihiro · A. Otsuka · Y. Okada · T. Sakamoto
Department of Mechanical Engineering, Faculty of Engineering,
Tokyo University of Science, Yamaguchi, 1-1-1 Daigaku-Dori,
Sanyo-Onoda 756-0884, Japan
e-mail: nagata@rs.tusy.ac.jp

T. Kusano
SOLIC Co. Ltd., 80-42 Yotsuyama-Machi, Omuta 836-0067, Japan

K. Watanabe
Graduate School of Natural Science and Technology, Okayama University,
3-1-1 Tsushima-naka, Kita-ku, Okayama 700-8530, Japan

M. K. Habib
American University in Cairo, AUC Avenue, 74, New Cairo 11835, Egypt

© Springer International Publishing AG, part of Springer Nature 2018
J. P. Davim (ed.), *Introduction to Mechanical Engineering*, Materials Forming,
Machining and Tribology, https://doi.org/10.1007/978-3-319-78488-5_11

1 Introduction

In manufacturing industries, there exist two representative systems for prototyping. One is the conventional removal manufacturing systems using router bits, such as computer numerical control (CNC) milling or CNC lathe machines which can precisely perform metalworking and woodworking. The other is the additive manufacturing systems, such as optical shaping apparatus or 3D printer which enables to quickly transform a design concept into a real model. As for removal machining, Lee introduced a machining automation using an industrial robot [1]. The robot had double parallel mechanism and consequently performed a large work space as well as a high stiffness to reduce deformation and vibration. Schreck et al. launched Hard Material Small-Batch Industrial Machining Robot (HEPHESTOS) project, where the focus of the objective was on developing robotic manufacturing methods in order to give rise to a cost-efficient solution in hard materials machining [2].

As one of examples of postprocessor in CAM system, CLS data written in ISO format produced by main processor in CAM system could be transformed into G-codes files (NC data) and an industrial five-axis machine tool with a nutating table could be actually controlled using the NC data [3]. However, it is not easy but complicated for the CAM system to generate each robot language according to different industrial robot makers.

The authors developed an industrial machining robotic system for foamed polystyrene materials as shown in Fig. 1a [4]. In the machining robot, the developed robotic CAM system called the direct servo system provided a simple interface for NC data and CLS ones, without the need to use any robot language between operators and the machining robot [5]. In other words, a desired trajectory consisting of numerous discrete position and orientation components could be calculated from CLS data for the robot. The direct servo system simplified the machining process without a postprocessor as shown in Fig. 1a. However, a CAM process to generate CLS data after the design process has to be still passed through in order to machine the designed model using the robot. Although the authors surveyed related papers, e.g., [6, 7] to make the process easier, any suitable system was not seen.

In this chapter, a smart machining system using preprocessor, postprocessor, and interpolation techniques is introduced [8]. The robotic preprocessor is used for the machining robot to directly convert STL data into CLS data as illustrated by the proposed process two shown in Fig. 1b, and this helps to remove the need for having the conventional CAM process. The STL originally means Stereolithography which is a file format proposed by 3D systems and recently is supported by many CAD/CAM software. It is also known as Standard Triangulated Language in Japan. The STL is widely used for rapid prototyping with a 3D printer which is a typical additive manufacturing system [9]. The STL deals with a triangular representation of a 3D surface geometry [10, 11]. The robotic preprocessor allows the machining robot to be controlled along zigzag path or spiral ones generated based on STL data. Then, a smart spline interpolation method, which is easily implemented, is proposed for smoothing coarse CLS data. The effectiveness and potential of this unique machining

Fig. 1 Comparing the proposed two processes for robotic machining

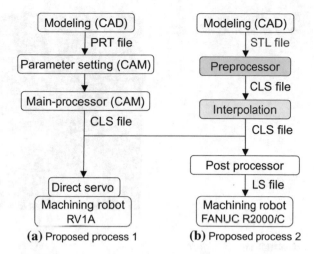

(a) Proposed process 1   (b) Proposed process 2

system with a smart spline interpolation are demonstrated through machining experiments of a foamed polystyrene and a wood.

2   Machining Robot with Robotic CAM System

A roboticpostprocessor shown in Fig. 1b was already proposed to enhance the affinity between FANUC industrial robots and 3D CAD/CAM systems [12]. Figure 2 shows the machining scene of an impeller model using FANUC industrial robot R2000iC whose robot language is called LS. A ball-end mill attached to the arm tip can be controlled as to follow the position and orientation components in LS data. The postprocessor is able to generate LS data directly from CLS ones. Our current interest is to enable the industrial machining robot to run through STL data that consist of numerous unstructured triangulated patches as shown in Fig. 3. $\mathbf{v}_{1i}$, $\mathbf{v}_{2i}$, $\mathbf{v}_{3i}$, and $\mathbf{n}_i$ are the three position vectors and a normal vector in ith triangular patch, respectively. The right in Fig. 3 shows an example of a STL file including a curved surface consisting of numerous triangle patches.

In this chapter, a preprocessor is proposed to convert STL data into two kinds of CLS data and it is integrated with the developed industrial machining robotic system to execute an assigned machining task using CLS data. Consequently, the system can implement the task and control its sequence of machining actions based on STL data.

Fig. 2 Machining robot
based on FANUC R2000iC

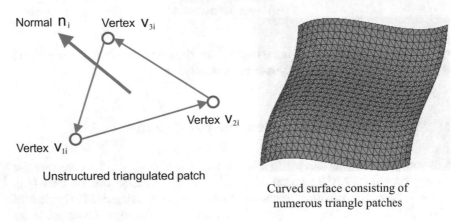

Unstructured triangulated patch

Curved surface consisting of
numerous triangle patches

Fig. 3 Unstructured triangulated patch and an example of STL file

3 Preprocessor by Analyzing Triangle Patches

3.1 Automatic Dimension Extraction of STL Data

In the early developed system, CLS data were basically generated along the original
STL data consisting of many triangle patches [13]. This subsection introduces an
advanced process, in which the CLS data for robot control are generated along a
zigzag path or spiral one by intelligently analyzing STL data. The significant
advantage of this approach is to eliminate the need to use any commercially provided CAM software, and accordingly, 3D printer-like data interface can be smartly
realized. First of all, dimensions of the STL data are extracted by retrieving all
patches in the STL file and they are set to two constants $\mathbf{v}_{min} = [x_{min}y_{min}z_{min}]^T$ and
$\mathbf{v}_{max} = [x_{max}y_{max}z_{max}]^T$. From the next subsection, two types of base paths viewed
in xy-plane are designed considering the extracted dimensions. They are a zigzag
path and a spiral one.

3.2 Automatic Dimension Extraction of STL Data

This subsection explains how a base zigzag path is designed. Two effective machining parameters, i.e., pick feed and step, are respectively set to p_f and s_p by referring the dimensions. s_p is a constant pitch viewed in xy-plane between two adjacent points $\mathbf{c}_j = [c_{xj} c_{yj}\ 0]^T$ and $\mathbf{c}_{j+1} = [c_{x(j+1)} c_{y(j+1)}\ 0]^T$ on a zigzag path. One pass point $\mathbf{c}_j$ is appended into a CLS file as a "GOTO" statement. $j(1 \leq j \leq m)$ is the number of the pass points generated from STL data. m is the total number of triangle patches in the STL file. Figure 4 illustrates an example of a base zigzag path drawn within STL data consisting of multiple triangulated patches. As can be seen, c_{xj} and c_{yj} are located just on the zigzag path, so that remained height c_{zj} has only to be determined by analyzing triangle patches in the STL data.

3.3 Generation of CLS Data Along Base Zigzag Path

Figure 5 shows an example of generation of pass points $\mathbf{c}_j = [c_{xj} c_{yj} c_{zj}]^T$ in a patch (2). The number of pass points within the patch depends on the length of the step illustrated in Fig. 4. The dashed line in the upper figure shows one section of the base zigzag paths $\mathbf{c}_j = [c_{xj} c_{yj}\ 0]^T$ viewed in xy-plane, and the chained line in the lower figure draws the generated path $\mathbf{c}_j = [c_{xj}\ c_{yj}\ c_{zj}]^T$ along the triangulated patch viewed in yz-plane. The pass points for constructing CLS data are generated along the chained line. The pass points in other patches such as (1), (3), (4), and (5) can be similarly obtained.

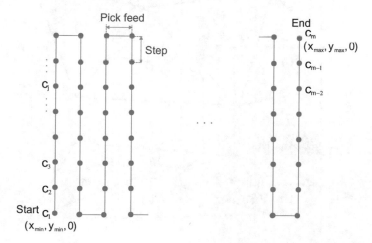

Fig. 4 Base zigzag path $\mathbf{c}_j = [c_{xj} c_{yj}\ 0]^T$ designed along STL data consisting of multiple triangulated patches

To realize the preprocessor based on STL data without using any commercially provided CAM system, the z-component c_{zj} must be calculated just on a triangulated patch. In order to calculate c_{zj}, first of all, a triangulated patch, in which the point $\mathbf{c}_j$ viewed in xy-plane is included, is searched in the target STL data. Figure 6 shows the scene where the pass point $\mathbf{c}_j$ is located within a patch. Note that $\mathbf{c}_{j-1}$ and $\mathbf{c}_{j+1}$ may be also within the triangle patch as shown in Fig. 5. Whether $\mathbf{c}_j$ is located in the patch or not can be known by checking the following outer products.

$$(\mathbf{v}_{2i} - \mathbf{v}_{1i}) \times (\mathbf{c}_j - \mathbf{v}_{1i}) \tag{1}$$

$$(\mathbf{v}_{3i} - \mathbf{v}_{2i}) \times (\mathbf{c}_j - \mathbf{v}_{2i}) \tag{2}$$

$$(\mathbf{v}_{1i} - \mathbf{v}_{3i}) \times (\mathbf{c}_j - \mathbf{v}_{3i}) \tag{3}$$

If $\mathbf{c}_j$ is located within the patch, then the above three equations have the same sign. After finding the first triangle satisfying this condition, the equation of the plane including the triangle is determined by the perpendicular condition to the normal vector $\mathbf{n}_i$, which leads to

$$n_{xi}(x - x_{1i}) + n_{yi}(y - y_{1i}) + n_{zi}(z - z_{1i}) = 0 \tag{4}$$

By respectively substituting c_{xj} and c_{yj} into x and y in Eq. (4), if $n_{zi} \neq 0$, then z-directional component c_{zj} can be calculated by

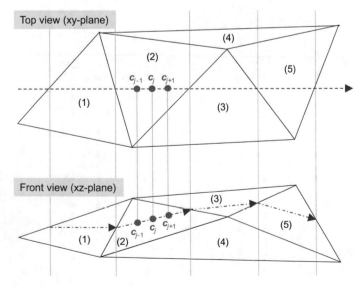

Fig. 5 Generation of pass points. Upper and lower figures show the top and front views of five adjacent triangulated patches, respectively

Fig. 6 Pass point $\mathbf{c}_j$ is
located within a triangle patch
viewed in xy-plane

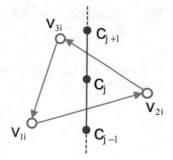

$$c_{zj} = z_{1i} - \frac{1}{n_{zi}}\{n_{xi}(c_{xj} - x_{1i}) + n_{yi}(c_{yj} - y_{1i})\} \qquad (5)$$

where c_{xj} and c_{yj} are extracted from the base path shown in Fig. 4. Consequently, by repeating the above calculations, all pass points $\mathbf{c}_j = [c_{xj}c_{yj}c_{zj}]^{\mathrm{T}}(1 \leq j \leq m)$ can be obtained. Figure 7a shows a STL file of Chinese characters (214 mm × 36 mm × 5 mm) meaning "Tokyo University of Science, Yamaguchi" made by using Adobe Illustrator and Photoshop; lower figure shows the regular and precise zigzag path (CLS data) generated from the STL file by the preprocessor, in which the values of step and pick feed are set to 0.4 mm.

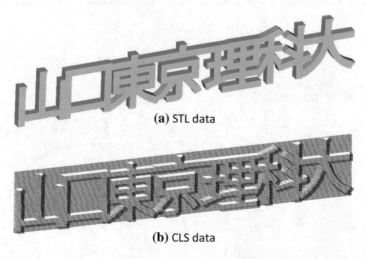

(a) STL data

(b) CLS data

Fig. 7 Upper figure shows a STL file made by using Adobe Illustrator and Photoshop; lower figure shows the regular and precise zigzag path (CLS data) generated by the preprocessor

3.4 Generation of CLS Data Along Base Circular Spiral Path

The base spiral $[c_{xj}c_{yj}\ 0]^T$ viewed in xy-plane for generating CLS data $[c_{xj}c_{yj}c_{zj}]^T$ is designed by

$$c_{xj} = \frac{r\theta_j}{2\pi}\cos\theta_j \tag{6}$$

$$c_{yj} = \frac{r\theta_j}{2\pi}\sin\theta_j \tag{7}$$

$$\frac{dc_{xj}}{d\theta_j} = \frac{r}{2\pi}(\cos\theta_j - \theta_j\sin\theta_j) \tag{8}$$

$$\frac{dc_{yj}}{d\theta_j} = \frac{r}{2\pi}(\sin\theta_j + \theta_j\cos\theta_j) \tag{9}$$

When pass points $[c_{xj}c_{yj}\ 0]^T$ viewed in xy-plane are generated while drawing a spiral path, two effective parameters need to be considered as shown in Fig. 8. One is the movement l_j [mm] from a pass point $\mathbf{c}_{j-1}$ to next one $\mathbf{c}_j$ along the spiral, i.e., arc length. The other is the pitch r [mm] along the radius vector. Operators can design a preferred base spiral path only by giving l_j and r.

The rate between the length l_j of an arc on the spiral and the varying angle θ_j is calculated by

Fig. 8 Base spiral path $\mathbf{c}_j = [c_{xj}c_{yj}\ 0]^T$ for generating CLS data

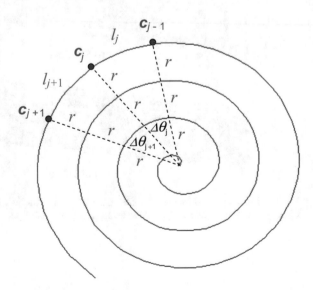

$$\frac{\mathrm{d}l_j}{\mathrm{d}\theta_j} = \sqrt{\left(\frac{\mathrm{d}c_{xj}}{\mathrm{d}\theta_j}\right)^2 + \left(\frac{\mathrm{d}c_{yj}}{\mathrm{d}\theta_j}\right)^2} = \frac{r}{2\pi}\sqrt{1+\theta_j^2} \tag{10}$$

To produce better surface quality with less cusp heights in actual removal machining using a ball-end mill, l_j shown in Fig. 8 must be a constant value 1. Hence, the angle θ_j needs to be varied as

$$\theta_j = \theta_{j-1} + \Delta\theta_j \tag{11}$$

$$\Delta\theta_j = \frac{l}{r}\frac{2\pi}{\sqrt{1+\theta_{j-1}^2}} \tag{12}$$

This means θ_j has only to be adjusted by Eqs. (11) and (12) in order that l_j is fixed to a constant value 1. Figure 9 shows a result of circular spiral path (CLS data: 215 mm × 215 mm × 5 mm) generated by the preprocessor viewed in Cartesian space, in which l and r are set to 0.7 mm. Note that c_{zj} can be calculated by Eq. (5).

3.5 Removal Machining Experiment

In the earlier subsections, the preprocessor that can generate two kinds of regular and precise tool paths has been proposed. In this subsection, two machining experiments were conducted using a machining robot and a desktop NC machine tool, in which two tool paths (CLS data) generated by the preprocessor from STL

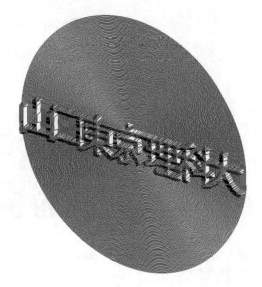

Fig. 9 Regular and precise circular spiral path (CLS data) $\mathbf{c}_j = [c_{xj}c_{yj}c_{zj}]^{\mathrm{T}}$ viewed in Cartesian space generated by the preprocessor

data shown in Figs. 3 and 10 are given to the systems. Figure 11 shows the successful machining scene of the machining robot using the CLS data made along a zigzag path (left photo) and the resultant surface (right photo). The material is a foamed polystyrene. Also, Figs. 10 and 12 show another example of conversion from STL data to spiral-based CLS data and its machining scene using a desktop NC machine tool, respectively. The material is a chemical wood. The feasibility and effectiveness were confirmed from the actual experiments using the machining robot and the desktop NC machine tool.

4 Smart Spline Interpolation of CLS Data

When the density of CLS data is not high but coarse, some interpolation method is effective to smoothly reconstruct the trajectory with preferable scaling. Figure 13 illustrates a third-order spline curve consisting of three cubic curves $x_{i-1}(t)$, $x_i(t)$ and $x_{i+1}(t)$ just passing on four points $p_{i-1}, p_i, p_{i+1}, p_{i+2}$ $(\in \Re^{3\times1})$. The third-order spline curve $x_i(t) = [x_i(t)\ y_i(t)\ z_i(t)]^{\mathrm{T}}$ in Cartesian space is written as

$$x_i(t) = a_i t^3 + b_i t^2 + c_i t + d_i \tag{13}$$

where $a_i = [a_{xi} a_{yi} a_{zi}]^{\mathrm{T}}$, $b_i = [b_{xi} b_{yi} b_{zi}]^{\mathrm{T}}$, $c_i = [c_{xi} c_{yi} c_{zi}]^{\mathrm{T}}$, $d_i = [d_{xi} d_{yi} d_{zi}]^{\mathrm{T}}$ are the coefficient vectors; t is the normalized variable. When the x-directional component is considered, the following relations are obtained from Fig. 13.

$$x_i(1) = p_{x(i1)} = a_{xi} + b_{xi} c_{xi} + d_{xi} \tag{14}$$

$$x_i(0) = p_{xi} = d_{xi} \tag{15}$$

$$x_i(1) = p_{x(i+1)} = a_{xi} + b_{xi} + c_{xi} + d_{xi} \tag{16}$$

$$x_i(2) = p_{x(i+2)} = 8a_{xi} + 4b_{xi} + 2c_{xi} + d_{xi} \tag{17}$$

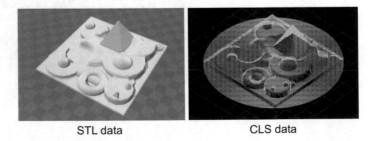

STL data   CLS data

Fig. 10 Another example of conversion from STL to spiral-based CLS data using the preprocessor

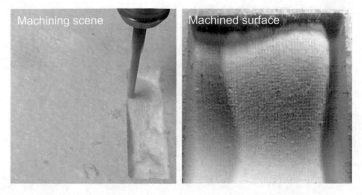

Fig. 11 Machining scene of machining robot RV1A using zigzag-based CLS data shown in Fig. 7 (left); resultant surface (right)

Fig. 12 Machining scene using a desktop NC machine tool, where the spiral-based CLS data shown in Fig. 10 generated by preprocessor are given

Fig. 13 Spline interpolation using third-order functions

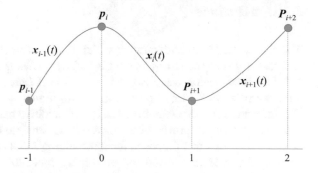

d_{xi} is fixed as p_{xi}, so that the remaining coefficients can be calculated by

$$\begin{pmatrix} a_{xi} \\ b_{xi} \\ c_{xi} \end{pmatrix} = \begin{pmatrix} -1/6 & 1/2 & -1/6 \\ -1/2 & 1/2 & 0 \\ 1/3 & 1 & -1/6 \end{pmatrix} \begin{pmatrix} p_{x(i-1)} - p_{xi} \\ p_{x(i+1)} - p_{xi} \\ p_{x(i+2)} - p_{xi} \end{pmatrix} \tag{18}$$

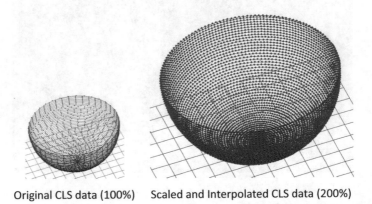

Original CLS data (100%)   Scaled and Interpolated CLS data (200%)

Fig. 14 Original CLS data (left), scaled and interpolated CLS data (right) illustrated in perspective view

Coefficients in other two directions are similarly calculated, so that the section between p_i and p_{i+1} can be simply interpolated with $x_i(t)$ $(0 \leq t \leq 1)$.

An experiment of scaling interpolation of a spiral path was conducted with the third-order spline curve. The left side of Fig. 14 shows the points in the original CLS data. Also, the right side of Fig. 14 draws the points in the CLS data interpolated with the third-order spline curve, which is scaled up with 200%. The number of interpolated points within a section is set to three. As can been, the coarse original CLS data were desirably interpolated, smoothed, and scaled up.

5   Conclusions

The STL means Stereolithography which is a file format proposed by 3D systems and recently is supported by many CAD/CAM software. It is known that the STL interface specification was designed in 1989 for fabbers. Fabbers means specialists who can perform 3D rapid prototyping from digital data, e.g., using a 3D printer. The 3D printer is also recognized as a typical additive manufacturing system. In this chapter, a robotic preprocessor to convert STL data into CLS data forming a zigzag or spiral path was first proposed for a machining robot of foamed polystyrene. The robotic preprocessor allowed the machining robot to be controlled along a continuous zigzag or spiral path smartly calculated from STL data. Then, a third-order spline interpolation method, which is easily implemented, was proposed for smoothing originally coarse CLS data. The effectiveness and promise of this unique machining system are demonstrated through actual machining experiments and interpolation tests. The noteworthy point is that the proposed preprocessor realized a promising data interface available with STL data like 3D printers and a smoother with variable scaling. Needless to say, the preprocessor can be applied to CNC machine tools that can deal with CLS data and/or NC data.

Questions

1. What is the problem in teaching process of an industrial robot using a teach pendant?

Answer: When the desired trajectory includes free-formed curve, a large number of points have to be taught. Accordingly, the teaching task is getting complicated and time-consuming.

2. Give application examples of industrial robot.

Answer: Industrial robots have been applied to the automation processes of welding, painting, handling, assembling, sanding, polishing, etc. The processes have been successfully rationalized.

3. What are the main parameters in CAM process?

Answer: Type of tool path such as zigzag, whirl, and contour line; tool shape (e.g., flat or ball end), tool length, tool diameter and/or radius, rotational speed; tolerance to a designed model, pick feed, feed rate, and so on.

4. Explain the main processor and postprocessor of CAM system.

Answer: The main processor generates cutter location source data called CLS data. Then, the postprocessor converts the CLS data into numerical control data called NC data according to various kinds of CNC machine tools.

5. Give an example of contents in CLS data.

Answer:
GOTO/0.0000,-100.0000,-50.0000,0.0000000,0.0000000,1.0000000
GOTO/4.0217,-100.0000,-50.0324,0.0160868,0.0000000,0.9998706
GOTO/8.0424,-100.0000,-50.1294,0.0321695,0.0000000,0.9994824
GOTO/12.0610,-100.0000,-50.2911,0.0482438,0.0000000,0.9988356
GOTO/16.0764,-100.0000,-50.5174,0.0643057,0.0000000,0.9979302
GOTO/20.0877,-100.0000,-50.8083,0.0803509,0.0000000,0.9967666
In this example, the former three elements in each GOTO statement describe a position vector, also the latter three elements form a normal vector at the position.

6. Give an example of contents in NC data.

Answer:
G01X44.3743Y7.1264Z0.2466B-32.2166C40.7584F3000.0
X48.4211Y0.0000Z-0.7536B-32.2166C40.7584
X50.0000Y0.0000Z0.0000B-32.2166C40.7584
X50.0428Y0.0403Z-0.0809B111.3599C-79.2415
X50.0000Y0.0000Z0.0000B111.3599C-79.2415
X48.4211Y0.0000Z-0.7536B111.3599C-79.2415

In this example, G01 and F3000.0 specify a linear motion with a feed rate of 3000 mm/min. B and C are rotational angles (deg.) around y-axis and z-axis, respectively, for a five-axis CNC machine tool.

7. Write the structure of STL data including numerous triangulated patches.

Answer:
UINT8[80]//header area
UINT32//number of triangulated patches
for each triangle//This block is repeated as many as the number of triangles
REAL32[3]//normal vector
REAL32[3]//vertex 1
REAL32[3]//vertex 2
REAL32[3]//vertex 3
UINT16//not used
End

8. Explain each advantage of removal machining using a CNC machine tool and additive machining using a 3D printer.

Answer: Various kinds of materials can be machined using an end mill in removal machining processes; however, there is a limitation about shapes to what the removal machining can produce. On the other hand, almost all shapes can be produced in additive machining using a 3D printer; however, there is a restriction related to the type of materials that the nozzle of 3D printer can discharge.

9. Give the examples of resin materials used for 3D printers.

Answer: ABS means a copolymer consisting of Acrylonitrile, Butadiene, and Styrene. PLA is the abbreviation of Polylactic Acid.

10. Explain the difference between the data interface for CNC machine tools and that for industrial robots.

Answer: Almost all CNC machine tools can be uniformly controlled based on standard NC data. However, the data interface of industrial robots has not been sufficiently standardized yet, so that engineers at user side have to learn different robot languages provided by each robots' maker.

Acknowledgements This work was supported by JSPS KAKENHI Grant Number 25420232 and 16K06203.

Glossary

CAD is the abbreviation of computer-aided design. Various shapes of 3D models can be easily and precisely designed and typical data formats such as Drawing

Interchange Format (DXF) and Initial Graphics Exchange Specification (IGES) can be outputted from the models

CAM is the abbreviation of computer-sided manufacturing, which includes main and post processes. The main process generates tool paths called cutter location source (CLS) data. Then, the post process further generates numerical control (NC) data according to the type of NC machine tool actually used for machining

STL means Stereolithography which is a file format proposed by 3D systems and recently is supported by not only many CAD, CAD/CAM software but also design tools such as Photoshop and Illustrator. It is also known as Standard Triangulated Language in Japan. The STL is widely spread to the rapid proto-typing mainly using a 3D printer which is a typical additive manufacturing system

Industrial robots are automated, programmable, and flexible due to the serial link structure with five or more axes, so that they can work as skilled workers or dexterous arms. Typical applications of industrial robots include welding, painting, pick and place, palletizing, assembly, packaging and labeling, product inspection, and so on. Those tasks can be accomplished with high endurance, speed, and accuracy

References

1. Lee, M. K. (1995). Design of a high stiffness machining robot arm using double parallel mechanisms. In *Proceedins of 1995 IEEE International Conference on Robotics and Automation*, Vol. 1, pp. 234–240.
2. Schreck, D. S., & Krueger, J. (2014). HEPHESTOS: Hard material small-batch industrial machining robot. In *Proceedings of 41st International Symposium on Robotics (ISR/Robotik 2014)*, pp. 1–6.
3. My, C. A. (2010) Integration of CAM systems into multi-axes computerized numerical control machines. In *Proceedings of 2010 Second International Conference on Knowledge and Systems Engineering (KSE)*, pp. 119–124.
4. Nagata, F., Otsuka, A., Watanabe, K., & Habib, M.K. (2014). Fuzzy feed rate controller for a machining robot. In *Proceedings of the 2014 IEEE International Conference on Mechatronics and Automation (IEEE ICMA 2014)*, pp. 198–203.
5. Nagata, F., Yoshitake, S., Otsuka, A., Watanabe, K., & Habib, M. K. (2013). Development of CAM system based on industrial robotic servo controller without using robot language. *Robotics and Computer-Integrated Manufacturing,29*(2), 454–462.
6. Al-Ahmari, A., & Moiduddin, K. (2014). CAD issues in additive manufacturing. In *Comprehensive Materials Processing, Vol. 10: Advances in Additive Manufacturing and Tooling*, pp. 375–399.
7. Matta, A. K., Raju, D. R., & Suman, K. N. S. (2015). The integration of CAD/CAM and rapid prototyping in product development: A review. *Materials Today: Proceedings,2*(4/5), 3438–3445.
8. Nagata, F., Okada, Y., Sakamoto, T., Kusano, T. M., Habib, K., & Watanabe, K. (2017). Preprocessor with spline interpolation for converting stereolithography into cutter location

source data. In *IOP Conference Series: Earth and Environmental Science*, Vol. 69, Conf. 1, 9 pages. DOI:https://doi.org/10.1088/1755-1315/69/1/012115.

9. Brown, A. C., & Beer, D.D. (2013). Development of a stereolithography (STL) slicing and G-code generation algorithm for an entry level 3-D printer. *In Proceedings of IEEE African Conference 2013*, pp. 1–5.

10. Szilvasi-Nagy, M., & Matyasi, G. (2003). Analysis of STL files. *Mathematical and Computer Modeling,38*(7/9), 945–960.

11. Iancu, C., Iancu, D., & Stancioiu, A. (2010). From CAD model to 3D print via STL file format. *Fiability & Durability,1*(5), 73–81.

12. Nagata, F., Yamane, Y., Okada, Y., Kusano, T., Watanabe, K., & Habib, M. K. (2017). Post processor for an industrial robot FANUC R2000*i*C, In *Proceedings of 22nd International Symposium on Artificial Life and Robotics*, pp. 634–637.

13. Nagata, F., Takeshita, K., Watanabe, K., & Habib, M. K. (2016). Generation of triangulated patches smoothed from original point cloud data with noise and its application to robotic machining. In *Proceedings of the 2016 IEEE International Conference on Mechatronics and Automation (ICMA 2016)*, pp. 535–540.

Chapter 12
Comparison of Non-conventional Intelligent Algorithms for Optimizing Sculptured Surface CNC Tool Paths

Nikolaos A. Fountas, Nikolaos M. Vaxevanidis, Constantinos I. Stergiou and Redha Benhadj-Djilali

Abstract The optimization of process parameters referring to sculptured surface tool path planning increases efficiency and enhances product quality; thus, it is for the major research subject for many noticeable studies. Optimization for process parameters is usually conducted by working with a two-phase scheme; regression modeling based on the results obtained by a design of experiments, and optimization by employing an intelligent algorithm. Currently, new artificial algorithms have been developed and deployed to address different kinds of problems in engineering. In the present work, six new intelligent algorithms have been tested to sculptured surface tool path optimization problems, namely particle swarm optimization (PSO), invasive weed optimization (IWO), shuffled frog-leaping algorithm (SFLA), shuffled complex evolution (SCE), teaching–learning-based optimization (TLBO), and virus-evolutionary genetic algorithm (VGA). Except from the VGA which has been developed from scratch, the rest of the algorithms have been adopted from the literature whilst the case studies the algorithms are applied to have been established using design of machining simulation experiments on benchmark sculptured surfaces. The results obtained from case studies are compared with each other to investigate the capabilities of the aforementioned algorithms in terms of their application to the sculptured surface machining problem.

N. A. Fountas · N. M. Vaxevanidis (✉)
Faculty of Mechanical Engineering-Laboratory of Manufacturing
Processes and Machine Tools (LMProMaT), School of Pedagogical
and Technological Education (ASPETE), 141 21 N. Heraklion, Athens, Greece
e-mail: vaxev@aspete.gr

N. A. Fountas
e-mail: fountasnikolaos@hotmail.com

C. I. Stergiou
Mechanical Engineering Department, Piraeus University
of Applied Sciences (PUAS), 122 44 Aigaleo, Greece

N. A. Fountas · R. Benhadj-Djilali
Faculty of Science, Engineering and Computing (SEC), School of Mechanical
and Automotive Engineering—Roehampton Vale Campus, Kingston University,
London SW15 3DW, UK

© Springer International Publishing AG, part of Springer Nature 2018
J. P. Davim (ed.), *Introduction to Mechanical Engineering*, Materials Forming,
Machining and Tribology, https://doi.org/10.1007/978-3-319-78488-5_12

Keywords Virus-evolutionary genetic algorithm · Tool path optimization
Sculptured surface machining · Machining simulation

1 Introduction

Mechanical engineering is a broad field that stems from the need to design and produce almost everything in terms of goods, small-to-large-scale products as well as devices and structures. The major scope of mechanical engineering is to capture an original idea and deliver it as a final product for the market. Several prerequisites are required to accomplish this such as skills and specialized knowledge of engineering fundamentals depending the area of interest, i.e., solar energy, materials, thermodynamics, product design, manufacturing.

The process of converting raw materials to final products/finished goods is found in the area of manufacturing which deals either with small-scale or large-scale production under predetermined requirements in terms of accuracy, functionality, quality, and aesthetics [1]. Manufacturing involves a wide range of processes distinguished to metal-forming and material-removal operations. In metal-forming processes, the material (usually metal) being handled is plastically deformed to take the shape of a desired geometry. To plastically deform a metal, a force must be applied with a magnitude such that yield strength of the material is exceeded. Representative manufacturing processes of this category include bending, forming, deep drawing, rolling, etc. [2]. Material removal is the major characteristic in metal cutting (machining) processes where a large amount of volume has to be removed from the stock to leave the final shape of the product that should correspond to its original engineering drawing [2, 3]. Material removal is achieved by employing different types of cutting tools that ought to maintain their properties during machining; hence, a large body of high-strength materials and coatings are used for their production. Most representative operations that fall in this category are turning, milling, and drilling [4–7]. The aforementioned manufacturing processes found to these categories are deployed using conventional or non-conventional technology.

Regardless of the selection of a single or multiple manufacturing processes to manufacture high-quality products, process parameter optimization is an essential task so as to cope with the trade-off between productivity and quality. The contradictory nature of quality objectives has led several researchers to investigate parameters for the aforementioned manufacturing processes and try to optimize them so that cost, production rate, quality, and accuracy are simultaneously met. Sculptured surface machining is the material-removal process extensively utilized to produce products with free-form surfaces. Hence, as it occurs to the rest of the aforementioned manufacturing processes, it is imperative to determine optimal tool path parameters for achieving qualitative and efficient sculptured surface machining operations [8]. Machining modeling for tool path generation is a major industrial practice for producing functional and aesthetical parts. This is usually conducted by testing several numerical values for tool path parameters that belong to applicable intervals being previously validated by laboratory experiments. During the last decade, several research works concerning intelligent optimization for engineering problems have either selected or developed artificial intelligence systems to respond to pressing industrial needs for quality and cost. In order to provide viable industrial solutions, a considerable body of researchers has integrated empirical models based on regression or neural networks to optimization modules such as genetic algorithms.

Palanikumar et al. [9] experimented on the machining of glass fiber-reinforced plastic using polycrystalline diamond cutting tool. Their aim was to minimize tool flank wear, surface roughness and maximize the material-removal rate given the speed, the feed, and the cutting depth. After the generation of three discrete regression models based on their experimental results, the multi-objective optimization problem was handled by a non-dominated sorting genetic algorithm-II (NSGA-II). In order for Garg et al. [10] to optimize wire-electric-discharge machining process, a series of experiments were performed to develop mathematical models for cutting speed and surface roughness as major optimization criteria. The same authors utilized the mathematical models as objective functions for an NSGA-II to optimize the aforementioned non-conventional machining process. Fountas et al. [11] predicted cutting forces for the case of turning the PA66 GF-30 GFRP polyamide by soft computing approaches. Bhavsar et al. [12] performed experiments for obtaining results necessary to build regression models referring to material-removal rate and surface roughness during focused ion beam micro-milling process. An NSGA-II was employed to optimize material-removal rate and surface roughness regression models as objective functions for maximization and minimization, respectively. Rao et al. [13] collected all aforementioned studies and retrieved their objective functions to apply a non-dominated sorting teaching–learning-based optimization (TLBO) algorithm for multi-objective process optimization. Zain et al. [14] applied a genetic algorithm to optimize the end-milling process using different cutting tools. Their study involved statistics to exploit the optimal regression equation obtained by results for different cutting tools, and the most trustworthy was taken as the fitness function for their genetic algorithm.

Despite the fact that tool path generation to machine sculptured surfacesis a well-established research field, further requirements for artificial systems involving convergence speed, high exploration–exploitation, and global optimal point guarantee impose the increasing development of new heuristics to improve the quality of solutions. Some distinguished meta-heuristics among others are invasive weed optimization (IWO), shuffled complex evolution (SCE), shuffled frog-leaping algorithm (SFLA), teaching–learning-based optimization algorithm (TLBO), and virus-evolutionary genetic algorithm (VGA). IWO is a meta-heuristic which simulates the spreading strategy of weeds and is heavily based on r/K-selection theory. First the artificial weeds employ the r-selection scheme to initialize their potential solutions whilst they gradually change to K-selection scheme to formulate more robust solutions as convergence process to global optimum continues [15]. SCE is a heuristic based on many sub-populations known as complexes. In each iteration, the internal optimization module of SCE evolves the complex. After an entire cycle (i.e., a generation in genetic algorithms) of all complexes, a recombination is performed to restructure the main population. Thereby, segmentation and partitioning operators are applied to shuffle the population and complexes [16]. SFLA is a memetic meta-heuristic that is designed to seek a global optimal solution by performing an informed heuristic search using a heuristic function. It is based on evolution of memes carried by interactive individuals and a global exchange of information among the population. SFLA progresses by transforming "frogs" in a memetic evolution. More information about SFLA can be found in Eusuff et al. [17] and Gomez-Gonzalez and Jurado [18]. Fountas et al. [19] performed experiments in a machining simulation system to create regression models and optimize surface machining tool paths with the use of an artificial immune algorithm. Virus-evolutionary genetic algorithm is a two-population heuristic that utilized the ability of viruses to overwrite their genetic material to that of their hosts. By considering the viruses as "solution carriers" robust individual schemes for the main population of individuals are developed during the evolution process. The first artificial model of a virus-evolutionary genetic algorithm was pioneered by Kubota et al. [20], and one of its enhanced forms was first developed and employed to tool path optimization by Fountas et al. [21, 22].

The aim of the present study is to compare the results obtained from the aforementioned Intelligent Algorithms on optimal tool path generation experiments conducted to a manufacturing software environment. Tool path generation experiments have been established using benchmark sculptured surfaces found in literature. Since, the ultimate scope is to find out which of these systems deserve to be integrated into an automated global tool path optimization methodology for sculptured surface machining [23]. The algorithms were tested using only few candidate solutions as well as generations under the prism that an algorithmic iteration could represent an entire machining simulation varying from few seconds to a considerable fraction of time in minutes or even hours. Problem formulation for representing the design space for these experiments has been conducted using response surface methodology (RSM) so as to provide a continuous optimization domain for the algorithms to search in.

2 Application of Intelligent Algorithms for Sculptured Surface Tool Path Optimization

Machining simulation experiments have been designed by taking into account successfully implemented 5-axis tool path generation strategies found in literature so as to formulate the optimization criteria. Since, criteria worthy of optimization involve part quality and process efficiency key attributes characterizing them have been assessed to generate the objective functions. Such attributes were selected to be the number of cutter locations given a particular sculptured surface and machining error as the combined effect of scallop height (remaining material among radial tool passes) and chord error (deviation among the theoretical surface and the one achieved by cutting interpolation). The independent variables taken under consideration were the type of cutting tool, the stepover, the lead and tilt angles for 5-axis machining, and the maximum discretization (forward to feed direction) step.

2.1 Design of Experiments

Based on the benchmark surfaces selected for machining simulation, design of experiments has been conducted to establish the design space for algorithmic evaluations. Applicable intervals for tool path planning parameters have been taken into account whilst the ranges for numerical values correspond to those previously implemented by known 5-axis tool path planning strategies [24, 25]. In all experiments, full quadratic regression models were created to correlate the independent tool path planning parameters to the responses representing efficiency and quality. It should be mentioned here that the capability of regression models in terms of their response prediction does not affect the quality of algorithmic operations; it only affects the quality of final solution. Should the coefficient of determination (R^2) is below 85%, a regression model is deemed rather unable to predict a response's result and therefore should not be trusted as an empirical model. Since this study aims at assessing algorithmic performance using regression models without the necessity of finding the real optimal result for the quality objectives of benchmark surfaces tested, regression models have been considered as objective functions for the algorithms nonetheless. On the other hand, the ability of algorithms to converge to the lowest (minimization) or the largest (maximization) possible result for objective function despite its prediction capability is of great interest though. The five crucial tool path planning parameters mentioned above were examined according to the optimization problem's needs and the selected references for adopting their operational intervals. That is, some of the independent variables were kept fixed to shift the research interest toward the rest of them. For the experimental simulations, only filleted end-mills were used owing to their efficiency and high surface finish against flat end and ball end mills.

2.2 Parameter Settings for PSO, IWO, SCE, SFLA, TLBO, and VGA

Owing to the unique characteristics and functionality of each algorithm, it is tedious to perform rigorous comparisons among results obtained by dedicated experiments. Note that each of the configurations among non-conventional heuristics imposes its corresponding settings in terms of the algorithm-specific parameters. These parameters refer to the intelligent operators for heuristics and are vital attributes for their successful performance. Even though some general guidelines do exist

Table 1 Parameters and settings for the intelligent algorithms selected

Algorithm	Parameters			
PSO	Inertia weight	Inertia weight damping ratio	Personal learning coefficient	Global learning coefficient
	$W_1 = 1$	$W_2 = 0.99$	$C_1 = 1.5$	$C_2 = 2$
IWO	Min. number of seeds	Max. number of seeds	Var. reduction exponent	Init. value of St. Dev.	Final value of St. Dev.
	$S_{min} = 1$	$S_{max} = 5$	$V_{re} = 2$	0.5	0.001
SCE			Complex size	Number of complexes	
			$C_s = 5$	$N_c = 1$	
SFLA	Number of memeplexes		Memeplex size	Step size	
	$N_{mem} = 1$		$M_s = 5$	$S = 2$	
TLBO	Teaching factor (TF)				
	$TF = 1$				
VGA	No viruses	Infection rate (min/max)		Reduction rate	
	3	60–100%		0.001	

concerning the proper selection of these parameters, it has been suggested that optimal settings strongly depend on the problem formulation. Therefore, the most common parameter settings recommended by the relevant literature have been determined for each of the selected intelligent algorithms. The algorithm-specific parameters and their corresponding settings are presented in Table 1.

Since intelligent algorithms have stochastic nature, each execution will often result to different outputs. A common practice is to apply the algorithms several times and study their behavior whilst keeping the best result. Despite the issue of being totally in agreement with this important guideline, it was decided to try a different concept that may represent more accurately real-world applications where no time for experiments and trial-and-error efforts may be available. According to the concept adopted, each of the algorithms was run only once and the solution was recorded. However to avoid jeopardizing the algorithms' actual capabilities in converging to optimal solution, several preliminary experiments were conducted revealing that they all were able to converge to the global minimum result for every objective function tested according to the benchmark surfaces' properties.

2.3 Sculptured Surface Tool Path Problems Formulation

Optimization problems that deal with machining suggest either the maximization or minimization of an objective function, usually reflecting a performance metric under constraints. The problem might be expressed as it is shown in the following equation.

$$\text{Min/Max}(f(x)) \text{ subject to :} \begin{array}{l} g_i(x) \le 0, \quad i = 1 \ldots n \\ x_j^l \le x_j \le x_j^u, \quad j = 1 \ldots m \end{array} \tag{1}$$

where x is a vector of tool path parameters; $f(x)$ is the objective function for either minimization or maximization; $g_i(x)$ is the ith constraint, and x_j^l, x_j^u are the lower and upper values for parameters x referring to their applicable intervals. Note that in multi-objective optimization, several non-dominated solutions exist in a Pareto front from which the user is prompted to select from. To ease the efforts when it comes to this case is to formulate a linear expression among objectives and their weights of importance to obtain only a single optimal solution from the Pareto front.

Such linear expressions have been developed according to the objectives of case studies examined in current experiments. Macro commands were developed and operated in CAM software automation environment to parse cutting data files; isolate tool axis coordinates and evaluate machining data for the objectives such as chord error computations between interpolated tool positions and effective radius owing to tool inclination angles affecting the scallop height. Since the objectives involved to tool path assessment for machining sculptures surfaces are of different magnitudes, they were normalized to 0–1 interval so as to alter the shift tending toward the largest magnitude.

3 Machining Simulation Tests on Benchmark Sculptured Surfaces

The machining simulation tests conducted to assess the aforementioned intelligent algorithms are presented in the following sub-sections. The benchmarksculptured surfaces were selected according to the availability of their design configurations so as to reproduce them exactly as they were studied by previous researchers. Another reason for selecting these surfaces was their generalized complexity to capture all possible tool axis variations for convex, concave, and saddle surface regions.

3.1 Benchmark Sculptured Surface 1: Multi-curvature Bi-cubic Bezier Patch

A bi-cubic Bezier surface has been designed in CAD environment for establishing the machining experiments. The control points for this sculptured surface may be found in Gray et al. [24]. The aim was to minimize the number of 5-axis tool positions by reducing the number of cutting data (CL data) and maintain as lowest machining error as possible. As far as machining error is concerned, the average chord error and scallop height given via effective radius of the tool have been considered. The benchmark surface as well as the tool path applied is shown in Fig. 1.

Response surface methodology (RSM) has been adopted to generate and conduct the machining simulations in CAM software for this sculptured surface. The experimental design suggested 20 runs with a default alpha equal to 1.682. A toroidal cutting tool Ø37.4 with 6 mm corner radius (Rc) has been used. According to this tool, the lower bound for stepover was 10% of its diameter (equal to 3.74 mm) whilst the upper bound was 35% (equal to 13.09 mm). Lead angle was set to take values between 10° and 25°; tilt angle between 0° and 7°; and maximum discretization step was kept to the constant value of 1.397 mm. Cutting tolerance was determined equal to 0.025 mm. A common criterion for assessing the response of tool path parameters was formulated and computed by an automated macro per each experimental run as follows:

$$C = \left\{ \text{AvgChordErr} + \frac{1}{R_{\text{eff}}} + \text{NoCLs} \times 10^{-3} \right\} \tag{2}$$

where AvgChordErr, R_{eff}, NoCLs, and C are the average chord error, the effective radius, the number of cutting data, and the criterion, respectively. Table 2 summarizes the values for parameters and the results obtained for benchmark sculptured surface 1.

According to the analysis of variance (ANOVA) corresponded to the aforementioned experimental design, the full quadratic regression equation with R^2 = 65.4% has been generated and is as follows:

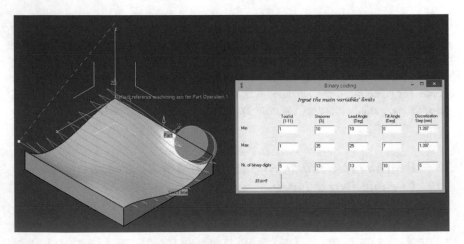

Fig. 1 Tool path applied on the benchmark sculptured surface 1 and corresponding bounds for testing optimal tool path parameter values

$$\begin{aligned} \text{ObjFun} = 11.58 &- 0.159 * x(1) - 0.096 * x(2) + 0.074 * x(3) \\ &- 0.00601 * x(1)^2 - 0.00238 * x(2)^2 - 0.0108 * x(3)^2 \\ &+ 0.01785 * x(1) * x(2) + 0.0076 * x(1) * x(3) - 0.00481 * x(2) * x(3) \end{aligned}$$

$$(3)$$

The algorithms were executed with 5 candidate solutions to be evolved for 5 iterations. Figure 2 shows the convergence diagrams obtained.

With reference to Fig. 2, Table 3 summarizes the optimal tool path parameter values recommended by the intelligent algorithms. Note that the global minimum for the objective function given in Eq. 3 has been found equal to 8.0288 for stepover = 3.74 mm; lead angle = 25°; and tilt angle = 7°, which is also the lowest result for the experimental design established for benchmark surface 1. However, it is mentioned that these results do not represent a generalized approach for solving the multi-objective optimization problem; they only cover this specific case whilst they are as much trustworthy as the coefficient of determination R^2 = 65.4% of the regression model taken as the objective function.

By considering the behavior of algorithms for optimizing the tool path creation for sculptured surface 1 (Fig. 2), the following is observed: VGA can arrive at the best possible solution even from early generations; TLBO is closely as efficient as VGA with similar convergence trend; SFLA produces a mixed "downward–upward" trend; PSO's convergence follows a "convex" trend; SCE follows a downward convergence accompanied with a small period of repeating the same objective function value (straight-line for iterations 3–4); and IWO appears to have the slowest convergence, yet with a noticeable downward trend.

Table 2 Parameters for tool path simulations and results obtained

a/a exp.	Stepover (mm)	Lead angle (°)	Tilt angle (°)	Criterion
1	3.74	10	0	10.047989
2	13.09	10	0	10.050489
3	3.74	25	0	08.085368
4	13.09	25	0	10.092768
5	3.74	10	7	10.047849
6	13.09	10	7	10.048149
7	3.74	25	7	07.082412
8	13.09	25	7	10.083612
9	0.553	17.5	3.5	10.066148
10	16.277	17.5	3.5	10.070248
11	8.415	4.886	3.5	10.027989
12	8.415	30.113	3.5	10.093778
13	8.415	17.5	−2.386	10.066967
14	8.415	17.5	9.386	10.067295
15–20	8.415	17.5	3.5	10.067948

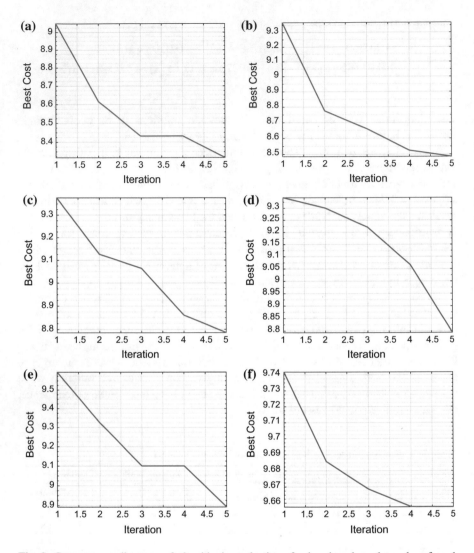

Fig. 2 Convergence diagrams of algorithmic evaluations for benchmark sculptured surface 1: **a** VGA; **b** TLBO; **c** SFLA; **d** PSO; **e** SCE; **f** IWO

3.2 Benchmark Sculptured Surface 2: Bi-cubic Bezier Mirrored Patches Connected with C_0 Continuity Edge

For the second machining simulation experiment, a benchmark part containing two mirrored bi-cubic patches connected by a C_0 continuous edge has been adopted. Challenge in this particular case is imposed by machining the surface having determined feed direction to be perpendicular to C_0 continuous edge even though

Table 3 Optimal tool path parameters for machining the benchmark sculptured surface 1

Algorithm	Stepover (mm)	Lead (°)	Tilt (°)	ObjFun
VGA	4.595	24.978	6.758	8.319361
TLBO	3.740	24.478	4.369	8.486351
SFLA	4.110	21.958	6.025	8.786061
PSO	4.796	24.42	4.118	8.793665
SCE	4.820	23.946	3.586	8.892553
IWO	6.958	20.192	5.037	9.657611

good practices suggest that such a surface would be machined with feed direction parallel to C_0 edge. The same surface has been investigated by Gray et al. [25] where a Ø50.8 Rc6.35 mm toroidal end-mill was applied to machine it, whereas two fixed values for determining the forward step were set. The first forward step value was set equal to 2.0 mm to interpolate most of the regions of the benchmark surface, and the second forward step value was set to 0.762 mm to interpolate those to the vicinity of C_0 continuous edge. Results referring to surface finish were scallop height and maximum undercut with the values of 0.1 and 0.07 mm, respectively, and under the critical cutting tolerance of 0.1 mm. The benchmark surface as well as the tool path applied is shown in Fig. 3.

Response surface methodology (RSM) has been adopted to generate and conduct the machining simulations in CAM software for this sculptured surface. The experimental design suggested 20 runs with a default alpha equal to 1.682. A toroidal cutting tool Ø50.8 with 6.35 mm corner radius (Rc) has been used. According to this tool, the lower bound for stepover was 10% of its diameter (equal to 5.08 mm) whilst the upper bound was 15% (equal to 7.62 mm). Lead angle was

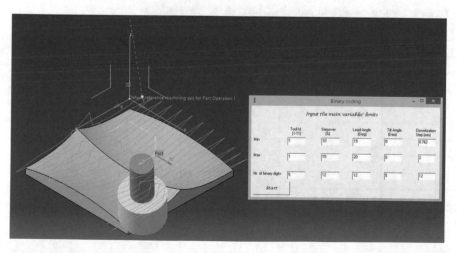

Fig. 3 Tool path applied on the benchmark sculptured surface 2 and corresponding bounds for testing optimal tool path parameter values

set to take values between 15° and 20°; tilt angle was kept constant to 0°; and maximum discretization step was varied from 0.762 to 2.0 mm. Cutting tolerance was determined equal to 0.05 mm. In this case, the common criterion for assessing the response of tool path parameters was formulated by utilizing CAM software utilities for examining remaining material in the form of scallop magnitude and the macro deployed in first case study to extract mean chord error and number of cutting data for superficial quality and tool path efficiency, respectively.

The results obtained for remaining material in the form of scallop, mean chord error, and number of cutting data were normalized in a "0–1" interval. The values for the final 3D Pareto criterion that statistically analyzed to generate its regression equation were computed using the relation shown in Eq. 4.

$$C = \sqrt{AvgChordErr^2 + h^2 + NoCLs^2} \tag{4}$$

Table 4 summarizes the values for parameters and the results obtained for benchmark sculptured surface 2.

According to the analysis of variance (ANOVA) corresponded to the aforementioned experimental design, the full quadratic regression equation with $R^2 = 63.81\%$ has been generated and is as follows:

$$
\begin{aligned}
ObjFun = {} & 1.56 + 0.35 * x(1) - 0.051 * x(2) - 0.19 * x(3) \\
& - 0.0063 * x(1)^2 + 0.00133 * x(2)^2 + 0.0237 * x(3)^2 \\
& - 0.00099 * x(1) * x(2) + 0.0225 * x(1) * x(3) - 0.00104 * x(2) * x(3)
\end{aligned}
\tag{5}
$$

Table 4 Parameters for tool path simulations and results obtained

a/a exp.	Stepover (mm)	Lead angle (°)	MaxDstep (mm)	Criterion
1	5.08	10	0.762	0.9502
2	7.62	10	0.762	0.8552
3	5.08	25	0.762	0.8964
4	7.62	25	0.762	0.7474
5	5.08	10	2	1.0133
6	7.62	10	2	0.9730
7	5.08	25	2	0.9241
8	7.62	25	2	0.8620
9	4.21412	17.5	1.381	0.9786
10	8.48588	17.5	1.381	0.8671
11	6.35	4.8866	1.381	1.4380
12	6.35	30.1134	1.381	0.8879
13	6.35	17.5	0.33997	1.0755
14	6.35	17.5	2.42203	0.8787
15–20	6.35	17.5	1.381	0.8756

The algorithms were executed with 5 candidate solutions to be evolved for 5 iterations. Figure 4 shows the convergence diagrams obtained.

With reference to Fig. 4, Table 5 summarizes the optimal tool path parameter values recommended by the intelligent algorithms. Note that the global minimum for the objective function given in Eq. 5 has been found equal to 0.7270 for stepover = 8.486 mm; lead angle = 22.580°; and maximum discretization step = 0.498 mm. As it has been mentioned in the first experimental design for benchmark sculptured surface 1, the results do not represent a generalized approach for solving the multi-objective optimization problem and they correspond only to this specific case. The problem's solution cannot be more reliable than 63.81% which is the value of the regression model's coefficient of determination R^2.

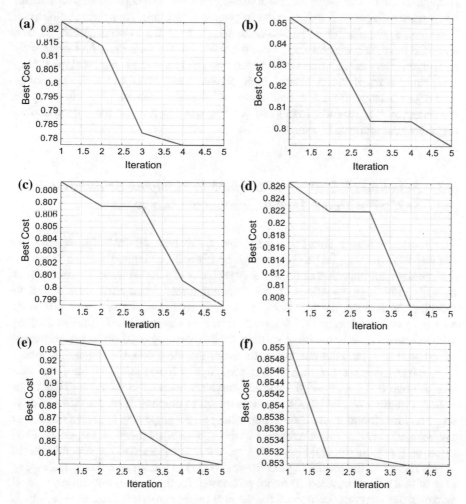

Fig. 4 Convergence diagrams of algorithmic evaluations for benchmark sculptured surface 2: **a** VGA; **b** TLBO; **c** SFLA; **d** PSO; **e** SCE; **f** IWO

Table 5 Optimal tool path parameters for machining the benchmark sculptured surface 2

Algorithm	Stepover (mm)	Lead (°)	MaxDstep (mm)	ObjFun
VGA	7.941	20.050	0.878	0.777870
TLBO	8.240	19.048	1.639	0.792076
SFLA	7.608	22.312	1.474	0.798637
PSO	7.426	21.717	1.394	0.806789
SCE	7.537	17.993	1.478	0.831343
IWO	7.232	27.871	1.406	0.852972

According to Fig. 4, the behavior of all algorithms in terms of their convergence is quite similar. The slope of the linear-downward convergence segment in all convergence diagrams is quite similar in genetic algorithms. An apparent difference is noticed to the number of iterations for which these slopes occur. VGA, TLBO, and SCE have their biggest slope from the second to the third iteration toward their convergence to final solution whilst SFLA and PSO have their biggest slope during third and fourth iteration toward their convergence. IWO shows a remarkable convergence slope from the first to second iteration representing efficiency in convergence rate in early iterations, yet the worst result for the corresponding objective function was obtained.

3.3 Benchmark Sculptured Surface 3: Complex Sculptured Surface with an Abrupt Curvature Change

The third machining simulation experiment involved an investigation for the optimal tool path applied to a complex benchmark sculptured surface having a surface segment imposing extremely abrupt changes in curvature and consecutively in tool positioning variations. Three different toroidal end-mills were tested to machine the surface: the first being a Ø8 mm, Rc2 mm; the second a Ø16 mm, Rc4 mm; and the third a Ø25.4 mm, Rc6 mm as suggested in Roman et al. [26] who examined exactly the same surface. Preliminary response surface experiments of categorical type were conducted to select the most suitable regression model form those generated per each cutting tool. It was found that Ø16 mm, Rc4 mm was the one that best fitted the surface curvature without violating it. Thereby, a constant number of tool passes was determined to emphasize on the effect of lead angle and maximum discretization step whilst no tilt angle was applied since it was not deemed necessary with reference to the surface's geometrical properties. Feed direction for the aforementioned cutting tools was determined to be perpendicular to X-axis regarding the machining axis system (G54) to introduce a more challenging case of tool path generation. The tool path tested for the benchmark surface as well as parameter intervals determined for the proposed approach is illustrated in Fig. 5.

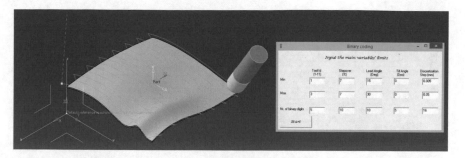

Fig. 5 Tool path applied to machine the complex sculptured surface and optimization intervals

The cut tolerance for the final experiments was set to 0.1 mm using the Ø16 mm, Rc4 mm.

RSM has been applied to build the experiments with two independent variables: lead angle and maximum discretization step. The former parameter was subjected to the bound [5, 45] whilst the latter to [0.01, 0.1]. Stepover was set to 45% of the Ø16 mm, Rc4 mm cutting tool (7.2 mm). According to the number of independent variables, a total of 13 machining simulation experiments were conducted with default alpha equal to 1.414. The multi-objective criterion formulated for this case study was consisted of mean chord error, maximum height of remaining material, number of cutting data, and maximum undercut. The results obtained for these objectives were normalized in a "0–1" interval. The values for the common Pareto criterion that statistically analyzed to generate its regression equation were computed using the relation shown in Eq. 6.

$$C = \sqrt{\text{AvgChordErr}^2 + h^2 + \text{NoCLs}^2 + u^2} \tag{6}$$

With h and u to represent the maximum height detected for remaining material in mm and the maximum undercut value in mm, respectively. Table 6 summarizes the values for parameters and the results obtained for benchmark sculptured surface 3.

Table 6 Parameters for tool path simulations and results obtained

a/a exp.	Lead angle (°)	MaxDstep (mm)	Criterion
1	5	10	0.9502
2	45	10	0.8552
3	5	100	0.8964
4	45	100	0.7474
5	−3.2843	55	1.0133
6	53.2843	55	0.9730
7	25	−8.64	0.9241
8	25	118.64	0.8620
9–13	25	55	0.9786

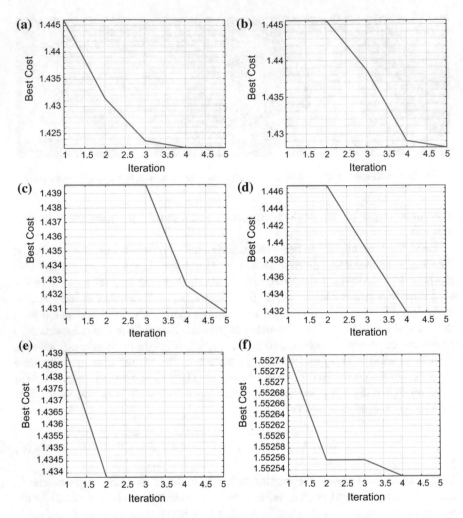

Fig. 6 Convergence diagrams of algorithmic evaluations for benchmark sculptured surface 3:
a VGA; **b** TLBO; **c** SFLA; **d** PSO; **e** SCE; **f** IWO

According to the analysis of variance (ANOVA) corresponded to the aforementioned experimental design, the full quadratic regression equation with
$R^2 = 83.34\%$ has been generated and is as follows:

$$\text{ObjFun} = 1.7841 - 0.02200 * x(1) - 0.00081 * x(2) + 0.000389 * x(1)^2$$
$$+ 0.000019 * x(2)^2 - 0.000041 * x(1) * x(2) \tag{7}$$

The algorithms were executed with 5 candidate solutions to be evolved for 5
iterations. Figure 6 shows the convergence diagrams obtained.

Table 7 Optimal tool path parameters for machining the benchmark sculptured surface 3

Algorithm	Lead angle (°)	MaxDstep (mm)	ObjFun
VGA	34.098	0.062	1.422339
TLBO	27.915	0.035	1.428145
SFLA	31.984	0.080	1.430669
PSO	33.466	0.081	1.431968
SCE	37.537	0.063	1.433817
IWO	49.514	0.098	1.552528

With reference to Fig. 6, Table 7 summarizes the optimal tool path parameter values recommended by the intelligent algorithms. Note that the global minimum for the objective function given in Eq. 5 has been found equal to 1.4189 for a lead angle value equal to 31.165° and for a maximum discretization step value equal to 0.0549 mm. These results cannot represent a generalized approach for solving the multi-objective optimization problem, and they correspond only to this specific case. The problem's solution cannot be more reliable than 83.34% which is the value of the regression model's coefficient of determination R^2 for the experiments conducted concerning benchmark sculptured surface 3.

According to Fig. 6, the following indications are observed in terms of the algorithms' convergence speed, rate, and efficiency. VGA starts to converge with a high convergence slope that gradually reduces to arrive finally in its minimum point which is the value of 1.422339 for the corresponding objective function. TLBO's behavior shows a steady-state convergence in early iterations (from 1 to 2) whilst a high convergence rate is reached during the second iteration until the fourth iteration. Its minimum point of 1.428145 is attained in the last iteration by producing a small slope from fourth to fifth iteration. SFLA is beneficial from the third to the fourth iteration since a high convergence rate is noticed. Form the first until the end of the third iteration, a steady-state (idle) convergence is observed. The final point for SFLA is equal to 1.430669. PSO's convergence behavior is quite similar to SFLA algorithm suggesting a beneficial convergence ratio from the second to the fourth iteration. The final point attained by PSO is equal to 1.431968. SCE algorithm appears to produce the highest slope in terms of its convergence speed from the first to the second iteration whilst no further optimization is attained until the fifth iteration which is the last one with minimum value for the objective function equal to 1.433817. IWO represents just as efficient convergence slope in early iterations (iterations 1 and 2) as SCE algorithm followed by no further improvement until the third iteration. IWO starts again to reduce its minimum point from the third until fourth iteration, and finally in fifth iteration, the minimum value for the corresponding objective function comes equal to 1.552528.

4 Discussion and Conclusions

Sculptured surface CNC machining is the most often-employed material-removal operation found in production chains worldwide owing to the requirements of functionality and aesthetics for new products. Such a material-removal operation imposes the usage of manufacturing software to model, simulate, and finally generate the CNC code. Since the process is heavily based on CAD models that imported as inputs to machining modeling environment, parameters affecting their geometrical accuracy should carefully be determined. Besides, feed rate and cutting speed parameters have extensively been examined for different materials and several cutting tools, yet the decision making concerning interpolation strategies, radial cutting depths, and 5-axis tool inclination angles has not globally addressed. These parameters directly affect the final part geometry and should simultaneously be optimized by an intelligent module.

Apparently to achieve a global handling of these parameters found in cutting-edge manufacturing software under an unbiased optimization concept, an intelligent system needs to be coupled to it so as to repetitively assess the randomly generated values corresponding to tool path parameters. In addition, evaluations for objectives need to be done by directly using the 3D model since this represents the part to be machined instead of any kind of mathematical expressions for the role of objective functions. Nevertheless, should such a concept is to be developed, one should examine the behavior of several artificial systems applied to the same problem and observe their strengths and weaknesses. Yet to accomplish it, a comparison among several artificial algorithms needs to be made in case-oriented experiments to simplify operational issues as well as observations.

In this work, an effort has been made to compare optimization solutions attained by six novel evolutionary algorithms deployed to address three case-oriented tool path generation problems using benchmarksculptured surfaces. The algorithms selected were a novel virus-evolutionary genetic algorithm (VGA), a particle swarm optimization algorithm (PSO), an invasive weed optimization algorithm (IWO), a shuffled frog-leaping algorithm (SFLA), a shuffled complex evolution (SCE) algorithm, and a teaching–learning-based optimization algorithm (TLBO). To simulate the real word's pressing demands for best possible results in time, few generations and candidate solutions have been determined for the evolutionary procedure toward convergence. This has been decided under the prism of considering iteration to be an entire tool path planning simulation. Should a simulation's duration may depend on the part's complexity, several minutes may be required to end up with the outcome. Therefore, it is of vital importance to choose an intelligent system capable of achieving high convergence rates even from early iterations.

In all three cases of tool path optimization experiments, VGA and TLBO were proven to be the most efficient in terms of their convergence result and fast exploration without getting trapped to local minima. This was observed by the convergence diagrams created based on the regression models from experiments which were considered as the objective functions. Without being concerned that much about the quality of these regression models in this work, both VGA and TLBO shown fast convergence to the lowest possible point according to the problem handled. Even though the rest of the algorithms have been shown quite promising, they did not obtain such minima as the ones obtained by VGA and TLBO. Not intending to degrade the algorithms, it is mentioned that all of them were capable of arriving to global minima for all experiments if more iterations and candidate solutions than five are to be determined. To emphasize on the basic optimization concept previously mentioned, VGA and TLBO ought to be preferred for system integration referring to special cases should solutions need to be quickly at hand. Unless optimal tool path planning does not fall into such particular cases, all of the algorithms investigated may be utilized to be operated as ad hoc with manufacturing software for globally optimal tool path planning.

Questions with answers (The correct answer is dictated with **bold**)

1. In metal-forming operations, the shape of the desired geometry for the product is obtained by applying processes such as milling drilling and turning (True/**False**)
2. One of the key processes in metal cutting is forging (True/**False**)
3. The final shape of the product must comply with specifications and requirements given in the engineering drawing (**True**/False)
4. Machining should be considered as a highly nonlinear optimization problem (**True**/False)
5. Material-removal rate is an optimization criterion reflecting surface finish (True/**False**)
6. Roughness parameters can be optimization criteria referring to quality (**True**/False)
7. Design of experiments can only be applied to analyze results for metal-forming operations (True/**False**)
8. Empirical models can be generated only to serve problem-solving referring to metal cutting (True/**False**)
9. Artificial intelligence techniques may be applied to all aspects of mechanical engineering for optimizing related problems (**True**/False)
10. Computer-aided manufacturing software is responsible for designing the product (True/**False**)

Glossary

Chordal deviation The type of machining error owing to cutting tool interpolation when applying CNC machining as a key metal cutting operation

Computer-aided manufacturing (CAM) Environment for modeling manufacturing processes with the aid of computers

Design of experiments The process of establishing experimental runs so as to investigate the influence of independent process parameters to one or more responses (dependent variables)

Empirical models Mathematical formulas (regression equations) relating independent variables and responses to be used for predicting crucial results prior to actual operations

Genetic algorithms Artificial intelligence heuristics used for optimizing one or more objectives related to an engineering problem

Metal cutting The field including the manufacturing processes where raw materials are turned to final products by removing the excess material by machining

Metal forming The field including the manufacturing processes where raw materials are shaped directly to final products

Objective function A mathematical relation expressing one or more criteria for optimization either with the use of artificial intelligence (genetic algorithms) or conventional engineering computing

Scallop height The remaining material a tool path leaves among adjacent tool passes when machining free-form products with CNC technology

Sculptured surface machining The metal cutting operation for producing free-form surfaces found in modern products, assisted by CNC machining technology

Tool path planning The process of computing the path to be followed by one or more cutting tools using the CAM environment

Virus-evolutionary genetic algorithm A special genetic algorithm following the principles of the virus theory of evolution

References

1. DeGarmo, E. P., Black, J. T., & Kohser, R. A. (2007). *Materials and processes in manufacturing* (10th ed.). New York: Wiley.
2. Kalpakjian, S., & Schmid, S. R. (2001). *Manufacturing engineering and technology* (4th ed.). New Jersey: Prentice-Hall.
3. Boothroyd, G., & Knight, W. A. (2006). *Fundamentals of machining and machine tools* (3rd ed.). Boca Raton: CRC.
4. Astakhov, V. P. (2006). *Tribology of metal cutting*. London: Elsevier.
5. Davim, J. P. (2003). Design of optimization of cutting parameters for turning metal matrix composites based on the orthogonal arrays. *Journal of Materials Processing Technology,132* (1), 340–344.
6. Davim, J. P., & Reis, P. (2005). Damage and dimensional precision of milling carbon fiber-reinforced plastics using design experiments. *Journal of Materials Processing Technology,160*(2), 160–167.

7. Davim, J. P., Reis, P., & Antonio, C. C. (2004). Experimental study of drilling glass fiber reinforced plastics (GFRP) manufactured by hand lay-up. *Computer Science Technology,64* (2), 289–297.
8. Choi, B. K., & Jerard, R. B. (1998). *Sculptured surface machining: theory and applications.* Dordrecht: Kluwer Academic Publishers.
9. Palanikumar, K., Latha, B., Senthilkumar, V. S., & Karthikeyan, R. (2009). Multiple performance optimization in machining of GFRP composites by a PCD tool using non-dominated sorting genetic algorithm (NSGA-II). *Metals and Materials International, 15*(2), 249–258.
10. Garg, M. P., Jain, A., & Bhushan, G. (2012). Modelling and multi-objective optimization of process parameters of wire electrical-discharge machining using non-dominated sorting genetic algorithm II. *Proceedings of the Institution of Mechanical Engineers Part B: Journal of Engineering Manufacture,226*(12), 1986–2001.
11. Fountas, N. A., Ntziantzias, I., Kechagias, J., Koutsomichalis, A., Davim, J. P., & Vaxevanidis, N. M. (2013). Prediction of cutting forces during turning PA66 GF-30 glass fiber reinforced polyamide by soft computing techniques. *Materials Science Forum,766,* 37–58.
12. Bhavsar, S. N., Aravindan, S., & Rao, P. V. (2015). Investigating material removal rate and surface roughness using multi-objective optimization for focused ion beam (FIB) micro-milling of cemented carbide. *Precision Engineering,40,* 131–138.
13. Rao, R. V., Rai Dhiraj, P., Balic, J. (2016). Multi-objective optimization of machining and micro-machining processes using non-dominated sorting teaching–learning-based optimization algorithm. *Journal of Intelligent Manufacturing,* 1–21. https://doi.org/10.1007/s10845-016-1210-5.
14. Zain, A. M., Haron, H., & Sharif, S. (2010). Application of GA to optimize cutting conditions for minimizing surface roughness in end milling machining process. *Expert System with Applications,37*(6), 4650–4659.
15. Sepehri Rad, H., & Lucas, C. (2007). A recommender system based on invasive weed optimization algorithm. *IEEE Congress Evolutionary Computer CEC,2007,* 4297–4304.
16. Duan, Q. Y., Gupta, V. K., & Sorooshian, S. (1993). Shuffled complex evolution approach for effective and efficient global minimization. *Journal of Optimization Theory and Applications,76*(3), 501–521.
17. Eusuff, M., Lansey, K., & Pasha, F. (2006). Shuffled frog-leaping algorithm: A memetic meta-heuristic for discrete optimization. *Eng Opt,38*(2), 129–154. https://doi.org/10.1080/03052150500384759.
18. Gomez-Gonzalez, M., & Jurado, F. (2015). Machining parameters selection for milling operations using shuffled frog-leaping algorithm. *International Journal of Emerging Technology and Advanced Engineering,5*(5), 50–59.
19. Fountas, N. A., Kechagias, J. D., Vaxevanidis, N. M. (2016). Artificial immune algorithm implementation for optimized multi-axis sculptured surface CNC machining. *IOP Conference Series: Material Science Engineering, 161*(1).
20. Kubota, N., Fukuda, T., & Shimojima, K. (1996). Virus-evolutionary algorithm for a self-organising manufacturing system. *Computer and Industrial Engineering,30*(4), 1015–1026.
21. Fountas, N. A., Benhadj-Djilali, R., Stergiou, C. I., & Vaxevanidis, N. M. (2017). An integrated framework for optimizing sculptured surface CNC tool paths based on direct software object evaluation and viral intelligence. *J Intell Manuf (Article in Press).* https://doi.org/10.1007/s10845-017-1338-y.
22. Fountas, N. A., Živković, S., Benhadj-Djilali, R., Stergiou, CI., Majstorovic, V. D., Vaxevanidis, N. M. (2017a). Intelligent dual curve-driven tool path optimization and virtual CMM inspection for sculptured surface CNC machining. In V. Majstorovic, & Z. Jakovljevic (Eds.), *Proceedings of 5th International Conference on Advanced Manufacturing Engineering and Technologies,* NEWTECH 2017, Lecture Notes in Mechanical Engineering, Berlin: Springer.

23. Fountas, N. A., Vaxevanidis, N. M., Stergiou, C. I., & Benhadj-Djilali, R. (2014). Development of a software-automated intelligent sculptured surface machining optimization environment. *International Journal of Advanced Manufacturing Technology,75*(5–8), 909–931.
24. Gray, P., Bedi, S., & Ismail, F. (2003). Rolling ball method for 5-axis surface machining. *Computer Aided Design,35,* 347–356.
25. Gray, P. J., Ismail, F., & Bedi, S. (2004). Graphics-assisted rolling ball method for 5-axis surface machining. *Computer Aided Design,36,* 653–663.
26. Roman, A., Barocio, E., huegel, J., & Bedi, S. (2015). Rolling Ball method applied to 3½½-axis machining for tool orientation and positioning and path planning. *Advances in Mechanical Engineering,7*(12), 1–12.

Index

© Springer International Publishing AG, part of Springer Nature 2018   371
J. P. Davim (ed.), *Introduction to Mechanical Engineering*, Materials Forming,
Machining and Tribology, https://doi.org/10.1007/978-3-319-78488-5

Printed in the United States
By Bookmasters